Biological Activities of Ribosome-Inactivating Proteins

Biological Activities of Ribosome-Inactivating Proteins

Editors

José Miguel Ferreras
Lucía Citores

MDPI • Basel • Beijing • Wuhan • Barcelona • Belgrade • Manchester • Tokyo • Cluj • Tianjin

Editors
José Miguel Ferreras
University of Valladolid
Valladolid
Spain

Lucía Citores
University of Valladolid
Valladolid
Spain

Editorial Office
MDPI
St. Alban-Anlage 66
4052 Basel, Switzerland

This is a reprint of articles from the Special Issue published online in the open access journal *Toxins* (ISSN 2072-6651) (available at: https://www.mdpi.com/journal/toxins/special_issues/biological_activities_RIPs).

For citation purposes, cite each article independently as indicated on the article page online and as indicated below:

LastName, A.A.; LastName, B.B.; LastName, C.C. Article Title. *Journal Name* **Year**, *Volume Number*, Page Range.

ISBN 978-3-0365-6822-5 (Hbk)
ISBN 978-3-0365-6823-2 (PDF)

Cover image courtesy of Lucía Citores

Contents

About the Editors

José Miguel Ferreras

José Miguel Ferreras (Prof., PhD) is a Full Professor (since 2016) in Biochemistry and Molecular Biology at the Department of Biochemistry and Molecular Biology and Physiology, University of Valladolid, Spain. His research activity is based on four main fields, concerning i) the study of protein synthesis inhibition, ii) the research on ribosome-inactivating proteins (RIPs) from plants, iii) the design and the in vitro and in vivo experimentation of immunotoxins and other conjugates in order to obtain molecules targeted specifically against cells responsible for a pathological state, and iv) the study of ribotoxins from fungi. Prof. Ferreras has co-authored more than one hundred scientific papers.

Lucía Citores

Lucía Citores (Prof., PhD) studied Chemistry at the University of Valladolid, Spain, and completed her PhD in Biochemistry at the same University. From 1997 to 2004 she worked at the Institute for Cancer Research in Oslo, Norway, as a postdoctoral researcher. During this time, she became interested in the endocytosis and intracellular transport of fibroblast growth factors and their receptors. She returned to Spain in 2004 as a postdoctoral researcher from the program Ramón y Cajal and in 2011 she was appointed as Associated Professor in Biochemistry and Molecular Biology at the University of Valladolid. Her research interests have focused on distribution, activity, structure and applications of ribosome-inactivating proteins and ribotoxins and how they enter and intoxicate mammalian and fungal cells.

Preface to "Biological Activities of Ribosome-Inactivating Proteins"

Since the discovery and purification of abrin and ricin in the late 19th century, ribosome-inactivating proteins (RIPs) have aroused great interest in many scientists due to their enormous potency in inhibiting protein synthesis. This activity has been used for noble purposes, such as fighting diseases like cancer or virosis, and for less noble purposes, such as murder or terrorism.

RIPs are ribosomal RNA N-glycosylases isolated mainly from plants that catalyze the hydrolysis of the N-glycosidic bond of a specific adenosine in the sarcin-ricin loop (SRL) of the major ribosomal RNA. Due to the fact that the SRL is crucial for anchoring translation elongation factors, RIPs cause irreversible inactivation of ribosomes. They have been classified into two types based on the presence (type 2 RIPs) or absence (type 1 RIPs) of a lectin chain (B chain) that can convert type 2 RIPs into potent toxins, such as ricin or abrin.

The exact biological role of these proteins is unknown, but they are thought to be a defense mechanism of some plants against pathogens and predators. As a consequence of their enzymatic action, RIPs show several biological activities, among which antiviral, antifungal, and antiproliferative activities stand out. The most promising applications of RIPs are related to their use as a component of immunotoxins, in which the RIPs are linked to antibodies that mediate their binding to and internalization by malignant cells. In agriculture, RIPs have been shown to increase resistance against viruses, fungi, and insects in transgenic plants.

The studies collected in this book offer the reader an overview of the most current and interesting lines of research in the field of RIPs and their applications in medicine and agriculture. The isolation and biological properties of some new RIPs, both type 1 and type 2, the mechanisms of toxicity of some previously reported RIPs, and two extensive reviews on the antiviral activity of RIPs and the strategies used to improve their pharmacological properties, are presented in this Issue.

In addition to our sincere thanks to all the authors and reviewers who have contributed to the success of this Special Issue, we also extend special thanks to the management team and editorial staff of MDPI for their valuable contribution and editorial support.

José Miguel Ferreras and Lucía Citores
Editors

toxins

MDPI

Editorial

Biological Activities of Ribosome-Inactivating Proteins

Lucía Citores * and José M. Ferreras *

Department of Biochemistry and Molecular Biology and Physiology, Faculty of Sciences, University of Valladolid, 47011 Valladolid, Spain
* Correspondence: lucia.citores@uva.es (L.C.); josemiguel.ferreras@uva.es (J.M.F.)

Citation: Citores, L.; Ferreras, J.M. Biological Activities of Ribosome-Inactivating Proteins. *Toxins* **2023**, *15*, 35. https://doi.org/10.3390/toxins15010035

Received: 19 December 2022
Accepted: 30 December 2022
Published: 1 January 2023

After more than 50 years of research, studies on the structure and biological activities of ribosome-inactivating proteins (RIPs) continue to provide a field of great interest within the scientific community, both for the health risks they pose and their applications in medicine and biotechnology. This Special Issue of *Toxins* offers a sample of the main research topics when studying these proteins. RIPs are ribosomal RNA N-glycosylases (EC 3.2.2.22), mainly isolated from plants, some bacteria, and fungi, that specifically catalyze the hydrolysis of the second N-glycosidic bond of the GAGA tetraloop located in the sarcin-ricin loop (SRL) of the major ribosomal RNA. Because SRL is crucial for anchoring elongation factors in the ribosome, the removal of adenine causes the irreversible inactivation of ribosomes, leading to cell death. In addition, RIPs usually demonstrate other enzymatic activities, including, most relevantly, their adenine polynucleotide glycosylase (APG) activity on all nucleic acid types; that is, some RIPs can remove adenines from both ribosomal and non-ribosomal RNA and DNA [1].

RIPs are structurally classified into two groups [1]: type 1 RIPs, consisting of a single polypeptide chain of approximately 30 kDa with enzymatic activity, and type 2 RIPs, of approximately 60 kDa, are formed by an enzymatically active A chain, similar to type 1 RIPs, which is linked through a disulfide bond to a B chain with lectin properties. With a strong affinity for cell surface sugars, the B chain can facilitate toxin entry into cells, thus conferring high toxicity to many type 2 RIPs in cells and animals.

This is the case of ricin and abrin: the first type 2 RIPs described [2]. In medicine, these proteins are mainly used for constructing immunotoxins directed against tumor cells but can also be used as chemical weapons. In this scenario, their most obvious use is in aerosols, which would cause lethal damage to the lungs. Sapoznikov et al. described the effects of abrin and ricin intoxication on the lungs following intranasal exposure in mice [3]. The results indicated that a lethal dose of abrin induced less pronounced damage to the pulmonary stroma and reduced deterioration of intercellular junction molecules compared to ricin, which could contribute to the higher level of protection achieved against abrin by postexposure antibody-mediated treatment.

Although ricin and abrin are the best-known and most used type 2 RIPs, they are not the most toxic. This distinction belongs to RIPs obtained from different species of the genus *Adenia*, such as volkensin, modeccin, lanceolins, and stenodactylin [4]. These proteins differentiate from ricin and abrin due to their unique aspect of being retrogradely transported along peripheral nerves and the central nervous system, providing interesting applications in neuroscience. Bortolotti et al. reported the purification of a new protein of this type from the caudex of *Adenia kirkii* Engl. [5]. Kirkiin is a RIP characterized by high cytotoxicity toward neuronal cell lines, making it a promising candidate for pharmacological purposes.

Type 2 RIPs obtained from species of the genus *Sambucus* are peculiarly hundreds of thousands of times less toxic than ricin and abrin. In the case of elderberry (*Sambucus nigra* L.), more than 20 RIPs and related lectins have been isolated and characterized from its flowers, seeds, fruits, and bark, making it a unique species for studying proteins of this type [6]. The work of Iglesias et al. has expanded our knowledge on the family of RIPs and RIP-related lectins produced by *S. nigra*; their purification and characterization of eight new proteins found in the leaves include one type 2 RIP and two related lectins specific for galactose, four type 2 RIPs with deficient sugar-binding domains, and one type 1 RIP. Several of these proteins are homologous to others found elsewhere in the plant [7].

Type 2 RIPs from *Sambucus* lack toxicity, mainly attributed to a reduced affinity for galactosides which could affect their cell binding, uptake, and the intracellular fate of RIPs. Iglesias et al. compared the binding, endocytosis mechanisms, and intracellular pathway followed by ebulin l (obtained from *Sambucus ebulus* L. leaves) with ricin [8]. The data showed that ebulin l binds to cells less than ricin and how, after binding, ebulin l was taken up by clathrin-dependent and clathrin-independent endocytosis into the endosomal/lysosomal system but not to the Golgi apparatus; importantly, ebulin l did not require clathrin or dynamin for intoxication.

Type 1 RIPs display lower toxicity, as they lack the lectin part and, therefore, cannot bind to cells, as type 2 RIPs demonstrate. The structure of type 1 RIPs is similar to the A-chain of type 2 RIPs, despite differences in the structure of various type 1 RIPs. While RIPs from cucurbits are comparable to the A-chain of type 2 RIPs, RIPs of the pokeweed, carnation, amaranth, and spurge families present greater differences [1]. Monocots, such as those from maize or rice, present the most contrasting RIPs. RIPs from the pokeweed (Phytolaccaceae), carnation (Caryophyllaceae), amaranth (Amaranthaceae), and spurge (Euphorbiaceae) families have been subject to much interest because of their antiviral properties and usefulness for the construction of immunotoxins [9,10]. Four examples of these type 1 RIPs are described in this Special Issue: the curcins from the euphorbiaceous *Jatropha curcas* L. [11], the sodins from the amaranthaceous *Salsola soda* L. [12], the sapovaccarins from the caryophyllaceous *Vaccaria hispanica* (Mill.) Rauschert (= *Saponaria vaccaria* L.) [13], and the PD-Ls from the phytolaccaceaous *Phytolacca dioica* L. [14].

Qin et al. studied the reason behind the differing toxicity of curcin and curcin C on the U20S osteosarcoma cell line [11] and found that curcin C cytotoxicity is higher because, unlike curcin, it is endocytosed by clathrin-dependent endocytosis mediated by LRP1 (low-density lipoprotein receptor-related protein 1): an abundant receptor in this type of cell.

Landi et al. isolated a new type 1 RIP from the seeds, edible leaves, and roots of *Salsola soda* [12]. Sodins showed APG activity and induced apoptosis in Hela and COLO 320 cell lines. Of note, sodin 5, from *S. soda* seeds, and quinoin, from *Chenopodium quinoa* Willd (another amaranthaceous species) seeds showed potent antifungal activity against *Penicillium digitatum* (Pers.) Sacc. [12], making them good candidates for obtaining transgenic plants resistant to fungi.

Schlaak et al. isolated a new type 1 RIP from *S. vaccaria* seeds homologous to type 1 RIPs from other caryophyllaceous that, similar to dianthin 30 and saporin-S6, are used for the construction of immunotoxins [13]. Compared to other type 1 RIPs, they exhibited greater thermostability, suggesting that they would be optimal candidates for targeted cancer therapy.

The exact biological role undertaken by RIPs remains unknown, though it has been considered to mirror a plant defense mechanism against pathogens and predators [1].

Notably, RIPs demonstrate antiviral activity. Their antiviral properties have been investigated for over four decades. However, the emergence of new viruses, giving rise to infectious diseases, has caused interest in these proteins to increase due to the difficulty of treating viral infections. On the other hand, a growing need to control crop diseases without the use of environmentally harmful phytosanitary products has resulted, in this regard, in proving RIPs to be promising tools for obtaining transgenic plants that are resistant to viruses. Citores et al., in this Special Issue, review the research studies addressing this topic, with special emphasis on the latest findings and mechanisms of action proposed [10].

Bulgari et al., using the *Phaseolus vulgaris*-tobacco necrosis virus (TNV) pathosystem, demonstrated that PD-L1 and PD-L4 possess strong antiviral activity [14]. Their experiments suggest that this activity targets viral and ribosomal RNA, explaining the near-complete abolition of infections when the virus and RIP enter the cells together.

The most promising applications of RIPs in experimental medicine, especially in cancer therapy, relate to their use as immunotoxins [15], in which RIPs are linked to antibodies that mediate their binding to and internalization by malignant cells. The main obstacles to treatment with RIPs include their short plasma half-life, nonselective cytotoxicity, and antigenicity. Lu et al. reviewed the strategies used to improve their pharmacological properties and discussed prospects for future developments in the engineering of RIPs [16].

In conclusion, the studies collected in this Special Issue provide the reader with an overview of the most current and interesting lines of research in the field of RIPs, including their applications in medicine and agriculture. Further research on the biological activities of RIPs will allow a greater understanding of their biological role, their more efficient use in medicines, mainly for the treatment of cancer and viral diseases and in the fight against crop diseases caused by viruses, fungi, and insects.

Acknowledgments: In addition to our sincere gratitude to the contributors of the various articles, special thanks are also given to the peer reviewers for their rigorous evaluations of all submitted manuscripts.

Conflicts of Interest: The authors declare no conflict of interest.

References

1. Di Maro, A.; Citores, L.; Russo, R.; Iglesias, R.; Ferreras, J.M. Sequence comparison and phylogenetic analysis by the Maximum Likelihood method of ribosome-inactivating proteins from angiosperms. *Plant Mol. Biol.* **2014**, *85*, 575–588. [CrossRef] [PubMed]
2. Olsnes, S. The history of ricin, abrin and related toxins. *Toxicon* **2004**, *44*, 361–370. [CrossRef] [PubMed]
3. Sapoznikov, A.; Gal, Y.; Alcalay, R.; Evgy, Y.; Sabo, T.; Kronman, C.; Falach, R. Characterization of Lung Injury following Abrin Pulmonary Intoxication in Mice: Comparison to Ricin Poisoning. *Toxins* **2022**, *14*, 614. [CrossRef]
4. Battelli, M.G.; Scicchitano, V.; Polito, L.; Farini, V.; Barbieri, L.; Bolognesi, A. Binding and intracellular routing of the plant-toxic lectins, lanceolin and stenodactylin. *Biochim. Biophys. Acta* **2010**, *1800*, 1276–1282. [CrossRef] [PubMed]
5. Bortolotti, M.; Maiello, S.; Ferreras, J.M.; Iglesias, R.; Polito, L.; Bolognesi, A. Kirkiin: A New Toxic Type 2 Ribosome-Inactivating Protein from the Caudex of Adenia kirkii. *Toxins* **2021**, *13*, 81. [CrossRef] [PubMed]
6. Ferreras, J.M.; Citores, L.; Iglesias, R.; Jimenez, P.; Girbes, T. Use of ribosome-inactivating proteins from Sambucus for the construction of immunotoxins and conjugates for cancer therapy. *Toxins* **2011**, *3*, 420–441. [CrossRef] [PubMed]
7. Iglesias, R.; Russo, R.; Landi, N.; Valletta, M.; Chambery, A.; Di Maro, A.; Bolognesi, A.; Ferreras, J.M.; Citores, L. Structure and Biological Properties of Ribosome-Inactivating Proteins and Lectins from Elder (*Sambucus nigra* L.) Leaves. *Toxins* **2022**, *14*, 611. [CrossRef] [PubMed]
8. Iglesias, R.; Ferreras, J.M.; Llorente, A.; Citores, L. Ebulin l Is Internalized in Cells by Both Clathrin-Dependent and -Independent Mechanisms and Does Not Require Clathrin or Dynamin for Intoxication. *Toxins* **2021**, *13*, 102. [CrossRef] [PubMed]
9. Polito, L.; Bortolotti, M.; Mercatelli, D.; Battelli, M.G.; Bolognesi, A. Saporin-S6: A useful tool in cancer therapy. *Toxins* **2013**, *5*, 1698–1722. [CrossRef] [PubMed]
10. Citores, L.; Iglesias, R.; Ferreras, J.M. Antiviral Activity of Ribosome-Inactivating Proteins. *Toxins* **2021**, *13*, 80. [CrossRef] [PubMed]
11. Qin, S.; Wang, X.; Han, P.; Lai, Z.; Ren, Y.; Ma, R.; Cheng, C.; Wang, T.; Xu, Y. LRP1-Mediated Endocytosis May Be the Main Reason for the Difference in Cytotoxicity of Curcin and Curcin C on U2OS Osteosarcoma Cells. *Toxins* **2022**, *14*, 771. [CrossRef] [PubMed]
12. Landi, N.; Ragucci, S.; Citores, L.; Clemente, A.; Hussain, H.Z.F.; Iglesias, R.; Ferreras, J.M.; Di Maro, A. Isolation, Characterization and Biological Action of Type-1 Ribosome-Inactivating Proteins from Tissues of *Salsola soda* L. *Toxins* **2022**, *14*, 566. [CrossRef] [PubMed]
13. Schlaak, L.; Weise, C.; Kuropka, B.; Weng, A. Sapovaccarin-S1 and -S2, Two Type I RIP Isoforms from the Seeds of *Saponaria vaccaria* L. *Toxins* **2022**, *14*, 449. [CrossRef] [PubMed]
14. Bulgari, D.; Landi, N.; Ragucci, S.; Faoro, F.; Di Maro, A. Antiviral Activity of PD-L1 and PD-L4, Type 1 Ribosome Inactivating Proteins from Leaves of Phytolacca dioica L. in the Pathosystem Phaseolus vulgaris–Tobacco Necrosis Virus (TNV). *Toxins* **2020**, *12*, 524. [CrossRef]

15. Polito, L.; Djemil, A.; Bortolotti, M. Plant Toxin-Based Immunotoxins for Cancer Therapy: A Short Overview. *Biomedicines* **2016**, *4*, 12. [CrossRef] [PubMed]
16. Lu, J.Q.; Zhu, Z.N.; Zheng, Y.T.; Shaw, P.C. Engineering of Ribosome-inactivating Proteins for Improving Pharmacological Properties. *Toxins* **2020**, *12*, 167. [CrossRef]

toxins

Review

Engineering of Ribosome-inactivating Proteins for Improving Pharmacological Properties

Jia-Qi Lu [1,†], Zhen-Ning Zhu [1,†], Yong-Tang Zheng [2] and Pang-Chui Shaw [1,*]

1 School of Life Sciences, The Chinese University of Hong Kong, Shatin, N.T., Hong Kong 99077, China; lujq@link.cuhk.edu.hk (J.-Q.L.); janet.chuk@gmail.com (Z.-N.Z.)
2 Key Laboratory of Animal Models and Human Disease Mechanisms, National Kunming High level Biosafety Research Center for Non-human Primates, Kunming Institute of Zoology, Chinese Academy of Sciences, Kunming 650223, Yunnan, China; zhengyt@mail.kiz.ac.cn
* Correspondence: pcshaw@cuhk.edu.hk
† These authors contributed equally.

Received: 19 February 2020; Accepted: 6 March 2020; Published: 9 March 2020

Abstract: Ribosome-inactivating proteins (RIPs) are N-glycosidases, which depurinate a specific adenine residue in the conserved α-sarcin/ricin loop (α-SRL) of rRNA. This loop is important for anchoring elongation factor (EF-G for prokaryote or eEF2 for eukaryote) in mRNA translocation. Translation is inhibited after the attack. RIPs therefore may have been applied for anti-cancer, and anti-virus and other therapeutic applications. The main obstacles of treatment with RIPs include short plasma half-life, non-selective cytotoxicity and antigenicity. This review focuses on the strategies used to improve the pharmacological properties of RIPs on human immunodeficiency virus (HIV) and cancers. Coupling with polyethylene glycol (PEG) increases plasma time and reduces antigenicity. RIPs conjugated with antibodies to form immunotoxins increase the selective toxicity to target cells. The prospects for future development on the engineering of RIPs for improving their pharmacological properties are also discussed.

Keywords: ribosome inactivating protein; therapeutic applications; immunotoxin; anti-HIV; anti-cancer

Key Contribution: This review summarizes an update of knowledge of the anti-HIV and anti-cancer activities of representative RIPs and the engineering methods to improve their pharmacological properties.

1. Introduction

Ribosome inactivating proteins (RIPs) are a group of cytotoxic N-glycosidases. They are mostly found from plants and a few from bacteria [1]. RIPs are classified into three types according to the number of subunits and the organization of the precursor sequences [2]. Type 1 RIPs such as trichosanthin (TCS) and momorcharin (MMC) have a single polypeptide chain with catalytic activity. Type 2 RIPs such as ricin and abrin are heterodimeric, with an active A chain linked to a lectin-binding B chain by a disulfide bond. Type 3 RIPs such as maize ribosome-inactivating protein and barley jasmonate-induced RIP (JIP60) contain a region within the protein that is removed for activation [3,4].

In general, RIPs remove a specific adenine in the α-sarcin/ricin loop (α-SRL) of rRNA, resulting in blocking the binding of elongation factor [5]. The depurination of α-SRL loop causes the GTP binding site to lose the ability to activate GTP hydrolysis. Protein synthesis is thus impeded and the cell dies [6]. Because of lectin-binding properties, most of type 2 RIPs have higher rate of cell entry and hence cytotoxicity [7]. RIPs have also been found to exhibit other enzymatic activities, including superoxide dismutase (SOD) [8], phospholipase [9] and depurination of DNA, RNA and poly (A) [10]. Several RIPs such as Momordica anti-HIV protein (MAP30) and gelonium anti-HIV protein (GAP31) exhibit a topological activity on plasmid and viral DNA, for example HIV-1 long terminal repeats (LTRs) [11]. RIPs can also induce cell apoptosis by a mechanism independent from the depurination of the SRL loop [12]. Because of their cytotoxicity, several RIPs have been tested for anti-tumor, anti-viral, anti-bacterial and anti-fungal properties. Clinical trials, such as gelonin for treating myeloid malignancies [13], ricin for treating leukemia [14] and TCS and pokeweed antiviral protein (PAP) for treating human immunodeficiency virus (HIV) [15,16] have been carried out.

Despite RIPs showing great potential in clinical applications, side effects such as inducing immune responses, short plasma half-life and non-specificity have limited their uses. Over the years, a number of works have been carried out to modify RIPs to reduce these problems. Antibodies can be conjugated with RIPs to form immunotoxins (ITs). Because of the selective function of the antibodies for targeting, ITs can achieve higher efficacy and lower side effects. Polyethylene glycol (PEG) has been used to couple with RIPs. The complex has increased molecular size. As a result, renal clearance, proteolytic degradation, immunogenicity and antigenicity are reduced.

2. Anti-HIV activity of RIPs

2.1. Anti-HIV Activity of Representative RIPs

RIPs including MAP30, saporin, TCS, gelonium anti-HIV protein (GAP31) and α-momorcharin (α-MMC) possess anti-HIV activity and other anti-viral activities. It has been found that RIPs affect the life cycle of human immunodeficiency viruses (HIV) including reverse transcription, integration, replication and assembly (Figure 1), not only due to their N-glycosidase activity.

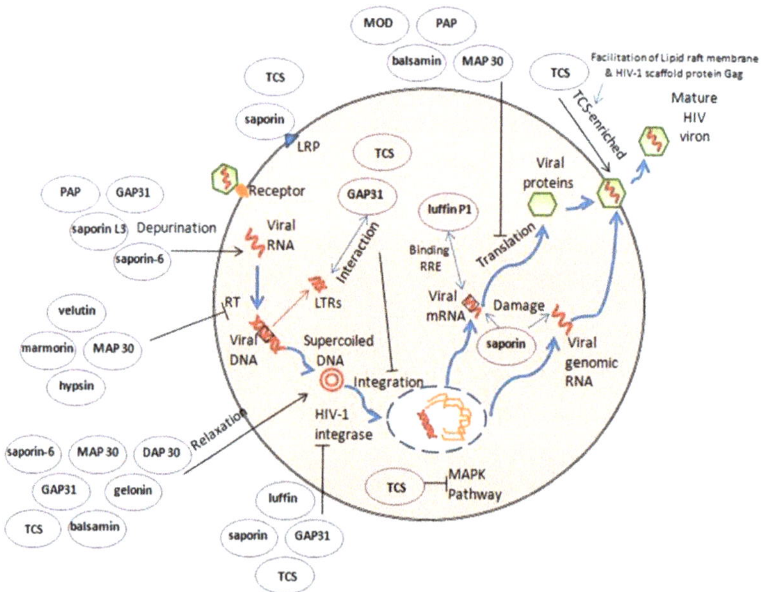

Figure 1. The cell cycle of HIV and the anti-HIV mechanism of representative RIPs. RIPs including TCS, GAP31, MAP30, PAP, marmorin and saporin can attack different steps of the life cycle of HIV and inhibit its growth. The mechanisms are not just due to rRNA depurination. TCS: trichosanthin; MAP30: momordica anti-HIV protein. GAP31: gelonium anti-HIV protein; PAP: pokeweed antiviral protein; MOD: maize ribosome-inactivating protein; DAP30: dianthus anti-HIV protein; MAPK: mitogen-activated protein kinase; HIV: human immunodeficiency virus.

Integration of the viral DNA plays a significant role in the replicative cycle of retroviruses. Saporin, TCS, GAP31 and luffin have all been reported possessing HIV-1 integrase inhibitory activities.

TCS is one of the most studied RIPs, regarding its anti-HIV activity. HIV-1 integration was also inhibited by TCS, which was attributed to the temporary interaction between TCS and HIV-1 long-terminal repeats (LTRs) [17]. Some TCS mutants without anti-HIV-1 activity still had depurination activity, which meant that N-glycosidase activity may be dissociated from the anti-HIV mechanism [18]. The anti-HIV activity of TCS was eliminated by c-Jun N-terminal kinase (JNK) inhibitor CEP-11004. Jun amino-terminal kinase (JNK) is a member of stress-activated protein kinases (SAPK), which belong to the mitogen-activated protein kinase (MAPK) family. TCS exhibited anti-HIV activity and it may through the MAPK signal pathway [19]. A similar result was found when HSV-1 infected cells were treated with TCS. The p38 MAPK and B-cell lymphoma 2 (Bcl-2) induced by HSV-1 were inhibited by TCS [20]. A singular TCS hijacking HIV-1 strategy was revealed. HIV-1 scaffold protein Gag and lipid raft membrane facilitated the formation of TCS-enriched virions. The infection ability of HIV-1 with TCS was reduced dramatically [21]. Chemokine (C-C motif) ligand 5 (RANTES) and alpha-stimulated chemotaxis by stromal cell-derived factor (SDF)-1 were significantly increased by TCS, while pertussis toxin-sensitive G proteins were activated simultaneously. The activation of chemokine receptors provoked by chemokine was strengthened by TCS [22]. The relaxed circular DNA can be broken into a linear DNA by TCS, which indicated that TCS has DNase-like activity [23], which may also be a possible mechanism. Trichobitacin isolated from the root tubers of *Trichosanthes kirilowii* transiently decreased the expression of HIV-1 p24 antigen [24].

Saporin was found to inhibit HIV integrase 3′ end processing activity (anti-HIV-1 integrase), induce viral apoptosis and suppress HIV propagation, which are unrelated to N-glycosidase activity [25]. Saporin-6 and saporin L3 exhibit classical depurination activity targeting the GAGA conserved sequence of RNA [26]. Saporin-6 was also reported to have DNA nuclease activity [27]. Isoform saporin-L1 can inhibit viral replication which may be related to the adenosine glycosidase activity on DNA, genomic RNA and mRNA [28]. However, the anti-HIV activity of saporin-6 is found independent of its RNA N-glycosidase activity, and may be related to apoptosis [25]. Like saporin, luffin also showed HIV integrase inhibitory activities on 3′ end processing and strand-transferring, which leads to anti-HIV-1 replication [29].

MAP30 (Momordica anti-HIV protein) displays DNA glycosylase activity contributing to HIV-1 integrase inhibition. Besides, MAP30, alpha- and beta-momorcharins depress HIV replication [30]. MAP30 is also able to relax supercoiled DNA [31]. MAP30 was found suppressing the expression of HIV core protein p24 and viral-related reverse transcriptase (RT) activity without cytotoxicity and cytostaticity [32]. MAP30 can assist other anti-HIV drugs including dexamethasone and indomethacin in achieving higher efficiency [30].

GAP31 (gelonium anti-HIV protein of 31 kDa) and MAP30 block the infection of HIV-1 in T lymphocytes and monocytes and viral replication [11]. They also show both anti-HIV and anti-HSV activity [33]. They manifest inhibitory activity on HIV-1 integrase attributed to the topological activity toward HIV-1 long-terminal repeats (HIV-1 LTRs) [11]. GAP31 interacts with 5′ overhanging adenosine ends, but not with blunt ends, which revealed that it acts like DNA adenosine glycosidase towards the accessible adenosine [34]. A 33-aa segment (KGATYITYVNFLNELRVKTKPEGNSHGIPSLRK) of GAP31, K10-K42, was shown to be the shortest peptide that elicits anti-HIV effect [35].

PAP (pokeweed antiviral protein) inhibits viral protein synthesis in HIV-1 infected cluster of differentiation 4 (CD4) + T cells [36]. Engineered nontoxic PAPs, FLP-102((151)AA (152)) and FLP-105((191)AA(192)), have the potency of nucleoside reverse transcriptase inhibition toward inhibitor-resistant HIV-1 with less cytotoxicity than native PAP [37]. PAP, MAP30 and GAP31 were nontoxic to human sperm, thereby they could be applied to inactivate infective viruses and virus-infected cells in semen [38].

DAPs 30 and 32 (dianthus anti-HIV proteins, 30 and 32 kDa), as well as GAP31, are able to relax supercoiled DNA and cleave double-stranded DNA into a linear one [39].

Balsamin, purified from *Momordica balsamina*, inhibits HIV-1 replication in both human T lymphocyte cell lines and human primary CD4+ T cells [40]. Balsamin is also capable of relaxing super-coiled DNA into the linear form [41].

2.2. Engineering of RIPs for Improving the Anti-HIV Efficacy

Acquired immune deficiency syndrome (AIDS) patients treated with TCS showed non-dose-related reversible mental status changes including dementia, and even coma [15]. GLQ223, an inhibitor of HIV replication, is a highly purified TCS formula to treat HIV patients with higher safety than TCS. A flu-like syndrome was the major adverse effect associated with GLQ223 [42]. Different strategies were used to remit the side-effect of RIPs, such as competitive binders and steric hindrance [43]. However, immunotoxins (ITs) are more effective because of their high specificity and selectivity. RIPs, especially PAP and ricin A chain (RTA), have been utilized to make ITs for therapeutic use [43].

Immunotoxins (ITs) are chimeric proteins that consist of RIPs or RIP fragments and moiety for targeting [43]. Targeting moiety includes antibodies, cytokines, growth factors, hormones and lectins. ITs were first designed with whole RIP linking to full length monoclonal antibody (mAb) by disulfide bond. Type 2 RIP has lectin-binding domain (B-chain) with multiple binding sites appearance maintained at a high level of non-specific internalization. To improve the performance of the IT, B-chain of type 2 RIP was removed, or its binding sites were blocked in the second generation of ITs. The third generation ITs are recombinant immunotoxins. Recombinant RIPs genetically fused to the targeting portion of mAb by a peptide linker increases homogeneity. A single chain Fv fragment (scFv) with retained targeting ability leads to smaller size of ITs, which may affect both cell penetration and clearance. Two major drawbacks of ITs are immunogenicity and vascular leak syndrome (VLS). Human or humanized antibody formats are applied to reduce immunogenicity in the fourth generation. Antigenic epitopes modification of RIPs is also applied. ITs can be used combining with other therapeutic agents to achieve synergic effect [44,45].

Most anti-HIV ITs were designed targeting the HIV envelope glycoprotein and surface antigens. Several RTA based ITs with different ligands targeting to an external envelope glycoprotein (gp120) of HIV and CD4 were tested for the anti-HIV efficacy; ligands included 0.5beta, anti-gp120 and mAb 924 [46–48]. It was later found not much improvement was observed, while anti-gp41 (mAb 7B2) together with soluble CD4 showed anti-HIV activity [46]. Pulchellin, was conjugated to mAb 924 and mAb 7B2 for recognizing gp120 and gp41; it showed similar characteristic with RTA ITs [49]. mAb 924 and 7B2 were conjugated to RTA and pulchellin by succinimidyl 6-[3(2-pyridyldithio) propionamido] hexanoate. The lysine and N-terminus on antibody and cysteine on RTA and pulchellin were involved in this conjugation [49].

RTA and Maize RIP variants were linked with HIV-1 protease recognition sequences to the C-terminal or internal inactivation region (Figure 2), which could be activated by HIV-1 protease [50,51]. Maize RIP has a 25-amino acid internal inactivation region, which is able to sterically block the interaction with ribosome. This provides a switching mechanism resulting in specific targeting to HIV-infected cells with low cytotoxicity to normal cells. The internal inactivation region was replaced by HIV-1 protease recognition site. Transcriptional activator Tat protein (TAT) sequence was fused to the N-terminus. The scheme of engineering is shown in Figure 2. Recombinant active maize RIP also exerted better anti-viral activity in vivo, with the decrease of plasma virial burden transiently in chimeric simian-human immunodeficiency virus (SHIV) 89.6-infected macaque model [52]. RTA linked with HIV-1 protease recognition sequences also exhibited a more specific activity towards HIV-1 infected cells [51].

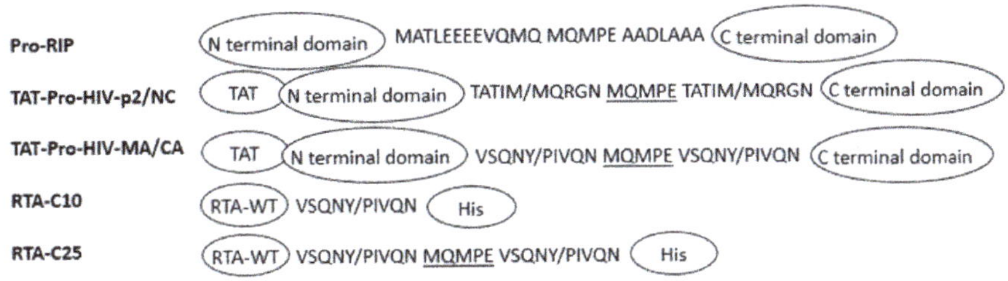

Figure 2. Schematic diagram of the RTA and maize RIP variants. The recombinant maize RIP precursor pro-RIP contains a 25 amino acids internal inactivation region. RTA: ricin A chain; HIV: human immunodeficiency virus; TAT: transcriptional activator Tat protein. TAT-Pro-HIV-P2/NC and TAT-Pro-HIV-MA/CA: N-termini of pro-RIP were fused with a TAT sequence. First and last 10 aa in internal inactivation region were replaced by the HIV-1 recognition p2/NC site (TATIM/MQRGN) and the MA/CA site (VSQNY/PIVQN), respectively. RTA-C10: MA/CA site fused to C-terminal of RTA. RTA-C25: two MA/CA sites were fused and separated by MQMPE (middle five residues of pro-RIP).

PAP was linked to mAbs targeting CD4, CD5 or CD7 antigens. The variants exerted anti-viral activity through inhibition of HIV-1 replication in HIV-1 infected CD4+ T cells and activating T cells from two asymptomatic HIV-1-seropositive patients [53].

TXU (anti-CD7)-PAP increased plasma half-life to 12.4 +/− 1.4 h and decreased systemic clearance to 2.7 +/− 0.7 mL/h/kg in adult HIV-infected patients. TXU-PAP had low toxicity. All six patients treated by 5 microg/kg dose level showed no adverse effects with viral burden reduction [16].

3. Anti-tumor activity of RIPs

3.1. Anti-Tumor Activity of Representative RIPs

TCS exerts anti-tumor activity to a wide spectrum of cancers by multiple mechanisms. The invasion, migration and epithelial-mesenchymal transition (EMT) of cervical cancer cells were inhibited, which might be relevant to the restriction of signal transducer and activator of transcription (STAT5))/C-myc signaling pathway activation by TCS. The level of B-cell lymphoma 2 (Bcl-2) and expression of antigen ki-67 associated with cellular proliferation and ribosomal RNA transcription and Phospho-c-Myc (P-C-myc) were decreased while the activation of caspase-3 was increased [54]. The apoptosis-inducing activity of TCS was attributed to the promotion of caspase-8 and caspase-9 pathways, along with the activation of caspase-3 and PARP cleavage in breast cancer cells [55]. TCS mediated Phosphoinositide 3-kinase (PI3K)/Protein kinase B (AKT) pathway and thus enhanced cytotoxicity and apoptosis-inducing activity of an anti-cancer therapy Gemcitabine against non-small cell lung cancer [56]. TCS also enhanced the cell penetration of Granzyme B leading to apoptosis of tumor cells [57]. It was shown that TCS incited autophagy in gastric cancer cells via increasing the level of autophagy protein 5 (ATG5), altering microtubule-associated protein 1A/1B-light chain 3 (LC3) I to LC3 II, inducing reactive oxygen species (ROS) and stimulating nuclear factor kappa-B (NF-κB)/Tumor protein p53 (p53) pathway [58]. ROS induction might be related to extracellular Ca^{2+}, which is involved in the apoptosis of human choriocarcinoma (JAR) cells [59]. TCS was able to inhibit angiogenesis in JAR cells through the reduction of vascular endothelial growth factor and inhibition of angiogenic signal, which contributed to the anti-cancer effect [60]. Smac (a mitochondrial protein) pathway was regulated by TCS in CaSki cervical cancer cells [61]. Low-density lipoprotein (LDL) receptor-related protein 1 (LRP1) is a receptor facilitating TCS to enter JAR cells, while no significant endocytosis of TCS was found in Hela cells [62]. Notch signal was downregulated by TCS in the nasopharyngeal carcinoma (NPC) cell line CNE2 [63]. Besides tumor cell apoptosis induction and antiproliferation ability, the host immune system mediated by TCS might be another pathway for inhibition. T cells such as interferon-gamma (IFN-γ) producing CD4(+) and CD8(+) T cells, were increased by TCS in the 3LL Lewis lung carcinoma tumor model.

TCS had upregulatory activity towards the expression of tumor suppressor in lung cancer 1 (TSLC1) and class I-restricted T cell-associated molecule (CRTAM) [64]. TCS exhibited antiproliferative activity on leukemia and lymphoma, which attributed to the induction of T-lymphocyte cell apoptosis and inhibition of B-lymphocyte cell growth by S-phase cell cycle arrest [65].

Ricin also exhibits an anti-tumor property. Ricin inhibited the growth of sarcomas in rats [66], and it increased the survival rate of Ehrlich ascites tumor-bearing mice [67]. It also shows promising effect on nude mice with human xenograft [68]. A phase I clinical study on 54 cancer patients with different kinds of tumors was taken and thus confirmed its properties [69]. The inhibition of protein synthesis was first considered attributing its anti-tumor activity. Cell apoptosis and the secretion of cytokine inflammatory mediators were shown to be the related mechanisms [70,71] (Figure 3). The treatment of ricin caused the activation of p38 and jun-N-terminal kinases (JNKs) [72]. Phosphoinositide 3-kinase (PI3K) and Janus kinase 2 (JAK2) were also involved in the activation of RAW264.7 mouse macrophage cells treated by ricin toxin-binding subunit B [73]. Ricin caused proinflammatory responses on human airway cells, which was related to stress-activated protein kinases and nuclear factor kappa-B (NF-kappaB) [74]. Both two main pathways, extrinsic (receptor mediated) and intrinsic (mitochondrial pathway), were involved in the ricin mediated cell apoptosis, following the activation of poly (ADP-ribose) polymerase (PARP) [75,76]. In addition, rapid release of cytochrome c was observed in ricin treated cells [75]. Ricin has been shown to induce the secretion of proinflammatory cytokines mediator such as tumor necrosis factor alpha (TNF-α) and interleukin-1 beta (IL-1β) [77,78].

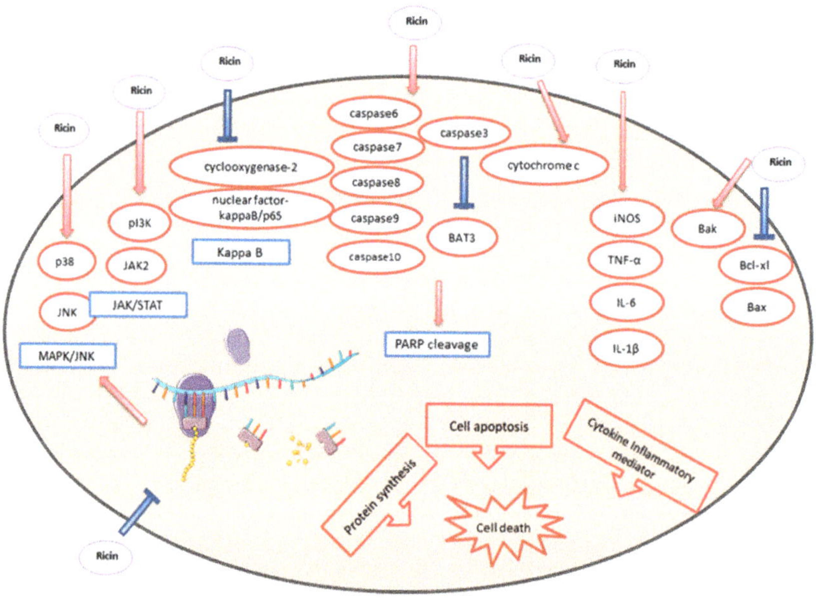

Figure 3. Ricin-induced cell death in the anti-tumor process. Arrows represent the activation of receptors and blunt arrows represent inhibition of receptors. Pathways involved are stated in blue boxes. The inhibition of protein synthesis, cell apoptosis and the release of cytokine inflammatory mediators are considered as the possible mechanism. p38: p38 mitogen-activated protein kinases; JNK: c-Jun N-terminal kinase; MAPK: mitogen-activated protein kinase; PI3K: phosphoinositide 3-kinases; JAK2: Janus kinase 2; STAT: signal transducer and activator of transcription; PARP: poly ADP-ribose polymerase; iNOS: inducible nitric oxide synthase; TNF-α: tumor necrosis factor-α; IL: Interleukin; Bcl-xl: B-cell lymphoma-extra large; Bax: Bcl-2-associated X protein.

Riproximin is a type 2 RIP that up-regulated the anti-cancer cytokine IL24/MDA-7 and ER-stress-related GADD genes; it also down-regulated the genes relating to migration (RHO GTPases), anti-apoptotic activities (BCL family), and cell cycle (cyclins) in selected human breast cancer cells MDA-MB-231 and MCF-7 [79]. Similar results were confirmed by using selected human and rat colorectal cancer (CRC) cell lines [80].

The anti-cancer effect of α-MMC was tested in human breast cancer cells MDA-MB-231 and MCF-7, but the relatively high cytotoxicity limited its therapeutic use [81]. After treated with α-MMC, cytochrome c was released, and intracellular free calcium concentration was increased, and calcium overloading led to cell death [82]. c-Jun N-terminal kinases (JNKs) signal pathway, which relates to cell apoptosis, was also triggered by α-MMC [83]. Many studies demonstrated that α-MMC involved in similar pathways with TCS, such as caspase-3 and 9 activations and interaction with low density lipoprotein receptor-related protein 1 (LRP1) [83,84]. LRP1 plays a vital role in the cytotoxicity mechanism of α-MMC because α-MMC mediated cytokine expression and MAPK pathway, which would be hindered by LRP1 silencing. α-MMC inhibited immune response through the inhibition of IL-1β, IL-2, IL-8, IL-9, IL-12, MIP-1α/β, MCP-1 and elevated the expression of IL-1ra and RANTES in human monocyte THP-1 cells. The regulation of cytokine release by α-MMC revealed that α-MMC might be applied to treat tumor-associated macrophages (M2 subtypes) [85].

Curcin, a type 1 RIP, could inhibit the growth of several tumor cell lines at 5 μg/mL, such as NCL-H446, SGC-7901 and S180 [86]. Curcin C, which shares highly conserved sequence with curcin, elicited inhibitory activity against the osteosarcoma cell line U20S with the half maximal inhibitory concentration (IC50) value 0.019 μM when IC50 of curcin is 0.27 μM [81].

Viscumarticulatum RIP (Articulatin-D), one of the mistletoe RIPs, could selectively inhibit acute T-cell leukemia. Caspase-8 and -3 were also involved. Early signals of apoptosis induction of Articulatin-D are exposure of phosphatidylserine and increased level of mitochondrial membrane potential [87]. Aviscumine and its native form mistletoe lectin-I increased the amount of cancer cell-specific T-cells resulting in more T-cell-mediated tumor cell lysis in a mouse glioma model. The level of the proteins associated with immune response was increased [88].

3.2. Engineering of RIPs for Improving the Anti-Cancer Efficacy

Most RIPs ITs have shown anti-cancer potential, particularly for hematological malignancies, which are easier accessed than solid tumors [45]. The presence of clusters of differentiation (CD) on hematological cells surface are considered as ideal targets for better design of ITs [89].

Sap-SO6 (the main isoform of saporin) was linked to CD2, CD7, CD19 and CD22 antigens found on human leukemia and lymphoma plasma membrane surface to make ITs [90]. The saporin ITs generated increased selective cytotoxicity at least 100-fold more than saporin alone [91]. Anti CD30-Saporin was reported reducing 60% tumor mass when used to treated refractory Hodgkin lymphoma patients [86]. However, it had transient hepatotoxicity when a single dose up to 0.8 mg/kg was applied [86].

RTA was also used in constructing ITs; some recent studies are reviewed below. A preliminary clinical study found that BCMab1-Ra, an IT consisting of RTA and BCMab1 (a novel monoclonal antibody that specifically recognized the aberrantly glycosylated Integrin a3b1 in bladder cancer), cured a patient with multiple tumors on the bladder and achieved no tumor recurrence in 3 years [92]. RTA conjugated with anti-HER2 scFv 4D5 and the endoplasmic reticulum-targeting peptide KDEL had 440-fold increase in anti-ovarian cancer cell activity compared to RTA alone. The specificity of this IT RTA-4D5-KDEL to HER2 was high so the toxicity to normal cells was low [93]. Combotox, a 1:1 combination of anti-CD19 and anti-CD22 immunotoxins, was conjugated to deglycosylated RTA, which showed higher efficacy than either IT in pediatric precursor B-lineage acute lymphoblastic leukemia (pre-B ALL) [94].

Besides plant RIPs, bacterial-originated toxins such as Pseudomonas exotoxin (PE) and diphtheria toxin (DT) were also used in ITs. Denileukin diftitox (Ontak) was the first immunotoxin approved by the Food and Drug Administration (FDA), which consists of Interleukin-2 ligand and DT [95]. A number of PE-based ITs have been under clinical trials. A recent example is the antimesothelin immunotoxin SS1(dsFv)PE38 (SS1P), which is the combination of PE38 (a modified Pseudomonas exotoxin A) and a murine antimesothelin variable antibody fragment. SS1P displayed high activity against malignant pleural mesothelioma in phase I clinical trial [96]. To achieve higher efficacy, researchers applied a tumor-seeking bacterial system by engineering *Salmonella typhimurium* to make it selectively expressed and released TGFα-PE38 (transforming growth factor alpha-PE38). The released TGFα-PE38 was then tested in mice with implanted colon or breast tumor cells, which expressed high levels of EGFR (epidermal growth factor receptor). Lower solid tumor growth rate was shown comparing to just intracerebral infusion of TGFα-PE38 [97].

Vascular leak syndrome is a major side effect of many RIP-based ITs. Ricin and T22, a ligand of the cell surface marker C-X-C motif chemokine receptor type 4 (CXCR-4), were assembled to form nanostructures, which exhibited specific anti-tumor activity and avoided VLS [98].

Besides targeting CD antigen, cell penetrate peptide (CPP) was adopted to improve specificity towards cancer cells. TCS fused with heparin-binding domain (HBD), a human derived cell-penetrating peptide CPP, could increase the apoptosis rate of HeLa cells when compared with treated TCS alone [99]. It offered an efficient delivery to cancer cells.

A co-delivery system of TCS and albendazole (ABZ) inhibited drug-resistant tumor cells (A549/T and HCT8/ADR) proliferation and tumor metastasis [100]. ABZ was covered by albumin-coated silver nanoparticles linked with low-molecular-weight protamine (a CPP) modified TCS; it could impair cytoskeleton.

4. Challenges in Therapeutic Applications

The immunogenicity of RIPs is an obstacle in usage. Although RIPs exhibit immunosuppressive activity [101], plant-originated RIPs readily stimulated immune system of patients, even causing allergic symptom [102]. Furthermore, the short plasma half-life of RIPs reduced drug exposure to targets, thus limiting clinical application. The plasma half-life of wild-type TCS was 9 min [103]. Repeated administration is needed to preserve the adequate level owing to renal insufficiency of small molecular weight RIPs [103]. However, repeating administration caused strong immune reaction [102]. Another side effect is neurotoxicity. TCS was examined without direct toxicity to the central nervous system (CNS). However, HIV-infected patients treated by TCS were reported to have adverse CNS reactions [104]. The HIV-infected macrophages might be altered by TCS treatment, which aggravated neurological symptoms [104]. Solid tumor mass is hard to access, leading to reduced efficacy. Intracavitary therapy with ITs might solve this problem [105]. Several studies have been carried out to enhance the pharmacological properties of RIPs.

5. Coupling with Polymer Polyethylene glycerol (PEG) and Dextran

PEGylation is a common strategy used to improve drug performance. PEG is biocompatible with high hydrophilicity, low toxicity and non-immunogenicity [106]. After coupling, the size and molecular weight are increased to avoid rapid renal clearance and proteolytic degradation for longer half-life. Antigenicity and immunogenicity can be decreased, while pharmacokinetics and pharmacodynamics can be improved. Moreover, permeation retention effect was increased by PEGylation, which helped to target tumoral tissues [107]. Non-specific PEGylation was applied at first. Although reactive amino acids such as cysteine, arginine and serine can be chosen [108], site-directed conjugation has provided better achievements. The antigenic sites are mapped and then antigenicity can be alleviated through PEGylation, while the original enzymatic activity could be least affected. For successful conjugation, a site extending from the protein surface is preferred [109]. Many RIPs PEGylations were attempted to advance their pharmacological properties (Table 1).

Table 1. Representative PEGylated RIPs.

RIP	Type	PEG	Disease	Site	Ref.
Gelonin (GAP31)	1	methoxypoly(ethylene glycol) (mPEG)2k/mPEG5k/mPEG succinimidyl succinate20k (SS-20PEG)	Tumor and HIV	Random	[110]
Alpha-Momorcharin (α-MMC)	1	mPEG-succinimidyl carbonate (mPEG-SC)10k	Tumor	N-terminal	[111]
α-MMC	1	20 kDa (mPEG)2-Lys-NHS	Tumor	Mono-, di-, tri-PEGylated	[112], [113], [114], [115]
α-MMC	1	20 kDa mPEG-butyraldehyde (mPEG-ALD)	/	N-terminal	[116]
Momordica anti-HIV protein (MAP30)	1	mPEG-SC10k	Tumor	N-terminal	[111]
MAP30	1	20 kDa (mPEG)2-Lys-NHS	Tumor	Mono-, di-, tri-PEGylated	[113]
Trichosanthin (TCS)	1	PEG5k	/	YFF81-83ACS/ KR173-174CG/[YFF81-83ACS, KR173-174CG]	[117]
TCS	1	PEG5k	HIV-1	Q219C/K173C/S7C	[118]
TCS	1	PEG5k/PEG20k	/	Q219C/K173C/S7C/[K173C, Q219C](KQ)	[119]
Ricin A Chain (RTA)	2	PEG2k/mPEG2k	/	Random	[99]
RTA	2	Monomethoxy-PEG hydrazide (mPEG-HZ)5k/monomethoxy-PEG succinimidyl propionate (mPEG-SPA)5k	Tumor	Carbohydrate/Amine-specifically	[120]
Maize RIP (MOD)	3	PEG5k/PEG20k	HIV, Chinese rhesus macaques	K78C/K264C	[109]

Gelonin (GAP31) was covalently coupled to methoxypoly (ethylene glycol) (mPEG) 2000, mPEG 5000 and mPEG succinimidyl succinate 20K (SS-20PEG). mPEG does not affect the positive charge of protein, while charges alteration may result in lowering biological activity. The plasma half-life time of all conjugations above was increased. PEGylation also decreased organ uptake. Coupled of mPEG retained immunogenicity, while SS-20PEG conjugated decreased cytotoxicity [110].

Site-directed PEGylation of TCS showed that PEG 20,000 is better than PEG 5000. Coupled to PEG 20000, the plasma half-life extended due to the enlarged size and resistance to proteolysis. Immunoglobulin G (IgG) level was also reduced because of decrease in immunogenicity [118]. The IgE level was reduced but the IgG level was maintained when conjugated to PEG 5000 [109,118]. Because tremendous liver uptake is through carbohydrate-mediated recognition, carbohydrate-directed PEGylation can improve the pharmacokinetics of RTA. Meanwhile, the circulation time was increased and antigenicity was reduced by masking the epitopes. This makes carbohydrate-directed PEG-RTA conjugate a potential anti-tumor drug [120]. Coupling PEG to maize RIP (MOD) gave similar results, including pharmacokinetics improvement and antigenicity reduction [109]. Mono-, di-, tri-PEGlyated α-MMC and MAP30 dramatically reduced the immunogenicity and maintained biological activities [112,113]. The enzymatic activity assay showed that mono-PEGylated α-MMC strongly inhibited the growth of human cervix adenocarcinoma cells [116]. Sun et al. provided a method to isolate high purity mono-mPEGylated MAP30 and α-MMC, which can be useful for further RIP drug examination [111]. In vivo study implied that the hepatic toxicity of α-MMC was reduced after PEGylation [115]. The plasma half-life of α-MMC was sharply increased from 6.2–7.5 h to 52–87 h [114]. The homogeneity of PEGlyated α-MMC was further improved by site-specific conjugation of mPEG-ALD [116]. Due to steric hindrance of active sites, PEGylation may lead to biological activity reduction, which can be compensated by longer plasma half-life [119].

Coupling dextran on RIPs was also studied to improve their performance on anti-cancer and anti-HIV. Similar to PEGylation, dextran can increase circulation time and reduce IgG or IgE responses [103,121,122]. Dextran may be used as a linker to connect monoclonal antibodies and RTA, and to improve selectivity toward target cells [123]. A lower rate of plasma clearance prolongs the plasma half-life of TCS after coupling to dextran T40 by a dialdehyde method [103]. The toxicity and potency were decreased [124]. TCS coupled with dextran had the IgE level reduced eight times compared to wild-type [121] To reduce the antigenicity, TCS was also coupled by bromodextran T20 [125]. To obtain better conjugate, potential antigenic site K173 of TCS was mutated to cysteine and coupled with dextran. Site-directed conjugate dextran-K173C had the hypersensitivity reaction and the level of IgG andIgE decreased [126].

6. Perspectives

The therapeutic applications of RIPs, in particular anti-HIV and anti-tumor potential were exploited in the last several decades [43]. The main obstacles of utilization include short plasma half-life, non-selective cytotoxicity and antigenicity. Most PEGylation increases plasma time and reduces antigenicity. RIPs conjugated with antibodies to form immunotoxins increase the selective toxicity to target cells.

Immunotherapy can assist chemotherapy to improve efficacy [127]. An IT is co-applied with a small molecular drug or another IT to achieve high efficacy and reduce side effects. Anti-CD3 and anti-CD7 were conjugated separately to RTA and mixed to treat steroid-refractory acute graft-versus-host-disease. The mixture exerted higher efficacy than one alone with low side effects in clinical trial I/II [128]. Saponin, which is generally classified as steroidal or triterpenoidal, can act as an endo/lysosomal escape enhancer. It was commonly used together with type I RIP to facilitate escape of RIP from endo/lysosomal degradation in order to increase efficacy of RIP or ITs [43].

"Cocktail therapy" is an effective strategy to increase the effectiveness of ITs [129]. The synergistic effect can be achieved when two or even more suitable ITs corporate. Lung cancer cells can easily generate resistance against tumor necrosis factor-related apoptosis-inducing ligand (TRAIL) [130]. TCS was found inducing TRAIL sensitivity of non-small cell lung cancer (NSCLC) by regulating death receptors and proteins involved in invasion and cell cycle [130].

A number of RIP-derived drugs reached clinical trial, but then failed due to severe side effects and little efficacy (Table 2). At present, an RIP diphtheria toxin (DT) derived drug denileukin diftitox (Ontak) has been approved by FDA for treating cutaneous T-cell lymphoma (CTCL).

Table 2. Recent (after 2000) representative ricin & RTA immunotoxins tested.

Immunotoxin	Toxin	Ligand	Target Antigen	Disease	Clinical Trial Status	Reason for Suspension	Ref.
anti-B4-bR	ricin	Anti-B4	CD19	B-cell lymphoma	III	no differences between event-free survival and overall survival	[14]
N901-bR	ricin	N901	CD56	small-cell lung cancer	II	vascular leak syndrome	[131]
Anti-CEA-bR	ricin	I-1	carcinoembryonic antigen	hepatic metastases	III	no obvious changes in the growth rate of injected lesions	[132]
Ki-4.dgA	RTA	Ki-4	CD30	refractory CD30+ Hodgkin's and non-Hodgkin's lymphoma	I	vascular leak syndrome; low tolerance	[133, 134]

With the rapid development of immunotoxin engineering technology, it is possible to acquire RIP-based drugs with high efficacy and low side effects. A list of endosomal escape enhancers such as saponin, TAT (transcriptional activator Tat protein), perforin and ricin B-chain were observed to facilitate RIP [43]. mAbs help increase the specificity; adequate combination of ITs with different mAbs can achieve better curative effect [96]. Novel technology has been tested, such as photochemical internalization (PCI). PCI is a light-based method and is employed to trigger the endosomal escape of RIP. Saporin linked with a photosensitizer functionalized CPP showed cytotoxicity augmentation in MC28 fibrosarcoma cells [135]. Another study also showed that the cytotoxicity of IT 225.28-saporin was strengthened by using PCI with a photosensitizer disulfonated tetraphenyl chlorin (TPCS2a) [136]. TCS was conjugated with an albumin-binding domain and a legumain-substrate peptide as a modified IT for better delivery efficiency, which can make use of the nutrient transporter pathway of albumin-binding proteins. Protease legumain at the tumor sites can dissociate TCS from an albumin-binding domain, which gives a new strategy for tumor therapy [137]. These new research findings provide RIP therapy a promising future.

Author Contributions: Conceptualization, P.-C.S.; writing, Z.-N.Z. and J.-Q.L.; review and editing: P.-C.S. and Y.-T.Z. All authors have read and agreed to the published version of the manuscript.

Funding: This research was funded by The Chinese University of Hong Kong: 3110130, The Research Grants Council of Hong Kong SAR: 14176617.

Conflicts of Interest: The authors declare no conflict of interest.

References

1. Olsnes, S. The history of ricin, abrin and related toxins. *Toxicon* **2004**, *44*, 361–370. [CrossRef] [PubMed]
2. De Virgilio, M.; Lombardi, A.; Caliandro, R.; Fabbrini, M.S. Ribosome-inactivating proteins: From plant defense to tumor attack. *Toxins (Basel)* **2010**, *2*, 2699–2737. [CrossRef] [PubMed]
3. Nielsen, K.; Boston, R.S. Ribosome-Inactivating Proteins: A Plant Perspective. *Annu. Rev. Plant. Physiol. Plant. Mol. Biol.* **2001**, *52*, 785–816. [CrossRef] [PubMed]
4. Zhu, F.; Zhou, Y.K.; Ji, Z.L.; Chen, X.R. The Plant Ribosome-Inactivating Proteins Play Important Roles in Defense against Pathogens and Insect Pest Attacks. *Front. Plant. Sci.* **2018**, *9*, 146. [CrossRef]
5. Hull, R. *Plant. virology*, 5th ed.; Elsevier/AP: Lodon, UK; Waltham, MA, USA, 2014; 1104p.
6. Choi, A.K.; Wong, E.C.; Lee, K.M.; Wong, K.B. Structures of eukaryotic ribosomal stalk proteins and its complex with trichosanthin, and their implications in recruiting ribosome-inactivating proteins to the ribosomes. *Toxins (Basel)* **2015**, *7*, 638–647. [CrossRef]
7. Weng, A.; Thakur, M.; von Mallinckrodt, B.; Beceren-Braun, F.; Gilabert-Oriol, R.; Wiesner, B.; Eichhorst, J.; Bottger, S.; Melzig, M.F.; Fuchs, H. Saponins modulate the intracellular trafficking of protein toxins. *J. Control. Release* **2012**, *164*, 74–86. [CrossRef]
8. Li, X.D.; Chen, W.F.; Liu, W.Y.; Wang, G.H. Large-scale preparation of two new ribosome-inactivating proteins–cinnamomin and camphorin from the seeds of Cinnamomum camphora. *Protein Expr. Purif.* **1997**, *10*, 27–31. [CrossRef]
9. Helmy, M.; Lombard, S.; Pieroni, G. Ricin RCA60: Evidence of its phospholipase activity. *Biochem. Biophys. Res. Commun.* **1999**, *258*, 252–255. [CrossRef]
10. Wang, S.; Li, Z.; Li, S.; Di, R.; Ho, C.-T.; Yang, G. Ribosome-inactivating proteins (RIPs) and their important health promoting property. *RSC Adv.* **2016**, *6*, 16791–16805. [CrossRef]
11. Lee-Huang, S.; Huang, P.L.; Huang, P.L.; Bourinbaiar, A.S.; Chen, H.C.; Kung, H.F. Inhibition of the integrase of human immunodeficiency virus (HIV) type 1 by anti-HIV plant proteins MAP30 and GAP31. *Proc. Natl. Acad. Sci. USA* **1995**, *92*, 8818–8822. [CrossRef]
12. Stirpe, F.; Battelli, M.G. Ribosome-inactivating proteins: Progress and problems. *Cell Mol. Life Sci.* **2006**, *63*, 1850–1866. [CrossRef] [PubMed]
13. Borthakur, G.; Rosenblum, M.G.; Talpaz, M.; Daver, N.; Ravandi, F.; Faderl, S.; Freireich, E.J.; Kadia, T.; Garcia-Manero, G.; Kantarjian, H.; et al. Phase 1 study of an anti-CD33 immunotoxin, humanized monoclonal antibody M195 conjugated to recombinant gelonin (HUM-195/rGEL), in patients with advanced myeloid malignancies. *Haematologica* **2013**, *98*, 217–221. [CrossRef] [PubMed]

14. Furman, R.R.; Grossbard, M.L.; Johnson, J.L.; Pecora, A.L.; Cassileth, P.A.; Jung, S.H.; Peterson, B.A.; Nadler, L.M.; Freedman, A.; Bayer, R.L.; et al. A phase III study of anti-B4-blocked ricin as adjuvant therapy post-autologous bone marrow transplant: CALGB 9254. *Leuk. Lymphoma* **2011**, *52*, 587–596. [CrossRef] [PubMed]
15. Byers, V.S.; Levin, A.S.; Waites, L.A.; Starrett, B.A.; Mayer, R.A.; Clegg, J.A.; Price, M.R.; Robins, R.A.; Delaney, M.; Baldwin, R.W. A phase I/II study of trichosanthin treatment of HIV disease. *AIDS* **1990**, *4*, 1189–1196. [CrossRef]
16. Uckun, F.M.; Bellomy, K.; O'Neill, K.; Messinger, Y.; Johnson, T.; Chen, C.L. Toxicity, biological activity, and pharmacokinetics of TXU (anti-CD7)-pokeweed antiviral protein in chimpanzees and adult patients infected with human immunodeficiency virus. *J. Pharmacol. Exp. Ther.* **1999**, *291*, 1301–1307.
17. Zhao, W.L.; Feng, D.; Wu, J.; Sui, S.F. Trichosanthin inhibits integration of human immunodeficiency virus type 1 through depurinating the long-terminal repeats. *Mol. Biol. Rep.* **2010**, *37*, 2093–2098. [CrossRef]
18. Ouyang, D.Y.; Chan, H.; Wang, Y.Y.; Huang, H.; Tam, S.C.; Zheng, Y.T. An inhibitor of c-Jun N-terminal kinases (CEP-11004) counteracts the anti-HIV-1 action of trichosanthin. *Biochem. Biophys. Res. Commun.* **2006**, *339*, 25–29. [CrossRef]
19. Bodmer, D.; Gloddek, B.; Ryan, A.F.; Huverstuhl, J.; Brors, D. Inhibition of the c-Jun N-terminal kinase signaling pathway influences neurite outgrowth of spiral ganglion neurons in vitro. *Laryngoscope* **2002**, *112*, 2057–2061. [CrossRef]
20. Huang, H.; Chan, H.; Wang, Y.Y.; Ouyang, D.Y.; Zheng, Y.T.; Tam, S.C. Trichosanthin suppresses the elevation of p38 MAPK, and Bcl-2 induced by HSV-1 infection in Vero cells. *Life Sci.* **2006**, *79*, 1287–1292. [CrossRef]
21. Zhao, W.L.; Zhang, F.; Feng, D.; Wu, J.; Chen, S.; Sui, S.F. A novel sorting strategy of trichosanthin for hijacking human immunodeficiency virus type 1. *Biochem. Biophys. Res. Commun.* **2009**, *384*, 347–351. [CrossRef]
22. Zhao, J.; Ben, L.H.; Wu, Y.L.; Hu, W.; Ling, K.; Xin, S.M.; Nie, H.L.; Ma, L.; Pei, G. Anti-HIV agent trichosanthin enhances the capabilities of chemokines to stimulate chemotaxis and G protein activation, and this is mediated through interaction of trichosanthin and chemokine receptors. *J. Exp. Med.* **1999**, *190*, 101–111. [CrossRef] [PubMed]
23. Li, M.; Yeung, H.; Pan, L.; Chan, S. Trichosanthin, a potent HIV-1 inhibitor, can cleave supercoiled DNA in vitro. *Nucleic Acids Res.* **1991**, *19*, 6309–6312. [CrossRef] [PubMed]
24. Zheng, Y.T.; Ben, K.L.; Jin, S.W. Anti-HIV-1 activity of trichobitacin, a novel ribosome-inactivating protein. *Acta Pharmacol. Sin.* **2000**, *21*, 179–182. [PubMed]
25. Yadav, S.K.; Batra, J.K. Mechanism of Anti-HIV Activity of Ribosome Inactivating Protein, Saporin. *Protein Pept. Lett.* **2015**, *22*, 497–503. [CrossRef] [PubMed]
26. Yuan, H.; Du, Q.; Sturm, M.B.; Schramm, V.L. Soapwort Saporin L3 Expression in Yeast, Mutagenesis, and RNA Substrate Specificity. *Biochemistry* **2015**, *54*, 4565–4574. [CrossRef]
27. Roncuzzi, L.; Gasperi-Campani, A. DNA-nuclease activity of the single-chain ribosome-inactivating proteins dianthin 30, saporin 6 and gelonin. *FEBS Lett.* **1996**, *392*, 16–20. [CrossRef]
28. Barbieri, L.; Valbonesi, P.; Righi, F.; Zuccheri, G.; Monti, F.; Gorini, P.; Samori, B.; Stirpe, F. Polynucleotide:Adenosine glycosidase is the sole activity of ribosome-inactivating proteins on DNA. *J. Biochem.* **2000**, *128*, 883–889. [CrossRef]
29. Au, T.K.; Collins, R.A.; Lam, T.L.; Ng, T.B.; Fong, W.P.; Wan, D.C. The plant ribosome inactivating proteins luffin and saporin are potent inhibitors of HIV-1 integrase. *FEBS Lett.* **2000**, *471*, 169–172. [CrossRef]
30. Puri, M.; Kaur, I.; Kanwar, R.K.; Gupta, R.C.; Chauhan, A.; Kanwar, J.R. Ribosome inactivating proteins (RIPs) from Momordica charantia for anti viral therapy. *Curr. Mol. Med.* **2009**, *9*, 1080–1094. [CrossRef]
31. Wang, Y.X.; Neamati, N.; Jacob, J.; Palmer, I.; Stahl, S.J.; Kaufman, J.D.; Huang, P.L.; Huang, P.L.; Winslow, H.E.; Pommier, Y.; et al. Solution structure of anti-HIV-1 and anti-tumor protein MAP30: Structural insights into its multiple functions. *Cell* **1999**, *99*, 433–442. [CrossRef]
32. Lee-Huang, S.; Huang, P.L.; Nara, P.L.; Chen, H.C.; Kung, H.F.; Huang, P.; Huang, H.I.; Huang, P.L. MAP 30: A new inhibitor of HIV-1 infection and replication. *FEBS Lett.* **1990**, *272*, 12–18. [CrossRef]
33. Bourinbaiar, A.S.; Lee-Huang, S. The activity of plant-derived antiretroviral proteins MAP30 and GAP31 against herpes simplex virus in vitro. *Biochem. Biophys. Res. Commun.* **1996**, *219*, 923–929. [CrossRef]

34. Li, H.G.; Huang, P.L.; Zhang, D.; Sun, Y.; Chen, H.C.; Zhang, J.; Huang, P.L.; Kong, X.P.; Lee-Huang, S. A new activity of anti-HIV and anti-tumor protein GAP31: DNA adenosine glycosidase–structural and modeling insight into its functions. *Biochem. Biophys. Res. Commun.* **2010**, *391*, 340–345. [CrossRef] [PubMed]
35. Lee-Huang, S.; Kung, H.F.; Huang, P.L.; Bourinbaiar, A.S.; Morell, J.L.; Brown, J.H.; Huang, P.L.; Tsai, W.P.; Chen, A.Y.; Huang, H.I.; et al. Human immunodeficiency virus type 1 (HIV-1) inhibition, DNA-binding, RNA-binding, and ribosome inactivation activities in the N-terminal segments of the plant anti-HIV protein GAP31. *Proc. Natl. Acad. Sci. USA* **1994**, *91*, 12208–12212. [CrossRef] [PubMed]
36. Zarling, J.M.; Moran, P.A.; Haffar, O.; Sias, J.; Richman, D.D.; Spina, C.A.; Myers, D.E.; Kuebelbeck, V.; Ledbetter, J.A.; Uckun, F.M. Inhibition of HIV replication by pokeweed antiviral protein targeted to CD4+ cells by monoclonal antibodies. *Nature* **1990**, *347*, 92–95. [CrossRef] [PubMed]
37. Uckun, F.M.; Rajamohan, F.; Pendergrass, S.; Ozer, Z.; Waurzyniak, B.; Mao, C. Structure-based design and engineering of a nontoxic recombinant pokeweed antiviral protein with potent anti-human immunodeficiency virus activity. *Antimicrob. Agents Chemother.* **2003**, *47*, 1052–1061. [CrossRef]
38. D'Cruz, O.J.; Uckun, F.M. Pokeweed antiviral protein: A potential nonspermicidal prophylactic antiviral agent. *Fertil. Steril.* **2001**, *75*, 106–114. [CrossRef]
39. Huang, P.L.; Chen, H.C.; Kung, H.F.; Huang, P.L.; Huang, P.; Huang, H.I.; Lee-Huang, S. Anti-HIV plant proteins catalyze topological changes of DNA into inactive forms. *Biofactors* **1992**, *4*, 37–41.
40. Kaur, I.; Puri, M.; Ahmed, Z.; Blanchet, F.P.; Mangeat, B.; Piguet, V. Inhibition of HIV-1 replication by balsamin, a ribosome inactivating protein of Momordica balsamina. *PLoS ONE* **2013**, *8*, e73780. [CrossRef]
41. Ajji, P.K.; Sonkar, S.P.; Walder, K.; Puri, M. Purification and functional characterization of recombinant balsamin, a ribosome-inactivating protein from Momordica balsamina. *Int. J. Biol. Macromol.* **2018**, *114*, 226–234. [CrossRef]
42. Kahn, J.O.; Gorelick, K.J.; Gatti, G.; Arri, C.J.; Lifson, J.D.; Gambertoglio, J.G.; Bostrom, A.; Williams, R. Safety, activity, and pharmacokinetics of GLQ223 in patients with AIDS and AIDS-related complex. *Antimicrob. Agents Chemother.* **1994**, *38*, 260–267. [CrossRef] [PubMed]
43. Gilabert-Oriol, R.; Weng, A.; von Mallinckrodt, B.; Melzig, M.F.; Fuchs, H.; Thakur, M. Immunotoxins Constructed with Ribosome-Inactivating Proteins and their Enhancers: A Lethal Cocktail with Tumor Specific Efficacy. *Curr. Pharm. Design* **2014**, *20*, 6584–6643. [CrossRef] [PubMed]
44. Rust, A.; Partridge, L.J.; Davletov, B.; Hautbergue, G.M. The Use of Plant-Derived Ribosome Inactivating Proteins in Immunotoxin Development: Past, Present and Future Generations. *Toxins (Basel)* **2017**, *9*, 344. [CrossRef] [PubMed]
45. Akbari, B.; Farajnia, S.; Ahdi Khosroshahi, S.; Safari, F.; Yousefi, M.; Dariushnejad, H.; Rahbarnia, L. Immunotoxins in cancer therapy: Review and update. *Int. Rev. Immunol.* **2017**, *36*, 207–219. [CrossRef]
46. Pincus, S.H.; Fang, H.; Wilkinson, R.A.; Marcotte, T.K.; Robinson, J.E.; Olson, W.C. In vivo efficacy of anti-glycoprotein 41, but not anti-glycoprotein 120, immunotoxins in a mouse model of HIV infection. *J. Immunol.* **2003**, *170*, 2236–2241. [CrossRef]
47. Matsushita, S.; Koito, A.; Maeda, Y.; Hattori, T.; Takatsuki, K. Selective killing of HIV-infected cells by anti-gp120 immunotoxins. *AIDS Res. Hum. Retrovir.* **1990**, *6*, 193–203. [CrossRef]
48. Pincus, S.H.; Wehrly, K.; Cole, R.; Fang, H.; Lewis, G.K.; McClure, J.; Conley, A.J.; Wahren, B.; Posner, M.R.; Notkins, A.L.; et al. In vitro effects of anti-HIV immunotoxins directed against multiple epitopes on HIV type 1 envelope glycoprotein 160. *AIDS Res. Hum. Retrovir.* **1996**, *12*, 1041–1051. [CrossRef]
49. Sadraeian, M.; Guimaraes, F.E.G.; Araujo, A.P.U.; Worthylake, D.K.; LeCour, L.J.; Pincus, S.H. Selective cytotoxicity of a novel immunotoxin based on pulchellin A chain for cells expressing HIV envelope. *Sci. Rep.* **2017**, *7*, 7579. [CrossRef]
50. Law, S.K.; Wang, R.R.; Mak, A.N.; Wong, K.B.; Zheng, Y.T.; Shaw, P.C. A switch-on mechanism to activate maize ribosome-inactivating protein for targeting HIV-infected cells. *Nucleic Acids Res.* **2010**, *38*, 6803–6812. [CrossRef]
51. Au, K.Y.; Wang, R.R.; Wong, Y.T.; Wong, K.B.; Zheng, Y.T.; Shaw, P.C. Engineering a switch-on peptide to ricin A chain for increasing its specificity towards HIV-infected cells. *Biochim. Biophys. Acta* **2014**, *1840*, 958–963. [CrossRef]
52. Wang, R.R.; Au, K.Y.; Zheng, H.Y.; Gao, L.M.; Zhang, X.; Luo, R.H.; Law, S.K.; Mak, A.N.; Wong, K.B.; Zhang, M.X.; et al. The recombinant maize ribosome-inactivating protein transiently reduces viral load in SHIV89.6 infected Chinese Rhesus Macaques. *Toxins (Basel)* **2015**, *7*, 156–169. [CrossRef] [PubMed]

53. Domashevskiy, A.V.; Goss, D.J. Pokeweed antiviral protein, a ribosome inactivating protein: Activity, inhibition and prospects. *Toxins (Basel)* **2015**, *7*, 274–298. [CrossRef] [PubMed]
54. Chen, Y.; Han, L.; Bai, L.; Tang, H.; Zheng, A. Trichosanthin inhibits the proliferation of cervical cancer cells and downregulates STAT-5/C-myc signaling pathway. *Pathol. Res. Pract.* **2019**, *215*, 632–638. [CrossRef]
55. Fang, E.F.; Zhang, C.Z.; Zhang, L.; Wong, J.H.; Chan, Y.S.; Pan, W.L.; Dan, X.L.; Yin, C.M.; Cho, C.H.; Ng, T.B. Trichosanthin inhibits breast cancer cell proliferation in both cell lines and nude mice by promotion of apoptosis. *PLoS ONE* **2012**, *7*, e41592. [CrossRef]
56. Tuya, N.; Wang, Y.; Tong, L.; Gao, W.; Yu, R.; Xue, L. Trichosanthin enhances the antitumor effect of gemcitabine in non-small cell lung cancer via inhibition of the PI3K/AKT pathway. *Exp. Ther. Med.* **2017**, *14*, 5767–5772. [CrossRef] [PubMed]
57. Li, C.; Zeng, M.; Chi, H.; Shen, J.; Ng, T.B.; Jin, G.; Lu, D.; Fan, X.; Xiong, B.; Xiao, Z.; et al. Trichosanthin increases Granzyme B penetration into tumor cells by upregulation of CI-MPR on the cell surface. *Oncotarget* **2017**, *8*, 26460–26470. [CrossRef]
58. Wei, B.; Huang, Q.; Huang, S.; Mai, W.; Zhong, X. Trichosanthin-induced autophagy in gastric cancer cell MKN-45 is dependent on reactive oxygen species (ROS) and NF-kappaB/p53 pathway. *J. Pharmacol. Sci.* **2016**, *131*, 77–83. [CrossRef]
59. Zhang, C.; Gong, Y.; Ma, H.; An, C.; Chen, D.; Chen, Z.L. Reactive oxygen species involved in trichosanthin-induced apoptosis of human choriocarcinoma cells. *Biochem. J.* **2001**, *355*, 653–661. [CrossRef]
60. He, D.; Jin, J.; Zheng, Y.; Bruce, I.C.; Tam, S.; Ma, X. Anti-angiogenesis effect of trichosanthin and the underlying mechanism. *Biochem. Biophys. Res. Commun.* **2013**, *430*, 735–740. [CrossRef]
61. Cui, L.; Song, J.; Wu, L.; Huang, L.; Wang, Y.; Huang, Y.; Yu, H.; Huang, Y.; You, C.C.; Ye, J. Smac is another pathway in the anti-tumour activity of Trichosanthin and reverses Trichosanthin resistance in CaSki cervical cancer cells. *Biomed. Pharmacother.* **2015**, *69*, 119–124. [CrossRef]
62. Jiao, Y.; Liu, W. Low-density lipoprotein receptor-related protein 1 is an essential receptor for trichosanthin in 2 choriocarcinoma cell lines. *Biochem. Biophys. Res. Commun.* **2010**, *391*, 1579–1584. [CrossRef] [PubMed]
63. Liu, F.; Wang, B.; Wang, Z.; Yu, S. Trichosanthin down-regulates Notch signaling and inhibits proliferation of the nasopharyngeal carcinoma cell line CNE2 in vitro. *Fitoterapia* **2012**, *83*, 838–842. [CrossRef] [PubMed]
64. Cai, Y.; Xiong, S.; Zheng, Y.; Luo, F.; Jiang, P.; Chu, Y. Trichosanthin enhances anti-tumor immune response in a murine Lewis lung cancer model by boosting the interaction between TSLC1 and CRTAM. *Cell. Mol. Immunol.* **2011**, *8*, 359–367. [CrossRef] [PubMed]
65. Wang, Y.Y.; Ouyang, D.Y.; Zheng, Y.T. Mechanism of trichosanthin against human leukemia/lymphoma cells in vitro. *Zhongguo Shi Yan Xue Ye Xue Za Zhi* **2007**, *15*, 729–732.
66. Mosinger, M. Necrosing or clastic effects of ricin on different organs and on experimental sarcomas. *C R Seances Soc. Biol. Fil.* **1951**, *145*, 412–415.
67. Lin, J.Y.; Chang, Y.C.; Huang, L.Y.; Tung, T.C. The cytotoxic effects of abrin and ricin on Ehrlich ascites tumor cells. *Toxicon* **1973**, *11*, 379–381. [CrossRef]
68. Fodstad, O.; Olsnes, S. Studies on the accessability of ribosomes to inactivation by the toxic lectins abrin and ricin. *Eur. J. Biochem.* **1977**, *74*, 209–215. [CrossRef]
69. Fodstad, O.; Kvalheim, G.; Godal, A.; Lotsberg, J.; Aamdal, S.; Host, H.; Pihl, A. Phase I study of the plant protein ricin. *Cancer Res.* **1984**, *44*, 862–865.
70. Griffiths, G.D.; Leek, M.D.; Gee, D.J. The toxic plant proteins ricin and abrin induce apoptotic changes in mammalian lymphoid tissues and intestine. *J. Pathol.* **1987**, *151*, 221–229. [CrossRef]
71. Kumar, O.; Sugendran, K.; Vijayaraghavan, R. Oxidative stress associated hepatic and renal toxicity induced by ricin in mice. *Toxicon* **2003**, *41*, 333–338. [CrossRef]
72. Iordanov, M.S.; Pribnow, D.; Magun, J.L.; Dinh, T.H.; Pearson, J.A.; Magun, B.E. Ultraviolet radiation triggers the ribotoxic stress response in mammalian cells. *J. Biol. Chem.* **1998**, *273*, 15794–15803. [CrossRef] [PubMed]
73. Xu, N.; Yuan, H.; Liu, W.; Li, S.; Liu, Y.; Wan, J.; Li, X.; Zhang, R.; Chang, Y. Activation of RAW264.7 mouse macrophage cells in vitro through treatment with recombinant ricin toxin-binding subunit B: Involvement of protein tyrosine, NF-kappaB and JAK-STAT kinase signaling pathways. *Int. J. Mol. Med.* **2013**, *32*, 729–735. [CrossRef] [PubMed]
74. Wong, J.; Korcheva, V.; Jacoby, D.B.; Magun, B.E. Proinflammatory responses of human airway cells to ricin involve stress-activated protein kinases and NF-kappaB. *Am. J. Physiol. Lung Cell. Mol. Physiol.* **2007**, *293*, L1385–L1394. [CrossRef] [PubMed]

75. Rao, P.V.; Jayaraj, R.; Bhaskar, A.S.; Kumar, O.; Bhattacharya, R.; Saxena, P.; Dash, P.K.; Vijayaraghavan, R. Mechanism of ricin-induced apoptosis in human cervical cancer cells. *Biochem. Pharmacol.* **2005**, *69*, 855–865. [CrossRef] [PubMed]
76. Komatsu, N.; Nakagawa, M.; Oda, T.; Muramatsu, T. Depletion of intracellular NAD(+) and ATP levels during ricin-induced apoptosis through the specific ribosomal inactivation results in the cytolysis of U937 cells. *J. Biochem.* **2000**, *128*, 463–470. [CrossRef]
77. Licastro, F.; Morini, M.C.; Bolognesi, A.; Stirpe, F. Ricin induces the production of tumour necrosis factor-alpha and interleukin-1 beta by human peripheral-blood mononuclear cells. *Biochem. J.* **1993**, *294*((Pt. 2)), 517–520. [CrossRef]
78. Gonzalez, T.V.; Farrant, S.A.; Mantis, N.J. Ricin induces IL-8 secretion from human monocyte/macrophages by activating the p38 MAP kinase pathway. *Mol. Immunol.* **2006**, *43*, 1920–1923. [CrossRef]
79. Pervaiz, A.; Zepp, M.; Adwan, H.; Berger, M.R. Riproximin modulates multiple signaling cascades leading to cytostatic and apoptotic effects in human breast cancer cells. *J. Cancer Res. Clin. Oncol.* **2016**, *142*, 135–147. [CrossRef]
80. Pervaiz, A.; Adwan, H.; Berger, M.R. Riproximin: A type II ribosome inactivating protein with anti-neoplastic potential induces IL24/MDA-7 and GADD genes in colorectal cancer cell lines. *Int. J. Oncol.* **2015**, *47*, 981–990. [CrossRef]
81. Zhang, Y.; Yang, Q.; Li, C.; Ding, M.; Lv, X.; Tao, C.; Yu, H.; Chen, F.; Xu, Y. Curcin C, a novel type I ribosome-inactivating protein from the post-germinating cotyledons of Jatropha curcas. *Amino Acids* **2017**, *49*, 1619–1631. [CrossRef]
82. Cao, D.; Sun, Y.; Wang, L.; He, Q.; Zheng, J.; Deng, F.; Deng, S.; Chang, S.; Yu, X.; Li, M.; et al. Alpha-momorcharin (alpha-MMC) exerts effective anti-human breast tumor activities but has a narrow therapeutic window in vivo. *Fitoterapia* **2015**, *100*, 139–149. [CrossRef] [PubMed]
83. Wang, L.; Shen, F.; Zhang, M.; He, Q.; Zhao, H.; Yu, X.; Yang, S.; Liu, Y.; Deng, N.; Zheng, J.; et al. Cytotoxicity mechanism of alpha-MMC in normal liver cells through LRP1 mediated endocytosis and JNK activation. *Toxicology* **2016**, *357–358*, 33–43. [CrossRef] [PubMed]
84. Manoharan, G.; Jaiswal, S.R.; Singh, J. Effect of alpha, beta momorcharin on viability, caspase activity, cytochrome c release and on cytosolic calcium levels in different cancer cell lines. *Mol. Cell. Biochem.* **2014**, *388*, 233–240. [CrossRef] [PubMed]
85. Deng, N.; Li, M.; Shen, D.; He, Q.; Sun, W.; Liu, M.; Liu, Y.; Zhou, Y.; Zheng, J.; Shen, F. LRP1 receptor-mediated immunosuppression of alpha-MMC on monocytes. *Int. Immunopharmacol.* **2019**, *70*, 80–87. [CrossRef] [PubMed]
86. Falini, B.; Bolognesi, A.; Flenghi, L.; Tazzari, P.L.; Broe, M.K.; Stein, H.; Durkop, H.; Aversa, F.; Corneli, P.; Pizzolo, G.; et al. Response of refractory Hodgkin's disease to monoclonal anti-CD30 immunotoxin. *Lancet* **1992**, *339*, 1195–1196. [CrossRef]
87. Mishra, R.; Das, M.K.; Singh, S.; Sharma, R.S.; Sharma, S.; Mishra, V. Articulatin-D induces apoptosis via activation of caspase-8 in acute T-cell leukemia cell line. *Mol. Cell. Biochem.* **2017**, *426*, 87–99. [CrossRef]
88. Schotterl, S.; Huber, S.M.; Lentzen, H.; Mittelbronn, M.; Naumann, U. Adjuvant Therapy Using Mistletoe Containing Drugs Boosts the T-Cell-Mediated Killing of Glioma Cells and Prolongs the Survival of Glioma Bearing Mice. *Evid. Based Complement. Alternat. Med.* **2018**, *2018*, 3928572. [CrossRef]
89. Polito, L.; Djemil, A.; Bortolotti, M. Plant Toxin-Based Immunotoxins for Cancer Therapy: A Short Overview. *Biomedicines* **2016**, *4*, 12. [CrossRef]
90. Flavell, D.J.; Warnes, S.L.; Noss, A.L.; Flavell, S.U. Anti-CD7 antibody and immunotoxin treatment of human CD7(+)T-cell leukaemia is significantly less effective in NOD/LtSz-scid mice than in CB.17 scid mice. *Br. J. Cancer* **2000**, *83*, 1755–1761. [CrossRef]
91. Giansanti, F.; Flavell, D.J.; Angelucci, F.; Fabbrini, M.S.; Ippoliti, R. Strategies to Improve the Clinical Utility of Saporin-Based Targeted Toxins. *Toxins (Basel)* **2018**, *10*, 82. [CrossRef]
92. Li, C.; Yan, R.; Yang, Z.; Wang, H.; Zhang, R.; Chen, H.; Wang, J. BCMab1-Ra, a novel immunotoxin that BCMab1 antibody coupled to Ricin A chain, can eliminate bladder tumor. *Oncotarget* **2017**, *8*, 46704–46705. [CrossRef] [PubMed]
93. Jiao, P.; Zhang, J.; Dong, Y.; Wei, D.; Ren, Y. Construction and characterization of the recombinant immunotoxin RTA-4D5-KDEL targeting HER2/neu-positive cancer cells and locating the endoplasmic reticulum. *Appl. Microbiol. Biotechnol.* **2018**, *102*, 9585–9594. [CrossRef] [PubMed]

94. Herrera, L.; Farah, R.A.; Pellegrini, V.A.; Aquino, D.B.; Sandler, E.S.; Buchanan, G.R.; Vitetta, E.S. Immunotoxins against CD19 and CD22 are effective in killing precursor-B acute lymphoblastic leukemia cells in vitro. *Leukemia* **2000**, *14*, 853–858. [CrossRef] [PubMed]
95. Allahyari, H.; Heidari, S.; Ghamgosha, M.; Saffarian, P.; Amani, J. Immunotoxin: A new tool for cancer therapy. *Tumour. Biol.* **2017**, *39*, 1010428317692226. [CrossRef] [PubMed]
96. Kontermann, R.E. Dual targeting strategies with bispecific antibodies. *MAbs* **2012**, *4*, 182–197. [CrossRef]
97. Lim, D.; Kim, K.S.; Kim, H.; Ko, K.C.; Song, J.J.; Choi, J.H.; Shin, M.; Min, J.J.; Jeong, J.H.; Choy, H.E. Anti-tumor activity of an immunotoxin (TGFalpha-PE38) delivered by attenuated Salmonella typhimurium. *Oncotarget* **2017**, *8*, 37550–37560. [CrossRef]
98. Diaz, R.; Pallares, V.; Cano-Garrido, O.; Serna, N.; Sanchez-Garcia, L.; Falgas, A.; Pesarrodona, M.; Unzueta, U.; Sanchez-Chardi, A.; Sanchez, J.M.; et al. Selective CXCR4(+) Cancer Cell Targeting and Potent Antineoplastic Effect by a Nanostructured Version of Recombinant Ricin. *Small* **2018**, *14*, e1800665. [CrossRef]
99. Hu, R.G.; Zhai, Q.W.; He, W.J.; Mei, L.; Liu, W.Y. Bioactivities of ricin retained and its immunoreactivity to anti-ricin polyclonal antibodies alleviated through pegylation. *Int. J. Biochem. Cell Biol.* **2002**, *34*, 396–402. [CrossRef]
100. Tang, Y.; Liang, J.; Wu, A.; Chen, Y.; Zhao, P.; Lin, T.; Zhang, M.; Xu, Q.; Wang, J.; Huang, Y. Co-Delivery of Trichosanthin and Albendazole by Nano-Self-Assembly for Overcoming Tumor Multidrug-Resistance and Metastasis. *ACS Appl. Mater. Interfaces* **2017**, *9*, 26648–26664. [CrossRef]
101. Benigni, F.; Canevari, S.; Gadina, M.; Adobati, E.; Ferreri, A.J.; Di Celle, E.F.; Comolli, R.; Colnaghi, M.I. Preclinical evaluation of the ribosome-inactivating proteins PAP-1, PAP-S and RTA in mice. *Int. J. Immunopharmacol.* **1995**, *17*, 829–839. [CrossRef]
102. Puri, M.; Kaur, I.; Perugini, M.A.; Gupta, R.C. Ribosome-inactivating proteins: Current status and biomedical applications. *Drug Discov. Today* **2012**, *17*, 774–783. [CrossRef] [PubMed]
103. Ko, W.H.; Wong, C.C.; Yeung, H.W.; Yung, M.H.; Shaw, P.C.; Tam, S.C. Increasing the plasma half-life of trichosanthin by coupling to dextran. *Biochem. Pharmacol.* **1991**, *42*, 1721–1728. [CrossRef]
104. Pulliam, L.; Herndier, B.G.; McGrath, M.S. Purified trichosanthin (GLQ223) exacerbation of indirect HIV-associated neurotoxicity in vitro. *AIDS* **1991**, *5*, 1237–1242. [CrossRef] [PubMed]
105. Yu, L.; Gu, F.; Zhang, C.; Xie, S.; Guo, Y. Targeted diagnosis and treatment of superficial bladder cancer with monoclonal antibody BDI-1. *Chin. Med. J.* **1998**, *111*, 404–407.
106. Ivanova, E.P.; Bazaka, K.; Crawford, R.J. Advanced synthetic polymer biomaterials derived from organic sources. In *New Functional Biomaterials for Medicine and Healthcare*; Horwood Publishing: Cambridge, UK, 2014; pp. 71–99. [CrossRef]
107. Milla, P.; Dosio, F.; Cattel, L. PEGylation of proteins and liposomes: A powerful and flexible strategy to improve the drug delivery. *Curr. Drug Metab.* **2012**, *13*, 105–119. [CrossRef]
108. Roberts, M.J.; Bentley, M.D.; Harris, J.M. Chemistry for peptide and protein PEGylation. *Adv. Drug Deliv. Rev.* **2002**, *54*, 459–476. [CrossRef]
109. Au, K.Y.; Shi, W.W.; Qian, S.; Zuo, Z.; Shaw, P.C. Improvement of the Pharmacological Properties of Maize RIP by Cysteine-Specific PEGylation. *Toxins (Basel)* **2016**, *8*, 298. [CrossRef]
110. Arpicco, S.; Dosio, F.; Bolognesi, A.; Lubelli, C.; Brusa, P.; Stella, B.; Ceruti, M.; Cattel, L. Novel poly(ethylene glycol) derivatives for preparation of ribosome-inactivating protein conjugates. *Bioconjug. Chem.* **2002**, *13*, 757–765. [CrossRef]
111. Sun, Y.; Sun, F.; Li, J.; Wu, M.; Fan, X.; Meng, Y.; Meng, Y. Mono-PEGylation of Alpha-MMC and MAP30 from Momordica charantia L.: Production, Identification and Anti-Tumor Activity. *Molecules* **2016**, *21*, 1457. [CrossRef]
112. Bian, X.; Shen, F.; Chen, Y.; Wang, B.; Deng, M.; Meng, Y. PEGylation of alpha-momorcharin: Synthesis and characterization of novel anti-tumor conjugates with therapeutic potential. *Biotechnol. Lett.* **2010**, *32*, 883–890. [CrossRef]
113. Meng, Y.; Liu, S.; Li, J.; Meng, Y.; Zhao, X. Preparation of an antitumor and antivirus agent: Chemical modification of alpha-MMC and MAP30 from Momordica Charantia L. with covalent conjugation of polyethyelene glycol. *Int. J. Nanomed.* **2012**, *7*, 3133–3142. [CrossRef]
114. Deng, N.H.; Wang, L.; He, Q.C.; Zheng, J.C.; Meng, Y.; Meng, Y.F.; Zhang, C.J.; Shen, F.B. PEGylation alleviates the non-specific toxicities of Alpha-Momorcharin and preserves its antitumor efficacy in vivo. *Drug Deliv.* **2016**, *23*, 95–100. [CrossRef] [PubMed]

115. Zheng, J.C.; Lei, N.; He, Q.C.; Hu, W.; Jin, J.G.; Meng, Y.; Deng, N.H.; Meng, Y.F.; Zhang, C.J.; Shen, F.B. PEGylation is effective in reducing immunogenicity, immunotoxicity, and hepatotoxicity of alpha-momorcharin in vivo. *Immunopharmacol. Immunotoxicol.* **2012**, *34*, 866–873. [CrossRef] [PubMed]
116. Sun, W.; Sun, J.; Zhang, H.; Meng, Y.; Li, L.; Li, G.; Zhang, X.; Meng, Y. Chemosynthesis and characterization of site-specific N-terminally PEGylated Alpha-momorcharin as apotential agent. *Sci. Rep.* **2018**, *8*, 17729. [CrossRef] [PubMed]
117. An, Q.; Lei, Y.; Jia, N.; Zhang, X.; Bai, Y.; Yi, J.; Chen, R.; Xia, A.; Yang, J.; Wei, S.; et al. Effect of site-directed PEGylation of trichosanthin on its biological activity, immunogenicity, and pharmacokinetics. *Biomol. Eng.* **2007**, *24*, 643–649. [CrossRef]
118. He, X.H.; Shaw, P.C.; Xu, L.H.; Tam, S.C. Site-directed polyethylene glycol modification of trichosanthin: Effects on its biological activities, pharmacokinetics, and antigenicity. *Life Sci.* **1999**, *64*, 1163–1175. [CrossRef]
119. He, X.H.; Shaw, P.C.; Tam, S.C. Reducing the immunogenicity and improving the in vivo activity of trichosanthin by site-directed pegylation. *Life Sci.* **1999**, *65*, 355–368. [CrossRef]
120. Youn, Y.S.; Na, D.H.; Yoo, S.D.; Song, S.C.; Lee, K.C. Carbohydrate-specifically polyethylene glycol-modified ricin A-chain with improved therapeutic potential. *Int. J. Biochem. Cell Biol.* **2005**, *37*, 1525–1533. [CrossRef]
121. Ko, W.H.; Wong, C.C.; Yeung, H.W.; Tam, S.C. Modulation of trichosanthin antigenicity by coupling to dextran. *Biochem. Int.* **1992**, *28*, 643–650.
122. Tam, S.C.; Blumenstein, J.; Wong, J.T. Soluble dextran-hemoglobin complex as a potential blood substitute. *Proc. Natl. Acad. Sci. USA* **1976**, *73*, 2128–2131. [CrossRef]
123. Printseva, O.; Faerman, A.I.; Maksimenko, A.V.; Tonevitsky, A.G.; Ilynsky, O.B.; Torchilin, V.P. Selective killing of smooth muscle cells in culture by the ricin A-chain conjugated with monoclonal antibodies to a cell surface antigen via a dextran bridge. *Experientia* **1985**, *41*, 1342–1344. [CrossRef] [PubMed]
124. Ko, W.H.; Yeung, H.W.; Tam, S.C. The biological activities of trichosanthin after coupling to dextran. *Gen. Pharmacol.* **1993**, *24*, 757–762. [CrossRef]
125. Yeung, H.W.; Ng, T.B. Properties of bromodextran-trichosanthin: A comparison with trichosanthin, an anti-AIDS protein. *Biochem. Int.* **1991**, *25*, 1051–1059. [PubMed]
126. Chan, W.L.; Shaw, P.C.; Li, X.B.; Xu, Q.F.; He, X.H.; Tam, S.C. Lowering of trichosanthin immunogenicity by site-specific coupling to dextran. *Biochem. Pharmacol.* **1999**, *57*, 927–934. [CrossRef]
127. Chandramohan, V.; Sampson, J.H.; Pastan, I.; Bigner, D.D. Toxin-based targeted therapy for malignant brain tumors. *Clin. Dev. Immunol.* **2012**, *2012*, 480429. [CrossRef]
128. Groth, C.; van Groningen, L.F.J.; Matos, T.R.; Bremmers, M.E.; Preijers, F.; Dolstra, H.; Reicherts, C.; Schaap, N.P.M.; van Hooren, E.H.G.; IntHout, J.; et al. Phase I/II Trial of a Combination of Anti-CD3/CD7 Immunotoxins for Steroid-Refractory Acute Graft-versus-Host Disease. *Biol. Blood Marrow Transplant.* **2019**, *25*, 712–719. [CrossRef] [PubMed]
129. Jain, R.K. Normalization of tumor vasculature: An emerging concept in antiangiogenic therapy. *Science* **2005**, *307*, 58–62. [CrossRef] [PubMed]
130. You, C.; Sun, Y.; Zhang, S.; Tang, G.; Zhang, N.; Li, C.; Tian, X.; Ma, S.; Luo, Y.; Sun, W.; et al. Trichosanthin enhances sensitivity of non-small cell lung cancer (NSCLC) TRAIL-resistance cells. *Int. J. Biol. Sci.* **2018**, *14*, 217–227. [CrossRef] [PubMed]
131. Fidias, P.; Grossbard, M.; Lynch, T.J., Jr. A phase II study of the immunotoxin N901-blocked ricin in small-cell lung cancer. *Clin. Lung Cancer* **2002**, *3*, 219–222. [CrossRef]
132. Zalcberg, J.R.; Pietersz, G.; Toohey, B.; Laird, J.; Huggins, R.; Zimet, A.S.; Hennessy, O.; McKenzie, A.; McKenzie, I.F.C. A phase III study of the intralesional injection of ricin-monoclonal antibody conjugates in patients with hepatic metastases. *Eur. J. Cancer* **1994**, *30*, 1227–1231. [CrossRef]
133. Schnell, R.; Staak, O.; Borchmann, P.; Schwartz, C.; Matthey, B.; Hansen, H.; Schindler, J.; Ghetie, V.; Vitetta, E.S.; Diehl, V.; et al. A Phase I study with an anti-CD30 ricin A-chain immunotoxin (Ki-4.dgA) in patients with refractory CD30+ Hodgkin's and non-Hodgkin's lymphoma. *Clin. Cancer Res.* **2002**, *8*, 1779–1786. [PubMed]
134. Schnell, R.; Borchmann, P.; Staak, J.O.; Schindler, J.; Ghetie, V.; Vitetta, E.S.; Engert, A. Clinical evaluation of ricin A-chain immunotoxins in patients with Hodgkin's lymphoma. *Ann. Oncol.* **2003**, *14*, 729–736. [CrossRef] [PubMed]

135. Yaghini, E.; Dondi, R.; Edler, K.J.; Loizidou, M.; MacRobert, A.J.; Eggleston, I.M. Codelivery of a cytotoxin and photosensitiser via a liposomal nanocarrier: A novel strategy for light-triggered cytosolic release. *Nanoscale* **2018**, *10*, 20366–20376. [CrossRef] [PubMed]
136. Eng, M.S.; Kaur, J.; Prasmickaite, L.; Engesaeter, B.O.; Weyergang, A.; Skarpen, E.; Berg, K.; Rosenblum, M.G.; Maelandsmo, G.M.; Hogset, A.; et al. Enhanced targeting of triple-negative breast carcinoma and malignant melanoma by photochemical internalization of CSPG4-targeting immunotoxins. *Photochem. Photobiol. Sci.* **2018**, *17*, 539–551. [CrossRef]
137. Chang, Y.; Yao, S.; Chen, Y.; Huang, J.; Wu, A.; Zhang, M.; Xu, F.; Li, F.; Huang, Y. Genetically-engineered protein prodrug-like nanoconjugates for tumor-targeting biomimetic delivery via a SHEATH strategy. *Nanoscale* **2019**, *11*, 611–621. [CrossRef]

toxins

MDPI

Review

Antiviral Activity of Ribosome-Inactivating Proteins

Lucía Citores †, Rosario Iglesias † and José M. Ferreras *

Department of Biochemistry and Molecular Biology and Physiology, Faculty of Sciences, University of Valladolid, E-47011 Valladolid, Spain; luciac@bio.uva.es (L.C.); riglesia@bio.uva.es (R.I.)
* Correspondence: josemiguel.ferreras@uva.es
† Authors contributed equally to this work.

Abstract: Ribosome-inactivating proteins (RIPs) are rRNA N-glycosylases from plants (EC 3.2.2.22) that inactivate ribosomes thus inhibiting protein synthesis. The antiviral properties of RIPs have been investigated for more than four decades. However, interest in these proteins is rising due to the emergence of infectious diseases caused by new viruses and the difficulty in treating viral infections. On the other hand, there is a growing need to control crop diseases without resorting to the use of phytosanitary products which are very harmful to the environment and in this respect, RIPs have been shown as a promising tool that can be used to obtain transgenic plants resistant to viruses. The way in which RIPs exert their antiviral effect continues to be the subject of intense research and several mechanisms of action have been proposed. The purpose of this review is to examine the research studies that deal with this matter, placing special emphasis on the most recent findings.

Keywords: adenine polynucleotide glycosylase; antiviral therapy; human virus; immunotoxin; ribosome-inactivating protein (RIP); rRNA glycosylase (EC 3.2.2.22); virus-resistant transgenic plant (VRTP)

Key Contribution: Ribosome-inactivating proteins might help in the fight against human and plant viruses.

Citation: Citores, L.; Iglesias, R.; Ferreras, J.M. Antiviral Activity of Ribosome-Inactivating Proteins. *Toxins* **2021**, *13*, 80. https://doi.org/10.3390/toxins13020080

Received: 22 December 2020
Accepted: 20 January 2021
Published: 22 January 2021

1. Introduction

One of the main efforts of virologists and molecular biologists is the search for antivirals that can help in the fight against viruses causing diseases in animals and especially in humans. Strategies are also being searched to tackle the challenge of plant viruses causing significant crop losses. This has led to the discovery of a number of antivirals with different chemical nature or proteins with different enzymatic activities [1,2]. The search for more effective and safer antivirals continues to be a field of intense investigation and plants are one of the most used sources, since they have evolved a variety of protein-based defense mechanisms to tackle viral infections [3]. Regarding ribosome-inactivating proteins (RIPs), it is worth noting the fact that one of the first RIPs to be purified was PAP (pokeweed antiviral protein) and although many RIPs have been purified as protein synthesis inhibitors, many others have been isolated as powerful antivirals. For many years, RIPs have been studied as potent inhibitors of protein synthesis that can be used for the construction of immunotoxins [4]. Since linked to a monoclonal antibody or a protein that specifically binds to a receptor, they can be used to specifically kill tumor cells [4,5]. RIPs have initially been studied as a family of proteins widely distributed among angiosperms although they have also been found in other taxons [6,7]. They irreversibly inactivate ribosomes inhibiting protein synthesis and thus causing cell death [6,7]. The first RIPs to be isolated, the extremely potent toxins ricin and abrin, were purified at the end of the nineteenth century and it was believed that their red cell agglutinating activity was the reason for the toxic effect [8,9]. In the early 1970s, it was reported that abrin, ricin, and PAP strongly inhibited protein synthesis in a cell-free rabbit reticulocyte system [8–10]; and Barbieri and Stirpe classified these and other related proteins as type 1 RIPs (a single polypeptide chain, such as PAP) and type 2 RIPs (two chains, an A chain similar to type 1 RIPs, and a B chain that possesses lectin activity, such as abrin and ricin) [4]. The enzymatic activity of ricin was discovered by Endo and colleagues, that is, RIPs are considered as 28S rRNA N-glycosylases (EC 3.2.2.22) that cleave the N-glycosidic bond between the adenine No. 4324 and its ribose in the 60S subunit of rat ribosomes [11] or the equivalent one in sensitive ribosomes from other organisms [12]. This adenine is located in the sarcin-ricin loop (SRL) that is crucial for anchoring the elongation factors EFG and EF2 on the ribosome during mRNA-tRNA translocation in prokaryotes and eukaryotes, respectively. This loop is also the target of ribotoxins such as α-sarcin, enzymes with rRNA endonuclease activity (EC 3.1.27.10) [13]. However, some RIPs are also able to remove more than an adenine from the rRNA [14] and many of them are able to deadenylate not only rRNA but also other polynucleotide substrates such as DNA, poly(A), mRNA, tRNA, and viral RNA [15], and because of this, the name of adenine polynucleotide glycosylase (or polynucleotide: adenosine glycosidase) was proposed for RIPs [15]. Additionally, other activities have been reported for RIPs, just as shown in Table 1.

Table 1. Proposed activities and other biological properties of ribosome-inactivating proteins (RIPs).

Activity	Example of RIP	References
Agglutinin	Ricin	[8]
Antiviral	PAP	[10]
rRNA N-glycosylase	Ricin	[11]
Adenine polynucleotide glycosylase	Saporin-L1	[15]
rRNA N-glycosylase/lyase	Gypsophilin/RALyase	[16]
RNase	BBAP1	[17]
DNase	BBAP1	[17]
Phosphatase	Trichosanthin	[18]
Superoxide dismutase	Camphorin	[19]
Phospholipase	Ricin	[20]
Chitinase	TKC 28-I	[21]
DNA nicking	BE27	[22]
Apoptosis induction	Stenodactylin	[4,23]
Necroptosis induction	Stenodactylin	[4,23]
Autophagia induction	Abrus Agglutinin	[24]
Senescence induction	JIP60	[25]
Plant tissue necrosis	JIP60	[26]

A convincing picture of the role played by these proteins in plants is not yet available. They seem to play different roles in different species, so antiviral, antifungal, plant defense, storage, programmed senescence, antifeedant, stress protection, and development regulation roles have been proposed for RIPs [7].

The need to find new antivirals has encouraged researchers to study the antiviral activity of RIPs. On the other hand, much research is underway, focused on the use of these proteins to obtain crops with resistance to viral pathogens. The aim of this review is to compile the advances that have been made within this field, placing special emphasis on the most recent findings.

2. Activity on Animal (Human) Viruses

Global health threats such as the emergence of human viruses resistant to commonly used antiviral drugs, has prompted the study of RIPs as possible tools for fighting these agents. Antiviral activity of RIPs against different animal viruses has been reported (Table 2).

Table 2. RIPs active against animal viruses. RIPs with antiviral activity, the families and species from which they have been obtained and the viruses in which this activity has been demonstrated are shown.

Species and RIP	Virus	References
POACEAE		
Zea mays L.		
Maize RIP	HIV, SHIV	[27,28]
EUPHORBIACEAE		
Ricinus communis L.		
Ricin A chain	HIV	[29]
Suregada multiflora (A.Juss.) Baill. (=*Gelonium multiflorum* A.Juss.)		
Gelonin	HIV, HPV, HSV, PICV,	[2,30–32]
GAP31	HIV	[33,34]
CUCURBITACEAE		
Trichosanthes kirilowii Maxim		
Trichosanthin (TCS)	HBV, HIV, HSV	[32,35–38]
TAP29	HIV	[36]
Trichobitacin	HIV	[36,39]
Momordica charantia L.		
Momordin (*M. charantia* inhibitor)	HPV, HSV	[30]
Alpha-momorcharin (α-MMC)	HBV, HIV, HSV	[2,32,40,41]
Beta-momorcharin	HIV	[2,32]
Momordica antiviral protein (MAP30)	DENV-2, HHV8, HBV, HIV, HSV	[35,42–46]
Momordica balsamina L.		
Balsamin	HIV	[47]
Luffa cylindrica (L.) M.Roem.		
Luffin	HIV	[32]
Bryonia cretica subsp. *dioica* (Jacq.) Tutin (=*Bryonia dioica* Jacq.)		
Bryodin	HIV	[48]
CARYOPHYLLACEAE		
Saponaria officinalis L.		
Saporin	HIV	[32,49,50]
Dianthus caryophyllus L.		
Dianthin 32 (DAP32)	HIV, HPV, HSV	[30,34]
Dianthin 30 (DAP30)	HIV	[34]
Agrostemma githago L.		
Agrostin	HIV	[2,32]
PHYTOLACCACEAE		
Phytolacca americana L.		
PAP (PAPI)	CHIKV, FLUV, HBV, HIV, HPV, HSV, HTLV, JEV, LCMV	[10,35,51–57]
PAPII	HIV	[57]
PAPIII	HIV	[57]
PAP-S	HSV, HPV, HBV	[30,56]

Virus name abbreviations: CHIKV (chikungunya virus), DENV (dengue virus), FLUV (human influenza virus), HBV (hepatitis B virus), HHV (human gammaherpesvirus), HIV (human immunodeficiency virus), HPV (human poliovirus), HSV (herpes simplex virus), HTLV (human T-cell leukemia virus), JEV (Japanese encephalitis virus), LCMV (lymphocytic choriomeningitis virus), PICV (Pichinde virus), SHIV (simian–human immunodeficiency virus).

RIPs with antiviral activity belong to the main types of RIPs found in angiosperms [7]: monocot type 1 RIPs (Poaceae), dicot type 1 RIPs (Euphorbiaceae, Caryophyllaceae, Phytolaccaceae), type 2 RIPs (ricin, Euphorbiaceae), and type 1 RIPs derived from type 2 RIPs (Cucurbitaceae); which suggests that all these proteins could have, to a greater or lesser extent, antiviral activity and that their main biological role could be precisely the defense of the plant against viruses. However, researchers have focused on the study of proteins obtained from species of the families Phytolaccaceae, Cucurbitaceae, Caryophyllaceae, and Euphorbiaceae; and the most studied RIPs are pokeweed antiviral protein (PAP), trichosanthin (TCS) and Momordica antiviral protein (MAP30), which have been the subject of recent reviews [10,35,36,38,58]. It is noteworthy that RIPs have shown to be active against viruses of very different nature: double-stranded (ds) DNA viruses (hepatitis B virus,

HBV; human gammaherpesvirus, HHV; human poliovirus, HPV; herpes simplex virus, HSV), retroviruses (human immunodeficiency virus, HIV; human T-cell leukemia virus, HTLV; simian–human immunodeficiency virus, SHIV), positive-sense single-stranded (ss) RNA viruses (Japanese encephalitis virus, JEV; dengue virus, DENV; chikungunya virus, CHIKV), and negative-sense (ss) RNA viruses (human influenza virus, FLUV; lymphocytic choriomeningitis virus, LCMV; Pichinde virus, PICV). Most of the viruses studied are enveloped viruses that infect humans, with the exceptions of the simian–human immunodeficiency virus (SHIV), the Pichinde virus (PICV), and the non-enveloped human poliovirus. This virus was the first in which activity against an animal virus was reported [59]. Results obtained with HEp-2 cells infected with human poliovirus or herpes simplex virus (HSV) showed that gelonin, momordin, dianthin 32, and PAP-S impaired viral replication by inhibiting protein synthesis in virus-infected cells, in which presumably they enter more easily than in uninfected cells [30], suggesting that antiviral activity could be a general property of RIPs.

2.1. *Activity on Human Immunodeficiency Virus*

The most studied virus is the human immunodeficiency virus (HIV). The lack of effective antivirals against this virus and its rapid spread around the world prompted studies on the activity of RIPs against this virus since 1989 [60]. At least 20 RIPs have shown activity against HIV (Table 2). Thus, several RIPs obtained from Euphorbiaceae and Caryophylaceae, but mostly from Cucurbitaceae and Phytolocaceae, inhibit the replication of HIV in vitro [35]. It has also been reported that maize RIP transiently reduces viral load in SHIV infected Chinese rhesus macaques [27]. The results obtained with RIPs promoted their use in clinical trials [61]. Although the development of specific HIV antivirals such as reverse-transcriptase and protease inhibitors have directed AIDS therapy to other treatments, these studies demonstrated the potential of RIPs for the treatment of virus-related diseases.

2.2. *Activity on Herpes Simplex Virus*

Another virus that has been targeted by RIPs is the herpes simplex virus (HSV). Currently, there is no treatment that completely eliminates HSV infection from the body, because once the virus enters an organism, it remains dormant until reactivated. This has encouraged researchers to study RIPs as candidates for HSV therapy. Gelonin, trichosanthin, dianthin 32, PAP, PAP-S, and several RIPs obtained from *Momordica charantia* have shown anti-HSV activity in vitro (Table 2).

2.3. *Activity on Other Animal Viruses*

Exposure of HepG2.2.15 cells to MAP30 [44], PAP-S [56], α-momorcharin [41], and an eukaryotic expression plasmid encoding PAP [56] inhibits the production of hepatitis B virus (HBV). Additionally, an extract from Radix Trichosanthis had a stronger inhibitive effect on expression of HBsAg and HBeAg in HepG2.2.15, and trichosanthin has been proposed as the main component of the aqueous extract responsible for the anti-hepatitis B viral effect [62].

On the other hand, it has also been reported that PAP inhibits replication of human T-cell leukemia (HTLV), human influenza, chikungunya (CHIKV), Japanese encephalitis (JEV), and lymphocytic choriomeningitis (LCMV) viruses, gelonin inhibits Pichinde virus replication, and MAP30 inhibits human gammaherpesvirus 8 (HHV8) and dengue virus [10,31,35,42,52–55].

2.4. *Citotoxicity of RIPs*

An important aspect to consider when working with antivirals is their cytotoxicity. In this sense, type 1 RIPs and type 2 RIPs can be distinguished. Type 1 RIPs consist of a polypeptide chain with rRNA N-glycosylase activity, while type 2 RIPs are constituted by two chains linked by a disulfide bond: The A chain (active) is equivalent to a type

1 RIP and the B chain (binding) is a lectin able to bind to membrane glycoproteins and glycolipids allowing endocytosis of RIP by cells. This is why RIPs such as ricin and abrin are extremely toxic showing IC_{50} (concentration that inhibits protein synthesis by 50%) values of 0.67–8 pM in cell cultures [63]. There are type 2 RIPs such as those from *Sambucus* which are much less toxic to cultured cells with IC_{50} values of 27–64 nM [64]. Type 1 RIPs are much less toxic and have highly variable IC_{50} values (0.2–10 μM) [63]. Due to the high cytotoxicity of type 2 RIPs, only type 1 RIPs or the ricin A-chain (which has a cytotoxicity similar to that of type 1 RIPs) [63] have been used as antiviral agents.

A good antiviral should display a substantial difference between the antiviral concentration and the cytotoxic concentration. Due to the diverse toxicities of type 1 RIPs, there are also differences in this regard, but the most commonly used proteins such as PAP, MAP30, or trichosanthin always show a substantial difference between toxic concentrations for cells (3–30 μM) [63,65,66] and concentrations that have antiviral activity (around 30 nM) [35].

Finally, it should be noted that some bacterial and fungal enzymes targeting the sarcin-ricin loop have also been reported to possess antiviral activity [2,67–73].

Therefore, RIPs have awakened over many years, and continue to do so, a keen interest as tools to fight viruses that cause diseases in humans. In fact, recently saporin and RTAM-PAP1 (a chimera constructed with ricin A-chain and PAP) have been proposed as candidates for therapy of COVID-19 [74,75].

3. Activity against Plant Viruses

To date, 39 RIPs have been described that display some type of activity against plant viruses (Table 3).

These RIPs have been found in 26 plant species belonging to one family of monocotyledons and ten families of dicotyledons, that are distributed throughout the phylogenetic tree of angiosperms in a similar way to the RIP-containing plants [7], thus suggesting that most RIPs could be active against plant viruses. As a matter of fact, only two type 2 RIPs from *Sambucus nigra* (SNAI and SNLRP) have been reported to fail to protect transgenic plants against viral infection [76].

Despite the fact that these antiviral proteins are distributed in a great variety of families, most of them (thirty one) belong to the orders Caryophyllales and Lamiales (families Caryophyllaceae, Amaranthaceae, Phytolaccaceae, Nyctaginaceae, Basellaceae, Lamiaceae), which are RIPs with well-defined structural and phylogenetic characteristics [7].

RIPs seem to be active against a wide range of viruses (Table 3), all of them belonging to different families of positive-sense single-stranded (ss) RNA viruses. The exception is the geminivirus ACMV (African cassava mosaic virus), which contains a single-stranded circular DNA genome. They seem to protect all kinds of plants and, although the most commonly used plant for testing has been *Nicotiana tabacum* L., RIPs have also shown ability to protect other species of the genus *Nicotiana* (*N. benthamiana* Domin and *N. glutinosa* L.) as well as other species commonly used in research or crops such as *Brassica rapa* L. (=*B. parachinensis* L.H.Bailey) (choy sum), *Cyamopsis tetragonoloba* (L.) Taub. (guar), *Crotalaria juncea* L. (sunn hemp). *Phaseolus vulgaris* L. (common bean). *Momordica charantia* L. (bitter melon), *Beta vulgaris* L. (sugar beet), *Cucurbita pepo* L. (squash), *Solanum tuberosum* L. (potato), *Carica papaya* L. (papaya), *Chenopodium quinoa* Willd. (quinoa), or *Lycopersicon esculentum* Mill. (tomato).

Table 3. RIPs active against plant viruses. RIPs with antiviral activity, the families and species from which they have been obtained and the viruses in which this activity has been demonstrated are shown.

Species and RIP	Virus	References
IRIDACEAE		
Iris x hollandica Tub.		
IRIP	TMV, TEV	[77]
IRAb	TMV, TEV	[77]
EUPHORBIACEAE		
Jatropha curcas L.		
Curcin 2	TMV	[78]
CUCURBITACEAE		
Trichosanthes kirilowii Maxim		
Trichosanthin	TuMV, CMV, TMV	[79,80]
Momordica charantia L.		
α-Momorcharin	CMV, ChiVMV, TMV, TuMV	[81,82]
LEGUMINOSAE		
Senna occidentalis (L.) Link (=*Cassia occidentalis* L.)		
Cassin	TMV	[83]
CARYOPHYLLACEAE		
Saponaria officinalis L.		
Saporin	BMV, TMV, AMV	[51]
Dianthus caryophyllus L.		
Dianthin 30	ACMV, TMV	[84,85]
Dianthin 32	TMV	[85]
AMARANTHACEAE		
Beta vulgaris L.		
BE27	TMV, AMCV	[86,87]
Amaranthus tricolor L.		
AAP-27	SHMV	[88]
Amaranthus viridis L.		
Amaranthin	TMV	[89]
Celosia argentea L. (=*Celosia cristata* L., =*Celosia plumosa* (Voss) Burv.)		
CCP 25	BMV, PMV, TMV, SHMV, ICRSV	[90–92]
CCP 27	TMV, SHMV, ICRSV	[92,93]
Chenopodium album L.		
CAP-I	TMV, SHMV	[94]
CAP-II	TMV, SHMV	[94]
CAP30	TMV	[95]
Salsola longifolia Forssk.		
SLP-32	BYMV, TNV	[96]
Spinacia oleracea L.		
VI (SoRIP2)	TMV	[97,98]

Table 3. *Cont.*

Species and RIP	Virus	References
PHYTOLACCACEAE		
Phytolacca insularis Nakai		
PIP	TMV, CMV, PVY, PVX, PLRV	[99]
Phytolacca dioica L.		
Dioicin 2	TMV	[87]
PD-S2	TMV	[87]
PD-L1	TNV	[100]
PD-L4	TMV, TNV	[87,100]
Phytolacca americana L.		
PAP (PAPI)	BMV, TMV, AMV, TBSV, SPMV, ZYMV CMV, PVY, PVX, TEV, SBMV	[51,58,101–105]
PAPII	TMV, PVX	[104]
PAP-S	AMCV	[105]
NYCTAGINACEAE		
Boerhaavia diffusa L.		
BDP-30	TMV	[106]
Mirabilis expansa (Ruiz & Pav.) Standl.		
ME1	TMV, BMV	[51]
Mirabilis jalapa L.		
MAP	TMV	[107]
Bougainvillea spectabilis Willd.		
Bouganin	ZYMV, AMCV	[105,108]
Bougainvillea buttiana Holttum & Standl.		
BBAP1	SHMV	[17]
BBP-24	TMV, SHMV	[109,110]
BBP-28	TMV, SHMV	[109,110]
BASELLACEAE		
Basella alba L. (=*Basella rubra* L.)		
RIP2	AMCV	[105]
LAMIACEAE		
Volkameria inermis L. (=*Clerodendrum inerme* (L.) Gaertn.)		
CIP-29	TMV, PRSV, SHMV	[111,112]
Volkameria aculeata L. (=*Clerodendrum aculeatum* (L.) Schltdl.)		
CA-SRI (CAP-34)	TMV, SHMV, PRSV	[113–115]
ADOXACEAE		
Sambucus nigra L.		
SNAI′	TMV	[116]
Nigrin b (SNAV)	TMV	[76]

Virus name abbreviations: ACMV (African cassava mosaic virus), AMCV (artichoke mottled crinkle virus), AMV (alfalfa mosaic virus), BMV (brome mosaic virus), BYMV (bean yellow mosaic virus), ChiVMV (Chilli veinal mottle virus), CMV (cucumber mosaic virus), ICRSV (Indian citrus ringspot virus = citrus ringspot virus, CRSV), PLRV (potato leafroll virus), PMV (pokeweed mosaic virus), PRSV (papaya ringspot virus), PVX (potato virus X), PVY (potato virus Y), SBMV (southern bean mosaic virus), SHMV (sunn-hemp mosaic virus = sunn-hemp rosette virus, SRV), SPMV (satellite panicum mosaic virus), TBSV (tomato bushy stunt virus), TEV (tobacco etch virus), TMV (tobacco mosaic virus), TNV (tobacco necrosis virus), TuMV (turnip mosaic virus), ZYMV (zucchini yellow mosaic virus).

It is difficult to compare the antiviral activity of the different RIPs because different criteria have been used to evaluate their antiviral capacity. In some cases, the putative antiviral character is based on their N-glycosylase activity on the virus genome [105]; all RIPs are able to release adenines from any kind of RNA or DNA, including viral genomes [4]. This adenine polynucleotide glycosylase activity has been detected by electrophoresis [87], or HPLC [103,105]. In many cases, the test has involved applying a RIP solution on the leaf surface of the plant together with the virus and comparing the result with the control that does not contain RIP. In some cases, the virus is applied simultaneously [86,92,113] and in others, sometime after the application of the RIP [90,115]. The evaluation of antiviral activity has been done by counting the number of lesions [88,93], the time of onset of symptoms [77,79], the number of infected plants [105], or the severity of the infection symptoms [78,115]. Virus levels have also been estimated by ELISA [99], Western blotting analysis [81], RT-PCR analysis [101], quantitative real-time PCR analysis [81,82], electron microscopy [92], or by determining the infection capacity of an extract from the infected plant [92]. Another approach has been the construction of virus-resistant transgenic plants [80,102]. The virus has been inoculated mechanically or by aphids [102] and the resistance has been determined by one of the methods listed above.

Other studies link RIPs to the defense of plants against viruses, especially studies of induction of RIPs through signaling compounds such as salicylic acid, hydrogen peroxide, or jasmonic acid, which are involved in the systemic acquired resistance (SAR) of plants against viruses and other pathogens. Thus, it has been reported that artichoke mottled crinkle virus (AMCV), salicylic acid, and hydrogen peroxide induce the expression of BE27 in both treated and untreated leaves of sugar beet plant [86,117]. On the other hand, it has been reported that alpha-momorcharin induces the generation of salicylic acid, jasmonic acid, and reactive oxygen species, which improve tobacco mosaic virus (TMV) tolerance [118]. Additionally, alpha-momorcharin induces the expression of the N gene [118], which encodes the N protein that recognizes the TMV replicase fragment and triggers signal transduction cascades, initiating a hypersensitive response (HR) and inhibiting the spread of TMV [118]. Other RIPs in which some type of elicitor activity has been reported are pokeweed antiviral protein II (PAPII) [104], CIP-29 [111], and CA-SRI [113,115]. By contrast, the antiviral activity of SNAI' [116], IRIP and IRAb [77], and nigrin b [76] is not accompanied by an induction of pathogenesis-related proteins. All this suggests that some, but not all RIPs, could be part of the SAR or/and HR to defend the plant against viral infections.

4. Antiviral Mechanisms of RIPs

RIPs have long been recognized as antiviral proteins in both plants and animals, but the mechanism responsible for this activity continues to be the subject of intense research today. The mechanism that triggers protection against viruses could have both common and different elements in plants and animals (Figure 1).

4.1. Antiviral Mechanisms of RIPs in Plants

4.1.1. Protein Synthesis Inhibition (rRNA N-glycosylase)

It has long been known that RIPs can inhibit protein synthesis in plants [119–122]. The mechanism is the same as that described for inhibition of protein synthesis in animals, i.e., RIPs act as N-glycosylases of the major rRNA by removing a specific adenine from the sarcin-ricin loop (SRL), which is highly conserved in animals and plants [120]. Moreover, it has been shown that some RIPs can inhibit protein synthesis carried out by ribosomes of the same plants that produce them [123] and in addition, in the case of some RIPs, a positive correlation between rRNA N-glycosylase activity on tobacco ribosomes and antiviral activity against TMV has been reported [124].

The fact that RIPs do not cause cell death in the absence of the virus and allow plant growth is due to the fact that, at least for type 1 RIPs from dicots, they are synthesized as preproteins with a leader peptide that directs them into the apoplastic space [125]. Viral infection is supposed to facilitate the entry of the RIP, which inactivates cell ribosomes, causing cell death and preventing the virus from using the cellular machinery to replicate and spread [125]. So far, the mechanism by which the virus facilitates the entry of RIPs has not been shown, although the ability of viruses to modify plasma membrane permeability is well-known [126].

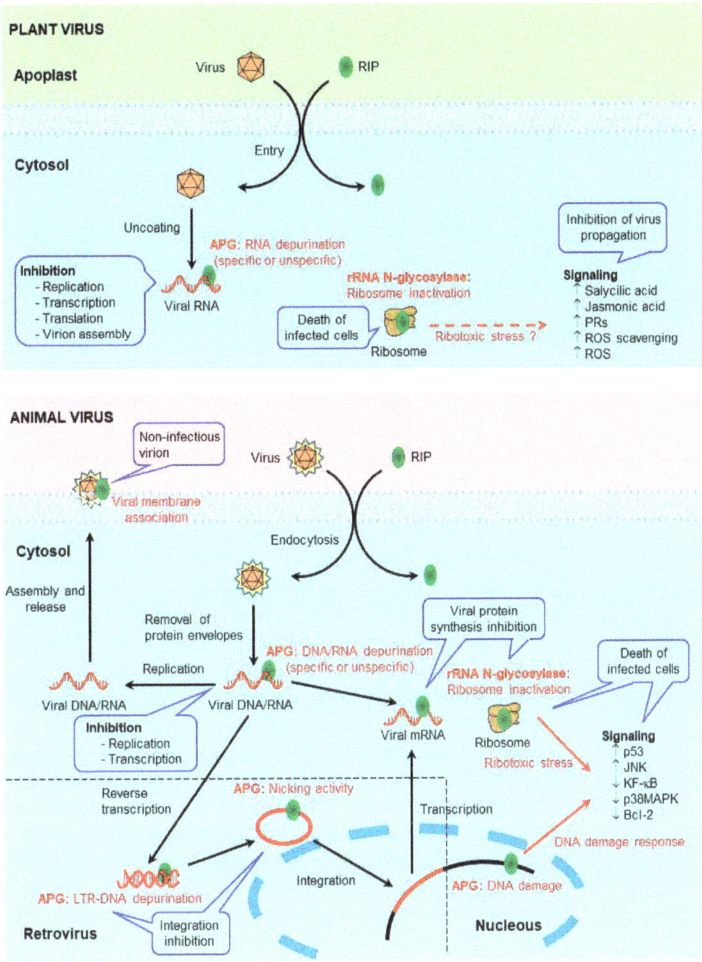

Figure 1. Proposed mechanisms for the antiviral activity of RIPs against plant viruses (upper panel), animal viruses (lower panel), and retroviruses (lower panel including dashed square). (**upper panel**) In plants, viral infection promotes the passage of the RIP from the apoplast to the cytosol. In the cytosol, it can inactivate ribosomes (rRNA glycosylase activity), causing the death of infected cells and thus preventing the spread of the virus. The RIP can also depurinate the viral RNA (adenine polynucleotide glycosylase, APG, activity), inhibiting its replication, transcription, translation, and assembly. It can also trigger antiviral defense signaling pathways, causing an increase in the levels of salicylic acid, jasmonic acid, pathogenesis-related (PR) proteins, and both reactive oxygen species (ROS) and ROS scavenging enzymes. (**lower panel**) In animal cells, the RIP can enter by pinocytosis or receptor-mediated endocytosis. RIP can inactivate ribosomes (rRNA glycosylase activity), causing the death of infected cells or inactivate the viral genome, DNA, or RNA (APG activity), preventing their replication, transcription, and translation. Some RIPs depurinate specific sequences (APG activity), blocking critical functions for the virus life cycle. In the case of retroviruses, the RIP can also depurinate the long terminal repeats (LTRs) (APG activity) or cleave the circular DNA (APG activity) preventing its integration into the cell genome. It can also be introduced into virions during budding (viral membrane association), making them less infective. Ribotoxic stress (rRNA glycosylase activity or APG activity on mRNA) and DNA damage (APG activity) caused by RIPs can trigger the activation of signaling pathways that cause infected-cell death preventing virus spreading.

4.1.2. Adenine Polynucleotide Glycosylase Activity

However, although some type 1 RIPs can inactivate ribosomes of some plants, they do not do so with those of others and usually act at much higher concentrations than in animal ribosomes [127]. In addition, mutants have been obtained from PAP that do not depurinate tobacco or reticulocyte lysate ribosomes but inhibit translation of brome mosaic virus (BMV) and potato virus X (PVX) [128].

The specificity of RIPs is highly variable, therefore some RIPs can act on other adenines in both animal [14] and plant [120,129] ribosomes. In addition, all RIPs release adenines from eukaryotic DNA and many of them also release adenines from other RNAs, including viral RNAs [15,22,87]. It has also been reported that some RIPs may have DNA nicking, DNase or RNase activities (Table 1). This can alter the life cycle of the virus, both its replication and transcription [130], translation [91], and assembly [131].

The adenine polynucleotide glycosylase activity on viral RNAs might be more specific. Thus, it has been reported that some RIPs can inhibit the translation of capped RNA by binding to the cap of viral RNAs and depurinating these RNAs downstream of the cap structure. For these RIPs, viral RNA depurination could be the main mechanism of their antiviral activity [51]. On the other hand, one of them (PAP) can also bind to translation initiation factors, allowing it to depurinate preferentially uncapped viral RNAs [103]. Viral capped RNA sequestration has also been proposed as an antiviral mechanism for MbRIP-1, a RIP from *Momordica balsamina* [132]. All this suggests that the antiviral mechanism of RIPs could be more complex than a simple and direct depurination of viral RNA.

4.1.3. Antiviral Protection through Signaling Pathways

The other proposed mechanism involves signaling molecules that defend the plant from viral infection. However, different results have been obtained depending on the RIP studied and the approach used. Thus, it has been reported that α-momorcharin (α-MMC), in *N. benthamiana* plants sprayed with a solution of the RIP, up-regulates the expression of reactive oxygen species (ROS) scavenging-related genes, modulating ROS homeostasis and conferring resistance to TMV, ChiVMV, and CMV infection [81,133]. Additionally, this RIP also up-regulates some salicylic acid-responsive defence-related genes [81]. By contrast, the same RIP sprayed in *M. charantia* plants increases plant resistance to CMV but by increasing jasmonic acid biosynthesis and inducing ROS without a relevant increase in salicylic acid [82]. It has also been reported that α-momorcharin induces an increase of both jasmonic acid and salicylic acid in tobacco plants, enhancing TMV resistance [118]. On the other hand, it has been postulated that PAP generates a signal that leads to the overexpression of pathogenesis-related proteins rendering transgenic tobacco plants resistant to virus infection in the absence of an increase in the salicylic acid levels [129,134,135]. Finally, it has been reported that the expression of IRAb and IRIP in transgenic tobacco plants provides a strong local protection against TMV and TEV but without induction of pathogenesis-related proteins [77]. The relationship between the enzymatic activity of RIPs and their ability to induce production of signaling molecules in plants has not been studied. In animals, the enzymes that exert their cytotoxic function through modification of the sarcin-ricin loop (SRL), such as ricin, α-sarcin, or Shiga toxin, strongly activate signaling pathways through the mitogen-activated protein kinases (MAPKs) p38 and JNK [136]. The trichothecenes deoxynivalenol (DON) and T-2 toxin inhibit protein synthesis and have been shown to induce activation of ERK1/2 and p38 MAP kinase in several animal and human cell lines followed by increased cytokine production [137]. This ribosome mediated activation of MAPKs is termed 'ribotoxic stress response' [137]. In Arabidopsis, DON and T-2 toxin led to the expression of MPK3 and MPK6 MAP kinases, implicated as positive regulators of the hypersensitive response via ethylene signaling and ROS [137]. Therefore, it would be possible that the generation of signaling compounds by plants was a response to ribotoxic stress produced by RIPs.

4.2. Antiviral Mechanisms of RIPs in Animals

4.2.1. Protein Synthesis Inhibition (rRNA N-glycosylase)

Early studies on the mechanism of antiviral action of RIPs in animal cells focused on their ability to inhibit protein synthesis [30]. Several type 1 RIPs (gelonin, *Momordica charantia* inhibitor, dianthin 32, and PAP-S) reduced viral production and plaque formation in HEp-2 cells infected with Herpes simplex virus-1 (HSV-1) or poliovirus I. In addition, the four RIPs inhibited protein synthesis more efficiently in cells infected with one of the two viruses than in uninfected cells, suggesting that RIPs inhibited viral replication by inhibiting protein synthesis of infected cells, presumably because they entered infected cells more easily than uninfected cells [30]. Although the mechanism by which viruses can facilitate the entry of RIPs is not established, it is known that type 1 RIPs can enter cells through pinocytosis or receptor-mediated endocytosis [138,139] and that both processes are stimulated by viruses [140,141].

4.2.2. Adenine Polynucleotide Glycosylase Activity

However, RIPs can inhibit virus replication without apparently inactivating ribosomes [34,52,142,143]. The adenine polynucleotide glycosylase activity on viral RNA [57] or DNA [33] is able to inactivate the viral genome and explains inhibition of virus replication [37,142,143]. In addition, RIPs can also depurinate viral mRNAs, thus avoiding the synthesis of proteins that are vital for its functions [52,144,145]. In the case of HIV, a strong inhibition of the integration of viral DNA into the host genome [32,45,50], caused by the adenine polynucleotide glycosylase activity on LTRs (long-terminal repeats) [33,146,147] and the nicking activity on the supercoiled DNA [148,149] of the virus, has been reported. Trichosanthin is also able to enter viral particles during budding, resulting in virions unable to infect other cells [150,151].

4.2.3. Antiviral Protection through Signaling Pathways

Finally, it has also been proposed that the antiviral activity of RIPs can be carried out through signaling pathways. Thus, it has been reported that RIPs promote p53 and c-Jun N-terminal kinase (JNK) activity [152,153] and block the activation of KF-κB, p38MAPK, and Bcl-2 [152,154,155] during viral infection. The modulation of these pathways would lead to the death of infected cells, thus preventing the spread of the virus. Cell DNA damage [152] or ribotoxic stress [153] caused by RIPs could trigger some of these signaling pathways. Ribotoxic stress response (RSR) is a response of cells to a variety of agents that affect the functions of ribosome, such as some antibiotics, alkaloids, mycotoxins, RIPs, ribotoxins, or ultraviolet radiation [136]. Ribotoxic stress is sensed by the MAP3K ZAKα that transduces the signal from ribosomes to activate MAP2K that in turn activates SAPKs. There are two SAPKs (stress-activated protein kinases) families in mammals: p38 and c-Jun N-terminal kinase (JNK). Activation of p38 induces cell-cycle arrest whereas activation of JNK promotes apoptosis [156], inducing both pro-survival and pro-apoptotic signaling. Additionally, mRNA damage by the adenine polynucleotide glycosylase activity of RIPs could trigger RSR as has been reported for ultraviolet radiation [156]. However, much research is still required to clarify how RIPs protect cells from viral infection through these pathways.

Therefore, RIPs can exert their antiviral effect through different mechanisms that could originate from their activity on the different nucleic acids from both the virus and the infected cell. Depending on the type of RIP, virus and infected cell, some mechanisms could predominate over others and more research is required to determine in each case which are the predominant ones.

5. Experimental Therapy

Because of its strong antiviral activity, RIPs have been used in experimental therapy, especially to treat the acquired immune deficiency syndrome (AIDS), but also against hepatitis, chikungunya, dengue, and lymphomas caused by the Epstein–Barr virus. Additionally, they have also been tested in vivo against viruses that infect animals, such as the murine cytomegalovirus, the Pichinde virus, or the simian–human immunodeficiency virus (Table 4).

5.1. RIPs and PEGylated RIPs

Trichosanthin (GLQ223) was used alone [61,157] or in combination with zidovudine (azidothymidine, AZT) [158] in clinical trials with AIDS patients. Trichosanthin infusions were safe and relatively well tolerated [157]. In patients, a decrease in serum p24 antigen [61] and an increase in $CD4^+$ and $CD8^+$ T cells [157,158] were observed. Recently, it has also been reported that maize RIP reduces the viral load of an HIV-related virus, the simian–human immunodeficiency virus in Chinese rhesus macaques [27].

Despite its potential as therapeutic agents, the strong immunogenicity, allergic reaction, and short half-life are the biggest barriers to their application as therapeutic agents. Polyethylene glycol (PEG) conjugation (PEGylation) can confer on these proteins, increasing plasma half-life, decreasing toxicity, and reducing immunogenicity and antigenicity. PEGylated alpha-momorcharin and MAP30 showed about 60%–70% antivirus activities against HSV-1, and at the same time decreased 50%–70% immunogenicity when compared with the non-PEGylated proteins [40].

5.2. Immunotoxins and Other Conjugates

RIPs have been used in medicine mainly as the toxic part of immunotoxins, that is, chimeric proteins consisting of an antibody specifically directed against a target, linked to a toxin of plant or bacterial origin. The design of immunotoxins has been improved over the past 40 years to minimize the off-target toxicity and immunogenicity [159,160]. Several types of antiviral immunotoxins have been constructed using either bacterial toxins (or their fragments) such as pseudomonas exotoxin A or diphteria toxin [161], and RIPs from plants (Table 4). The most commonly used RIP has been the ricin A-chain and the most studied virus the HIV. Viral proteins (gp41, gp120, or gp 160) or proteins from infected cells (CD4, CD25, or CD45RO) have been selected as targets. Despite the success of highly active antiretroviral therapy (HAART), antiviral immunotoxins continue to be developed in order to deplete persisting HIV-infected cell reservoirs [162]. Immunotoxins have also shown to be active in vitro against Epstein–Barr [163,164] and Pichinde [31] viruses and in vivo (in combination with the synthetic analogue of 2′-deoxy-guanosine ganciclovir) against the murine cytomegalovirus [165].

Targeting can also be carried out by conjugating RIPs with other proteins or peptides that specifically bound to viral proteins or proteins present only in infected cells [49,166].

5.3. Designed Antiviral Proteins and Nanocapsules

RIPs have also been used to design antiviral proteins. One of these engineered proteins contains an internal sequence that is recognized by the HIV protease and that is blocking the N-glycosylase activity of the RIP. This protein is activated in infected cells and has shown antiviral activity [28]. Similarly, variants of the ricin A-chain with the sequence recognized by the HIV protease in the C-terminus are activated in infected cells and show antiviral activity [29].

Another approach is to fuse the sequences of RIPs with antimicrobial peptides such as latarcin, thanatin, protegrin-1, and plectasin that are able to inhibit viral replication inside the infected cells, viral entry and replication, dengue NS2B-NS3 serine protease, and virus replication, respectively [42,53]. The aim is to target different stages of the viral life cycle. Thus, the peptide-fusion proteins Latarcin-PAP1-Thanatin and Protegrin1-MAP30-Plectasin inhibit virus replication in vitro and protect the virus-infected mice from chikungunya and dengue viruses, respectively [42,53]. Another fusion protein containing ricin A-chain and PAP-S displays antiviral activity in vitro against hepatitis B virus suggesting a synergistic activity of both proteins [167]. This has encouraged its authors to propose it as an anti-SARS-CoV-2 agent [75].

Table 4. Ribosome-inactivating proteins used in experimental antiviral therapy. RIPs have been used alone, PEGylated, or as part of immunotoxins, conjugates, engineered proteins, or nanocapsules.

Virus	Target	RIP	References
RIPs alone			
HIV	HIV infected cells	TCS	[61,157,158]
SHIV	SHIV infected cells	Maize RIP	[27]
PEGylated RIPs			
HSV-1	HIV infected cells	α-MMC	[40]
	HIV infected cells	MAP30	[40]
Immunotoxins			
HIV	gp 120	RAC, PAP-S, PAC, Gelonin	[168–172]
	gp 41	RAC, PAC, Gelonin	[170–175]
	gp 160	RAC	[173]
	CD45RO	RAC	[176]
	CD4	PAP	[143,177]
	CD25	RAC	[178]
PICV	PICV	Gelonin	[31]
EBV	CD30	Saporin 6	[163]
	EBV/C3d receptor	Gelonin	[164]
MCMV	MCMV	RAC	[165,179]
Conjugates			
HIV	gp 120	RAC	[166]
	$CD8^+$ T-cells	Saporin	[49]
Engineered proteins			
HIV	HIV protease	RAC	[29]
	HIV protease	Maize RIP	[28]
CHIKV	Viral life cycle	PAP	[53]
DENV	Viral life cycle	MAP30	[42]
HBV	HBV infected cells	RAC-PAP	[167]
Nanocapsules			
HIV	HIV infected cells	MAP30	[180]
	HIV protease	RAC	[181]

Virus name abbreviations: CHIKV (chikungunya virus), DENV (dengue virus), EBV (Epstein–Barr virus), HBV (hepatitis B virus), HIV (human immunodeficiency virus), HSV (herpes simplex virus), MCMV (murine cytomegalovirus), PICV (Pichinde virus), SHIV (simian–human immunodeficiency virus). RIP name abbreviations: MAP (Momordica antiviral protein), α-MMC (alpha-momorcharin), PAC (Pulchellin A-chain), PAP (pokeweed antiviral protein), RAC (ricin A-chain), TCS (trichosanthin).

The latest approach is the use of nanocapsules to deliver RIPs to virus-infected cells. Nanocapsules are vesicular objects in which the encapsulated compound is confined in an internal cavity surrounded by an outer membrane [182,183]. Nanocapsules containing MAP30 [180] or ricin A-chain [181] have shown antiviral activity in vitro against HIV. In the latter case, targeting has been achieved by using peptide crosslinkers that are sensitive to cleavage by HIV-1 protease [181].

5.4. Side Effects of RIP Therapy

Although trichosanthin was, in general, well tolerated in clinical trials when used in AIDS patients [157], some side effects were reported [61,157,158]. Clinical trials using RIPs as antivirals are scarce, but there are many clinical trials that have used RIPs as part of immunotoxins for the treatment of malignancies [9,64,184]. Side effects that may be mild or moderate like fever, nausea, vomiting, diarrhea, myalgia, edema, and hypoalbuminemia have been reported in these trials. Other effects are severe, such as immunogenicity and vascular leak syndrome (VLS), and could limit the therapeutic use of immunotoxins [64,184]. Immunogenicity may be the result of the formation of human anti-mouse antibodies (HAMA) or human anti-toxin antibodies (HATA). These antibodies can prevent repeated treatment cycles. The development of immunotoxins containing humanized antibodies or the use of part of antibodies containing only the variable domains can solve this problem [64,184]. To address the problem of the immunogenicity of RIPs,

PEGylation [40,184] and elimination of epitopes through genetic manipulation have been used [184]. Vascular leak syndrome, characterized by increased vascular permeability, is caused by the nonspecific binding of RIP to vascular endothelial cells. The identification and elimination of some peptides present in RIPs, nonessentials for RIP activity and responsible for this unspecific binding, have allowed the obtaining of less toxic recombinant RIPs [184].

6. Genetically Engineered Virus-Resistant Plants

Viruses cause epidemics in all major crops, representing a significant restriction on the yield and quality of agricultural production. As strict intracellular pathogens, they cannot be chemically controlled and prophylactic measures consist mainly in the destruction of infected plants and biocide applications to limit the population of vector organisms (arthropods, nematodes, and plasmodiophorids). A powerful alternative often used in agriculture is based on the use of crop genetic resistances, an approach that depends on mechanisms governing plant-virus interactions [185]. Several transgenic plants carrying virus resistance genes have been obtained by transferring virus-derived genes, including viral coat proteins, replicases, movement proteins, defective interfering RNAs, non-coding RNA sequences and proteases into susceptible plants, or non-viral genes including R genes, microRNAs, RIPs, protease inhibitors, dsRNAses, RNA modifying enzymes, and scFvs [186]. In recent years, transgenic plants carrying RIP genes that are resistant to fungi, insects and, above all, viruses have been reported. Thus, transgenic plants bearing RIP genes have been obtained that are resistant to a wide variety of viruses (Table 5).

Table 5. Transgenic plants bearing RIP genes. The degree of protection achieved is indicated as the percentage reduction of lesions, infected plants or detected virus levels, or as the delay in the onset of symptoms.

RIP	Host	Virus	Protection	Ref.
IRIP	*Nicotiana tabacum*	TMV, TEV	73% L.L.	[77]
IRAb	*Nicotiana tabacum*	TMV, TEV	54% L.L.	[77]
Curcin 2	*Nicotiana tabacum*	TMV	9 D.D.	[78]
Trichosanthin	*Nicotiana tabacum*	TuMV	100% L.L.	[79]
	Nicotiana tabacum	TMV, CMV	14 D.D.	[80]
	Lycopersicon esculentum	TMV, CMV	100% L.I.P.	[187]
Cassin	*Nicotiana tabacum*	TMV	13 D.D.	[83]
Dianthin 30	*Nicotiana benthamiana*	ACMV	100% L.L.	[84]
PIP	*Solanum tuberosum*	PVY, PYX, PLRV	98% L.V.L	[99]
PAP	*Nicotiana tabacum*	PVY, PYX, CMV	100% L.I.P.	[102,188]
	Nicotiana benthamiana	PVY	67% L.I.P.	[102]
	Solanum tuberosum	PVY, PYX	84% L.I.P.	[102]
PAPII	*Nicotiana tabacum*	TMV, PVX	89% L.L.	[104]
SNAI'	*Nicotiana tabacum*	TMV	59% L.L.	[116]
Nigrin b (SNAV)	*Nicotiana tabacum*	TMV	43% L.L.	[76]

Virus name abbreviations: ACMV (African cassava mosaic virus), CMV (cucumber mosaic virus), PLRV (potato leafroll virus), PVX (potato virus X), PVY (potato virus Y), TEV (tobacco etch virus), TMV (tobacco mosaic virus), TuMV (turnip mosaic virus). Protection abbreviations: L.L. (less lesions), D.D. (days of delay), L.I.P. (less infected plants), L.V.L. (less virus level).

Most of the times, tobacco has been transformed (*Nicotiana tabacum* L. and *N. benthamiana* Domin) but also potato (*Solanum tuberosum* L.) and tomato (*Lycopersicon esculentum* Mill.). *Agrobacterium tumefaciens* containing the plant transformation vectors has been used to transform either tobacco by the leaf disc co-cultivation method or potato (*S. tuberosum*) by the stem or tuber section co-cultivation method. The CaMV 35S promoter has always been used to express the RIPs, except in the case of dianthin 30 [84]. In the case of trichosanthin, tissue-specific promoters have also been used [80]. The CaMV 35S promoter is the most studied and most widely used plant promoter for transgenic expression [189], it is a very strong constitutive promoter that facilitates a high level of RNA transcription in a wide variety of plant species. For effective protection against viruses, it is preferable to achieve high levels of RIP expression since there is a direct correlation between expression level and resistance to viruses [78]. So, for example, in lines expressing small amounts of curcin 2, symptoms of TMV infection begin to appear after about 7 days, while lines that accumulate the highest level of curcin 2 (about 1.45 μg/mg) begin to develop symptoms after about 18 days.

Using the promoter CaMV 35S, plants with a RIP content of up to 2.7% of the total soluble protein have been obtained [80]. However, a high expression of RIP results in plants with an aberrant phenotype, which usually includes leaf mottling, extreme leaf discoloration, stunted leaf growth and/or excessive curvature, slow rooting and growth rates, and high plant mortality rates [80,188]. This could be because some RIPs can kill plant cells by inactivating their ribosomes [120–122]. Several approaches have been used to overcome this problem. One strategy might be to introduce the gene encoding for the preprotein [80], this allows the RIP to accumulate in the apoplasma instead of the cytosol, thus preventing access to the ribosomes. Transgenic tobacco plants expressing the preprotein of trichosanthin exhibited resistance to cucumber mosaic virus (CMV) and tobacco mosaic virus (TMV) but did not show an abnormal phenotype [80]. In the case of PAP, despite being the most widely used, it inhibits protein synthesis and is toxic to plant cells, but transgenic plants have been obtained with mutants that are not toxic to the plant maintaining the antiviral activity [188]. The lack of toxicity of these mutants has been attributed to a change in the location of the protein preventing contact with ribosomes [188]. PAP (PAPI) has also been replaced by PAPII in order to obtain virus-resistant plants [104]. The protein sequence of PAPII shows only 41% identity to PAPI. PAPII expressed in transgenic tobacco was correctly processed to the mature form and accumulated to at least 10-fold higher levels than wild-type PAP (up to 250 ng/mg PAPII). PAPII is less toxic than PAP and symptomless transgenic lines expressing PAPII were resistant to TMV and PVX [104]. Another approach is to use a promoter that is induced by viral infection, thus, the gene that encodes for dianthin 30 was introduced into *N. benthamiana* and expressed from the promoter ACMV virion-sense [84]. This promoter is induced specifically by the ACMV infection and transgenic plants displayed a normal phenotype and were resistant to ACMV [84].

Finally, it should be noted that some virus-resistant transgenic plants have been reported to be also resistant to fungi [78,104], which adds interest to this type of approach to improve crop resistance.

7. Conclusions

After decades of research, RIPs continue to be a topic of interest and a useful tool in many research fields. The new advances in plant molecular biology, virology, immunotherapy, and nanotechnology open new possibilities in the use of RIPs in medicine and agriculture in order to find solutions to the continuous challenge posed by viruses to human health and crop yields.

Author Contributions: Conceptualization, L.C. and J.M.F.; writing—original draft preparation, R.I. and J.M.F.; writing—review and editing, L.C. and R.I.; funding acquisition, J.M.F. All authors have read and agreed to the published version of the manuscript.

Funding: This research was funded by Consejería de Educación (Junta de Castilla y León) to the GIR ProtIBio, grant number VA033G19.

Institutional Review Board Statement: Not applicable.

Informed Consent Statement: Not applicable.

Data Availability Statement: Data are available upon request. Please, contact the contributing authors.

Conflicts of Interest: The authors declare no conflict of interest.

References

1. De Clercq, E. Looking Back in 2009 at the Dawning of Antiviral Therapy Now 50 Years Ago: An Historical Perspective. In *Advances in Virus Research*; Maramorosch, K., Shatkin, A.J., Murphy, F.A., Eds.; Elsevier Academic Press Inc.: San Diego, CA, USA, 2009; Volume 73, pp. 1–53.
2. Ng, T.; Cheung, R.C.F.; Wong, J.H.; Chan, W.-Y. Proteins, peptides, polysaccharides, and nucleotides with inhibitory activity on human immunodeficiency virus and its enzymes. *Appl. Microbiol. Biotechnol.* **2015**, *99*, 10399–10414. [CrossRef] [PubMed]
3. Musidlak, O.; Nawrot, R.; Goździcka-Józefiak, A. Which Plant Proteins Are Involved in Antiviral Defense? Review on In Vivo and In Vitro Activities of Selected Plant Proteins against Viruses. *Int. J. Mol. Sci.* **2017**, *18*, 2300. [CrossRef] [PubMed]
4. Bolognesi, A.; Bortolotti, M.; Maiello, S.; Battelli, M.G.; Polito, L. Ribosome-Inactivating Proteins from Plants: A Historical Overview. *Molecules* **2016**, *21*, 1627. [CrossRef] [PubMed]
5. Ferreras, J.M.; Citores, L.; Iglesias, R.; Jiménez, P.; Girbés, T. Use of Ribosome-Inactivating Proteins from Sambucus for the Construction of Immunotoxins and Conjugates for Cancer Therapy. *Toxins* **2011**, *3*, 420–441. [CrossRef] [PubMed]
6. Stirpe, F. Ribosome-inactivating proteins: From toxins to useful proteins. *Toxicon* **2013**, *67*, 12–16. [CrossRef] [PubMed]
7. Di Maro, A.; Citores, L.; Russo, R.; Iglesias, R.; Ferreras, J.M. Sequence comparison and phylogenetic analysis by the Maximum Likelihood method of ribosome-inactivating proteins from angiosperms. *Plant Mol. Biol.* **2014**, *85*, 575–588. [CrossRef] [PubMed]
8. Olsnes, S. The history of ricin, abrin and related toxins. *Toxicon* **2004**, *44*, 361–370. [CrossRef]
9. Polito, L.; Bortolotti, M.; Battelli, M.G.; Calafato, G.; Bolognesi, A. Ricin: An Ancient Story for a Timeless Plant Toxin. *Toxins* **2019**, *11*, 324. [CrossRef]
10. Domashevskiy, A.V.; Goss, D.J. Pokeweed Antiviral Protein, a Ribosome Inactivating Protein: Activity, Inhibition and Prospects. *Toxins* **2015**, *7*, 274–298. [CrossRef]
11. Endo, Y.; Tsurugi, K. The RNA N-glycosidase activity of ricin A-chain. The characteristics of the enzymatic activity of ricin A-chain with ribosomes and with rRNA. *J. Biol. Chem.* **1988**, *263*, 8735–8739. [CrossRef]
12. Iglesias, R.; Citores, L.; Ferreras, J.M. Ribosomal RNA N-glycosylase Activity Assay of Ribosome-inactivating Proteins. *Bio-Protocol* **2017**, *7*, e2180. [CrossRef]
13. Citores, L.; Ragucci, S.; Ferreras, J.M.; Di Maro, A.; Iglesias, R. Ageritin, a Ribotoxin from Poplar Mushroom (Agrocybe aegerita) with Defensive and Antiproliferative Activities. *ACS Chem. Biol.* **2019**, *14*, 1319–1327. [CrossRef] [PubMed]
14. Barbieri, L.; Ferreras, J.M.; Barraco, A.; Ricci, P.; Stirpe, F. Some ribosome-inactivating proteins depurinate ribosomal RNA at multiple sites. *Biochem. J.* **1992**, *286*, 1–4. [CrossRef]
15. Barbieri, L.; Valbonesi, P.; Bonora, E.; Gorini, P.; Bolognesi, A.; Stirpe, F. Polynucleotide: Adenosine glycosidase activity of ribosome-inactivating proteins: Effect on DNA, RNA and poly(A). *Nucleic Acids Res.* **1997**, *25*, 518–522. [CrossRef]
16. Ogasawara, T.; Sawasaki, T.; Morishita, R.; Ozawa, A.; Madin, K.; Endo, Y. A new class of enzyme acting on damaged ribosomes: Ribosomal RNA apurinic site specific lyase found in wheat germ. *EMBO J.* **1999**, *18*, 6522–6531. [CrossRef]
17. Choudhary, N.L.; Yadav, O.P.; Lodha, M.L. Ribonuclease, deoxyribonuclease, and antiviral activity of Escherichia coli-expressed Bougainvillea xbuttiana antiviral protein-1. *Biochemistry* **2008**, *73*, 273–277. [CrossRef] [PubMed]
18. Chen, H.; Wang, Y.; Yan, M.; Yu, M.; Yao, Q. 5′-AMP Phosphatase activity on trichosanthin and other single chain ribosome inactivating proteins. *Chin. Biochem. J.* **1996**, *12*, 125–130.
19. Li, X.; Chen, W.-F.; Liu, W.-Y.; Wang, G.-H. Large-Scale Preparation of Two New Ribosome-Inactivating Proteins—Cinnamomin and Camphorin from the Seeds of Cinnamomum camphora. *Protein Expr. Purif.* **1997**, *10*, 27–31. [CrossRef] [PubMed]
20. Helmy, M.; Lombard, S.; Piéroni, G. Ricin RCA60: Evidence of Its Phospholipase Activity. *Biochem. Biophys. Res. Commun.* **1999**, *258*, 252–255. [CrossRef]
21. Shih, N.; McDonald, K.A.; Jackman, A.P.; Girbés, T.; Iglesias, R. Bifunctional plant defence enzymes with chitinase and ribosome inactivating activities from Trichosanthes kirilowii cell cultures. *Plant Sci.* **1997**, *130*, 145–150. [CrossRef]
22. Iglesias, R.; Citores, L.; Di Maro, A.; Ferreras, J.M. Biological activities of the antiviral protein BE27 from sugar beet (Beta vulgaris L.). *Planta* **2014**, *241*, 421–433. [CrossRef] [PubMed]
23. Polito, L.; Bortolotti, M.; Pedrazzi, M.; Mercatelli, D.; Battelli, M.G.; Bolognesi, A. Apoptosis and necroptosis induced by stenodactylin in neuroblastoma cells can be completely prevented through caspase inhibition plus catalase or necrostatin-1. *Phytomedicine* **2016**, *23*, 32–41. [CrossRef] [PubMed]
24. Panda, P.K.; Behera, B.; Meher, B.R.; Das, D.N.; Mukhopadhyay, S.; Sinha, N.; Naik, P.P.; Roy, B.; Das, J.; Paul, S.; et al. Abrus Agglutinin, a type II ribosome inactivating protein inhibits Akt/PH domain to induce endoplasmic reticulum stress mediated autophagy-dependent cell death. *Mol. Carcinog.* **2016**, *56*, 389–401. [CrossRef] [PubMed]
25. Rustgi, S.; Pollmann, S.; Buhr, F.; Springer, A.; Reinbothe, C.; Von Wettstein, D.; Reinbothe, S. JIP60-mediated, jasmonate- and senescence-induced molecular switch in translation toward stress and defense protein synthesis. *Proc. Natl. Acad. Sci. USA* **2014**, *111*, 14181–14186. [CrossRef]
26. Przydacz, M.; Jones, R.; Pennington, H.G.; Belmans, G.; Bruderer, M.; Greenhill, R.; Salter, T.; Wellham, P.A.; Cota, E.; Spanu, P.D. Mode of Action of the Catalytic Site in the N-Terminal Ribosome-Inactivating Domain of JIP60. *Plant Physiol.* **2020**, *183*, 385–398. [CrossRef]
27. Wang, R.-R.; Au, K.-Y.; Zheng, H.-Y.; Gao, L.-M.; Zhang, X.; Luo, R.-H.; Law, S.K.-Y.; Mak, A.N.-S.; Wong, K.-B.; Zhang, M.-X.; et al. The Recombinant Maize Ribosome-Inactivating Protein Transiently Reduces Viral Load in SHIV89.6 Infected Chinese Rhesus Macaques. *Toxins* **2015**, *7*, 156–169. [CrossRef]

28. Law, S.K.-Y.; Wang, R.-R.; Mak, A.N.-S.; Wong, K.-B.; Zheng, Y.-T.; Shaw, P.C. A switch-on mechanism to activate maize ribosome-inactivating protein for targeting HIV-infected cells. *Nucleic Acids Res.* **2010**, *38*, 6803–6812. [CrossRef]
29. Au, K.-Y.; Wang, R.-R.; Wong, Y.-T.; Wong, K.-B.; Zheng, Y.-T.; Shaw, P.C. Engineering a switch-on peptide to ricin A chain for increasing its specificity towards HIV-infected cells. *Biochim. Biophys. Acta Gen. Subj.* **2014**, *1840*, 958–963. [CrossRef]
30. Foà-Tomasi, L.; Campadelli-Fiume, G.; Barbieri, L.; Stirpe, F. Effect of ribosome-inactivating proteins on virus-infected cells. Inhibition of virus multiplication and of protein synthesis. *Arch. Virol.* **1982**, *71*, 323–332. [CrossRef]
31. Barnett, B.B.; Burns, N.J.; Park, K.J.; Dawson, M.I.; Kende, M.; Sidwell, R.W. Antiviral immunotoxins: Antibody-mediated delivery of gelonin inhibits Pichinde virus replication in vitro. *Antivir. Res.* **1991**, *15*, 125–138. [CrossRef]
32. Au, T.K.; Collins, R.A.; Lam, T.L.; Ng, T.B.; Fong, W.P.; Wan, D. The plant ribosome inactivating proteins luffin and saporin are potent inhibitors of HIV-1 integrase. *FEBS Lett.* **2000**, *471*, 169–172. [CrossRef]
33. Li, H.-G.; Huang, P.L.; Zhang, D.; Sun, Y.; Chen, H.-C.; Zhang, J.; Huang, P.L.; Kong, X.-P.; Lee-Huang, S. A new activity of anti-HIV and anti-tumor protein GAP31: DNA adenosine glycosidase – Structural and modeling insight into its functions. *Biochem. Biophys. Res. Commun.* **2010**, *391*, 340–345. [CrossRef] [PubMed]
34. Lee-Huang, S.; Kung, H.-F.; Huang, P.L.; Huang, P.L.; Li, B.-Q.; Huang, P.; Huang, H.I.; Chen, H.-C. A new class of anti-HIV agents: GAP31, DAPs 30 and 32. *FEBS Lett.* **1991**, *291*, 139–144. [CrossRef]
35. Kaur, I.; Gupta, R.C.; Puri, M. Ribosome inactivating proteins from plants inhibiting viruses. *Virol. Sin.* **2011**, *26*, 357–365. [CrossRef]
36. Fang, E.F.; Ng, T.B.; Shaw, P.C.; Wong, R.N.S. Recent Progress in Medicinal Investigations on Trichosanthin and other Ribo-some Inactivating Proteins from the Plant Genus Trichosanthes. *Curr. Med. Chem.* **2011**, *18*, 4410–4417. [CrossRef]
37. He, D.-X.; Tam, S.-C. Trichosanthin affects HSV-1 replication in Hep-2 cells. *Biochem. Biophys. Res. Commun.* **2010**, *402*, 670–675. [CrossRef]
38. Shi, W.-W.; Wong, K.-B.; Shaw, P.C. Structural and Functional Investigation and Pharmacological Mechanism of Trichosanthin, a Type 1 Ribosome-Inactivating Protein. *Toxins* **2018**, *10*, 335. [CrossRef] [PubMed]
39. Zheng, Y.T.; Ben, K.L.; Jin, S.W. Anti-HIV-1 activity of trichobitacin, a novel ribosome-inactivating protein. *Acta Pharmacol. Sin.* **2000**, *21*, 179–182.
40. Meng, Y.; Liu, S.; Li, J.; Zhao, X. Preparation of an antitumor and antivirus agent: Chemical modification of α-MMC and MAP30 from Momordica Charantia L. with covalent conjugation of polyethyelene glycol. *Int. J. Nanomed.* **2012**, *7*, 3133–3142. [CrossRef]
41. Yao, X.; Li, J.; Deng, N.; Wang, S.; Meng, Y.; Shen, F. Immunoaffinity purification of α-momorcharin from bitter melon seeds (Momordica charantia). *J. Sep. Sci.* **2011**, *34*, 3092–3098. [CrossRef]
42. Rothan, H.A.; Bahrani, H.; Mohamed, Z.; Rahman, N.A.; Yusof, R. Fusion of Protegrin-1 and Plectasin to MAP30 Shows Significant Inhibition Activity against Dengue Virus Replication. *PLoS ONE* **2014**, *9*, e94561. [CrossRef]
43. Sun, Y.; Huang, P.L.; Li, J.J.; Huang, Y.Q.; Zhang, L.; Huang, P.L.; Lee-Huang, S. Anti-HIV Agent MAP30 Modulates the Expression Profile of Viral and Cellular Genes for Proliferation and Apoptosis in AIDS-Related Lymphoma Cells Infected with Kaposi's Sarcoma-Associated Virus. *Biochem. Biophys. Res. Commun.* **2001**, *287*, 983–994. [CrossRef]
44. Fan, J.M.; Zhang, Q.; Xu, J.; Zhu, S.; Ke, T.; Gao, D.F.; Xu, Y.B. Inhibition on Hepatitis B virus in vitro of recombinant MAP30 from bitter melon. *Mol. Biol. Rep.* **2007**, *36*, 381–388. [CrossRef]
45. Lee-Huang, S.; Huang, P.L.; Bourinbaiar, A.S.; Chen, H.C.; Kung, H.F. Inhibition of the integrase of human immunodeficiency virus (HIV) type 1 by anti-HIV plant proteins MAP30 and GAP31. *Proc. Natl. Acad. Sci. USA* **1995**, *92*, 8818–8822. [CrossRef]
46. Bourinbaiar, A.S.; Lee-Huang, S. The Activity of Plant-Derived Antiretroviral Proteins MAP30 and GAP31 against Herpes Simplex Virus Infectionin Vitro. *Biochem. Biophys. Res. Commun.* **1996**, *219*, 923–929. [CrossRef]
47. Kaur, I.; Puri, M.; Ahmed, Z.; Blanchet, F.P.; Mangeat, B.; Piguet, V. Inhibition of HIV-1 Replication by Balsamin, a Ribosome Inactivating Protein of Momordica balsamina. *PLoS ONE* **2013**, *8*, e73780. [CrossRef]
48. Wachinger, M.; Samtleben, R.; Gerhauser, C.; Wagner, H.; Erfle, V. Bryodin, a single-chain ribosome-inactivating protein, selectively inhibits the growth of HIV-1-infected cells and reduces HIV-1 production. *Res. Exp. Med.* **1993**, *193*, 1–12. [CrossRef]
49. Leitman, E.M.; Palmer, C.D.; Buus, S.; Chen, F.; Riddell, L.; Sims, S.; Klenerman, P.; Saez-Cirion, A.; Walker, B.D.; Hess, P.R.; et al. Saporin-conjugated tetramers identify efficacious anti-HIV CD8+ T-cell specificities. *PLoS ONE* **2017**, *12*, e0184496. [CrossRef]
50. Yadav, S.K.; Batra, J.K. Mechanism of Anti-HIV Activity of Ribosome Inactivating Protein, Saporin. *Protein Pept. Lett.* **2015**, *22*, 497–503. [CrossRef]
51. Vivanco, J.M.; Tumer, N.E. Translation Inhibition of Capped and Uncapped Viral RNAs Mediated by Ribosome-Inactivating Proteins. *Phytopathology* **2003**, *93*, 588–595. [CrossRef] [PubMed]
52. Mansouri, S.; Choudhary, G.; Sarzala, P.M.; Ratner, L.; Hudak, K.A. Suppression of Human T-cell Leukemia Virus I Gene Expression by Pokeweed Antiviral Protein. *J. Biol. Chem.* **2009**, *284*, 31453–31462. [CrossRef]
53. Rothan, H.A.; Bahrani, H.; Shankar, E.M.; Rahman, N.A.; Yusof, R. Inhibitory effects of a peptide-fusion protein (Latarcin–PAP1–Thanatin) against chikungunya virus. *Antivir. Res.* **2014**, *108*, 173–180. [CrossRef]
54. Ishag, H.Z.A.; Li, C.; Huang, L.; Sun, M.-X.; Ni, B.; Guo, C.-X.; Mao, X. Inhibition of Japanese encephalitis virus infection in vitro and in vivo by pokeweed antiviral protein. *Virus Res.* **2013**, *171*, 89–96. [CrossRef]
55. Uckun, F.M.; Rustamova, L.; Vassilev, A.O.; Tibbles, H.E.; Petkevich, A.S. CNS activity of Pokeweed Anti-viral Protein (PAP) in mice infected with Lymphocytic Choriomeningitis Virus (LCMV). *BMC Infect. Dis.* **2005**, *5*, 9. [CrossRef]

56. He, Y.-W.; Guo, C.-X.; Pan, Y.-F.; Peng, C.; Weng, Z.-H. Inhibition of hepatitis B virus replication by pokeweed antiviral protein in vitro. *World J. Gastroenterol.* **2008**, *14*, 1592–1597. [CrossRef]
57. Rajamohan, F.; Venkatachalam, T.K.; Irvin, J.D.; Uckun, F.M. Pokeweed Antiviral Protein Isoforms PAP-I, PAP-II, and PAP-III Depurinate RNA of Human Immunodeficiency Virus (HIV)-1. *Biochem. Biophys. Res. Commun.* **1999**, *260*, 453–458. [CrossRef]
58. Di, R.; Tumer, N.E. Pokeweed Antiviral Protein: Its Cytotoxicity Mechanism and Applications in Plant Disease Resistance. *Toxins* **2015**, *7*, 755–772. [CrossRef]
59. Ussery, M.A.; Irvin, J.D.; Hardesty, B. Inhibition of Poliovirus Replication by A Plant Antiviral Peptide. *Ann. N. Y. Acad. Sci.* **1977**, *284*, 431–440. [CrossRef]
60. McGrath, M.S.; Hwang, K.M.; Caldwell, S.E.; Gaston, I.; Luk, K.C.; Wu, P.; Ng, V.L.; Crowe, S.; Daniels, J.; Marsh, J. GLQ223: An inhibitor of human immunodeficiency virus replication in acutely and chronically infected cells of lymphocyte and mononuclear phagocyte lineage. *Proc. Natl. Acad. Sci. USA* **1989**, *86*, 2844–2848. [CrossRef]
61. Byers, V.S.; Levin, A.S.; Waites, L.A.; Starrett, B.A.; Mayer, R.A.; Clegg, J.A.; Price, M.R.; Robins, R.A.; Delaney, M.; Baldwin, R.W. A phase I/II study of trichosanthin treatment of HIV disease. *AIDS* **1990**, *4*, 1189–1196. [CrossRef]
62. Wen, D.; Wang, J.; Yan, H.; Chen, J.; Xia, K.; Liu, J.; Zhang, A. Effect of Radix Trichosanthis and Trichosanthin on Hepatitis B Virus in HepG2.2.15 Cells. *J. Nanosci. Nanotechnol.* **2015**, *15*, 2094–2098. [CrossRef] [PubMed]
63. Barbieri, L.; Battelli, M.G.; Stirpe, F. Ribosome-inactivating proteins from plants. *Biochim. Biophys. Acta Rev. Biomembr.* **1993**, *1154*, 237–282. [CrossRef]
64. Citores, L.; Iglesias, R.; Ferreras, J.M. Ribosome Inactivating Proteins from Plants: Biological Properties and their Use in Experimental Therapy. In *Antitumor Potential and Other Emerging Medicinal Properties of Natural Compounds*; Fang, E.F., Ng, T.B., Eds.; Springer: Dordrecht, The Netherlands, 2013; pp. 127–143.
65. Lv, Q.; Yang, X.-Z.; Fu, L.-Y.; Lu, Y.-T.; Lu, Y.-H.; Zhao, J.; Wang, F.-J. Recombinant expression and purification of a MAP30-cell penetrating peptide fusion protein with higher anti-tumor bioactivity. *Protein Expr. Purif.* **2015**, *111*, 9–17. [CrossRef] [PubMed]
66. Lin, B.; Yang, X.-Z.; Cao, X.-W.; Zhang, T.-Z.; Wang, F.-J.; Zhao, J. A novel trichosanthin fusion protein with increased cytotoxicity to tumor cells. *Biotechnol. Lett.* **2016**, *39*, 71–78. [CrossRef] [PubMed]
67. Ferens, W.A.; Hovde, C.J. Antiviral Activity of Shiga Toxin 1: Suppression of Bovine Leukemia Virus-Related Spontaneous Lymphocyte Proliferation. *Infect. Immun.* **2000**, *68*, 4462–4469. [CrossRef] [PubMed]
68. Shi, P.L.; Binnington, B.; Sakac, D.; Katsman, Y.; Ramkumar, S.; Gariépy, J.; Kim, M.; Branch, D.R.; Lingwood, C. Verotoxin A Subunit Protects Lymphocytes and T Cell Lines against X4 HIV Infection in Vitro. *Toxins* **2012**, *4*, 1517–1534. [CrossRef]
69. Yadav, S.K.; Batra, J.K. Ribotoxin restrictocin manifests anti-HIV-1 activity through its specific ribonuclease activity. *Int. J. Biol. Macromol.* **2015**, *76*, 58–62. [CrossRef]
70. Wong, J.H.; Wang, H.X.; Ng, T.B. Marmorin, a new ribosome inactivating protein with antiproliferative and HIV-1 reverse transcriptase inhibitory activities from the mushroom Hypsizigus marmoreus. *Appl. Microbiol. Biotechnol.* **2008**, *81*, 669–674. [CrossRef]
71. Wang, H.; Ng, T.B. Isolation and characterization of velutin, a novel low-molecular-weight ribosome-inactivating protein from winter mushroom (Flammulina velutipes) fruiting bodies. *Life Sci.* **2001**, *68*, 2151–2158. [CrossRef]
72. Lam, S.; Ng, T. First Simultaneous Isolation of a Ribosome Inactivating Protein and an Antifungal Protein from a Mushroom (Lyophyllum shimeji) Together with Evidence for Synergism of their Antifungal Effects. *Arch. Biochem. Biophys.* **2001**, *393*, 271–280. [CrossRef]
73. Yao, Q.Z.; Yu, M.M.; Ooi, L.S.; Ng, T.B.; Chang, S.T.; Sun, S.S.; Ooi, V.E. Isolation and Characterization of a Type 1 Ribosome-Inactivating Protein from Fruiting Bodies of the Edible Mushroom (Volvariella volvacea). *J. Agric. Food Chem.* **1998**, *46*, 788–792. [CrossRef]
74. Arslan, I.; Akgul, H.; Kara, M. Saporin, a Polynucleotide–Adenosine Nucleosidase, May Be an Efficacious Therapeutic Agent for SARS-CoV-2 Infection. *SLAS Discov. Adv. Life Sci.* **2020**. [CrossRef]
75. Hassan, Y.; Ogg, S.; Ge, H. Novel Binding Mechanisms of Fusion Broad Range Anti-Infective Protein Ricin A Chain Mutant-Pokeweed Antiviral Protein 1 (RTAM-PAP1) against SARS-CoV-2 Key Proteins in Silico. *Toxins* **2020**, *12*, 602. [CrossRef]
76. Vandenbussche, F.; Desmyter, S.; Ciani, M.; Proost, P.; Peumans, W.J.; Van Damme, E.J.M. Analysis of the in planta antiviral activity of elderberry ribosome-inactivating proteins. *Eur. J. Biochem.* **2004**, *271*, 1508–1515. [CrossRef]
77. Vandenbussche, F.; Peumans, W.J.; Desmyter, S.; Proost, P.; Ciani, M.; Van Damme, E.J.M. The type-1 and type-2 ribosome-inactivating proteins from Iris confer transgenic tobacco plants local but not systemic protection against viruses. *Planta* **2004**, *220*, 211–221. [CrossRef]
78. Huang, M.-X.; Hou, P.; Wei, Q.; Xu, Y.; Chen, F. A ribosome-inactivating protein (curcin 2) induced from Jatropha curcas can reduce viral and fungal infection in transgenic tobacco. *Plant Growth Regul.* **2007**, *54*, 115–123. [CrossRef]
79. Lam, Y.-H.; Wong, Y.-S.; Wang, B.; Wong, R.N.-S.; Yeung, H.-W.; Shaw, P.-C. Use of trichosanthin to reduce infection by turnip mosaic virus. *Plant Sci.* **1996**, *114*, 111–117. [CrossRef]
80. Krishnan, R.; McDonald, K.A.; Dandekar, A.M.; Jackman, A.P.; Falk, B. Expression of recombinant trichosanthin, a ribosome-inactivating protein, in transgenic tobacco. *J. Biotechnol.* **2002**, *97*, 69–88. [CrossRef]
81. Zhu, F.; Zhang, P.; Meng, Y.-F.; Xu, F.; Zhang, D.-W.; Cheng, J.; Lin, H.-H.; Xi, D. Alpha-momorcharin, a RIP produced by bitter melon, enhances defense response in tobacco plants against diverse plant viruses and shows antifungal activity in vitro. *Planta* **2012**, *237*, 77–88. [CrossRef]

82. Yang, T.; Meng, Y.; Chen, L.-J.; Lin, H.; Xi, D.-H. The Roles of Alpha-Momorcharin and Jasmonic Acid in Modulating the Response of Momordica charantia to Cucumber Mosaic Virus. *Front. Microbiol.* **2016**, *7*, 1796. [CrossRef]
83. Ruan, X.-L.; Liu, L.-F.; Li, H. Transgenic tobacco plants with ribosome inactivating protein gene cassin from Cassia occidentalis and their resistance to tobacco mosaic virus. *J. Plant Physiol. Mol. Biol.* **2007**, *33*, 517–523.
84. Hong, Y.; Saunders, K.; Hartley, M.R.; Stanley, J. Resistance to Geminivirus Infection by Virus-Induced Expression of Dianthin in Transgenic Plants. *Virology* **1996**, *220*, 119–127. [CrossRef] [PubMed]
85. Stirpe, F.; Williams, D.G.; Onyon, L.J.; Legg, R.F.; Stevens, W.A. Dianthins, ribosome-damaging proteins with anti-viral properties from Dianthus caryophyllus L. (carnation). *Biochem. J.* **1981**, *195*, 399–405. [CrossRef]
86. Iglesias, R.; Pérez, Y.; de Torre, C.; Ferreras, J.M.; Antolín, P.; Jiménez, P.; Ángeles Rojo, M.; Méndez, E.; Girbés, T. Molecular characterization and systemic induction of single-chain ribosome-inactivating proteins (RIPs) in sugar beet (Beta vulgaris) leaves. *J. Exp. Bot.* **2005**, *56*, 1675–1684. [CrossRef]
87. Iglesias, R.; Citores, L.; Ragucci, S.; Russo, R.; Di Maro, A.; Ferreras, J.M. Biological and antipathogenic activities of ribosome-inactivating proteins from *Phytolacca dioica* L. *Biochim. Biophys. Acta Gen. Subj.* **2016**, *1860*, 1256–1264. [CrossRef]
88. Roy, S.; Sadhana, P.; Begum, M.; Kumar, S.; Lodha, M.; Kapoor, H. Purification, characterization and cloning of antiviral/ribosome inactivating protein from Amaranthus tricolor leaves. *Phytochemistry* **2006**, *67*, 1865–1873. [CrossRef]
89. Kwon, S.-Y.; An, C.S.; Liu, J.R.; Paek, K.-H. A Ribosome–Inactivating Protein fromAmaranthus viridis. *Biosci. Biotechnol. Biochem.* **1997**, *61*, 1613–1614. [CrossRef]
90. Gholizadeh, A. Purification of a ribosome-inactivating protein with antioxidation and root developer potencies from Celosia plumosa. *Physiol. Mol. Biol. Plants* **2018**, *25*, 243–251. [CrossRef]
91. Baranwal, V.K.; Tumer, N.E.; Kapoor, H.C. Depurination of ribosomal RNA and inhibition of viral RNA translation by an antiviral protein of Celosia cristata. *Indian J. Exp. Biol.* **2002**, *40*, 1195–1197.
92. Balasubrahmanyam, A.; Baranwal, V.K.; Lodha, M.; Varma, A.; Kapoor, H. Purification and properties of growth stage-dependent antiviral proteins from the leaves of Celosia cristata. *Plant Sci.* **2000**, *154*, 13–21. [CrossRef]
93. Begam, M.; Kumar, S.; Roy, S.; Campanella, J.J.; Kapoor, H. Molecular cloning and functional identification of a ribosome inactivating/antiviral protein from leaves of post-flowering stage of Celosia cristata and its expression in E. coli. *Phytochemistry* **2006**, *67*, 2441–2449. [CrossRef] [PubMed]
94. Dutt, S.; Narwal, S.; Kapoor, H.C.; Lodha, M.L. Isolation and Characterization of Two Protein Isoforms with Antiviral Activity from Chenopodium album L Leaves. *J. Plant Biochem. Biotechnol.* **2003**, *12*, 117–122. [CrossRef]
95. Park, J.-S.; Hwang, D.-J.; Lee, S.-M.; Kim, Y.-T.; Choi, S.-B.; Cho, K.-J. Ribosome-inactivating activity and cDNA cloning of antiviral protein isoforms of Chenopodium album. *Mol. Cells* **2004**, *17*, 73–80. [PubMed]
96. Torky, Z.A. Isolation and characterization of antiviral protein from Salsola longifolia leaves expressing polynucleotide adenosine glycoside activity. *TOJSAT* **2012**, *2*, 52–58.
97. Straub, P.; Adam, G.; Mundry, K.-W. Isolation and Characterization of a Virus Inhibitor from Spinach (Spinacia oleracea L.). *J. Phytopathol.* **1986**, *115*, 357–367. [CrossRef]
98. Kawade, K.; Ishizaki, T.; Masuda, K. Differential expression of ribosome-inactivating protein genes during somatic embryogenesis in spinach (Spinacia oleracea). *Physiol. Plant.* **2008**, *134*, 270–281. [CrossRef]
99. Moon, Y.H.; Song, S.K.; Choi, K.W.; Lee, J.S. Expression of a cDNA encoding Phytolacca insularis antiviral protein confers virus resistance on transgenic potato plants. *Mol. Cells* **1997**, *7*, 807–815.
100. Bulgari, D.; Landi, N.; Ragucci, S.; Faoro, F.; Di Maro, A. Antiviral Activity of PD-L1 and PD-L4, Type 1 Ribosome Inactivating Proteins from Leaves of *Phytolacca dioica* L. in the Pathosystem *Phaseolus vulgaris*–Tobacco Necrosis Virus (TNV). *Toxins* **2020**, *12*, 524. [CrossRef]
101. Sipahioglu, H.M.; Kaya, I.; Usta, M.; Ünal, M.; Ozcan, D.; Özer, M.; Güller, A.; Pallás, V. Pokeweed (Phytolacca americana L.) antiviral protein inhibits Zucchini yellow mosaic virus infection in a dose-dependent manner in squash plants. *Turk. J. Agric. For.* **2017**, *41*, 256–262. [CrossRef]
102. Lodge, J.K.; Kaniewski, W.K.; Tumer, N.E. Broad-spectrum virus resistance in transgenic plants expressing pokeweed antiviral protein. *Proc. Natl. Acad. Sci. USA* **1993**, *90*, 7089–7093. [CrossRef]
103. Domashevskiy, A.V.; Williams, S.; Kluge, C.; Cheng, S.-Y. Plant Translation Initiation Complex eIFiso4F Directs Pokeweed Antiviral Protein to Selectively Depurinate Uncapped Tobacco Etch Virus RNA. *Biochemistry* **2017**, *56*, 5980–5990. [CrossRef] [PubMed]
104. Wang, P.; Zoubenko, O.; Tumer, N.E. Reduced toxicity and broad spectrum resistance to viral and fungal infection in transgenic plants expressing pokeweed antiviral protein II. *Plant Mol. Biol.* **1998**, *38*, 957–964. [CrossRef] [PubMed]
105. Bolognesi, A.; Polito, L.; Olivieri, F.; Valbonesi, P.; Barbieri, L.; Battelli, M.G.; Carusi, M.V.; Benvenuto, E.; Blanco, F.D.V.; Di Maro, A.; et al. New ribosome-inactivating proteins with polynucleotide:adenosine glycosidase and antiviral activities from Basella rubra L. and Bougainvillea spectabilis Willd. *Planta* **1997**, *203*, 422–429. [CrossRef] [PubMed]
106. Srivastava, S.; Verma, H.N.; Srivastava, A.; Prasad, V. BDP-30, a systemic resistance inducer from Boerhaavia diffusa L., suppresses TMV infection, and displays homology with ribosome-inactivating proteins. *J. Biosci.* **2015**, *40*, 125–135. [CrossRef] [PubMed]
107. Bolognesi, A.; Polito, L.; Lubelli, C.; Barbieri, L.; Parente, A.; Stirpe, F. Ribosome-inactivating and Adenine Polynucleotide Glycosylase Activities in Mirabilis jalapa L. Tissues. *J. Biol. Chem.* **2002**, *277*, 13709–13716. [CrossRef] [PubMed]

108. Güller, A.; Sipahioğlu, H.M.; Usta, M.; Durak, E.D. Antiviral and Antifungal Activity of Biologically Active Recombinant Bouganin Protein from Bougainvillea spectabilis Willd. *J. Agric. Sci.* **2018**, *24*, 227–237. [CrossRef]
109. Narwal, S.; Balasubrahmanyam, A.; Lodha, M.L.; Kapoor, H.C. Purification and properties of antiviral proteins from the leaves of Bougainvillea xbuttiana. *Indian J. Biochem. Biophys.* **2001**, *38*, 342–347.
110. Narwal, S.; Balasubrahmanyam, A.; Sadhna, P.; Kapoor, H.; Lodha, M.L. A systemic resistance inducing antiviral protein with N-glycosidase activity from Bougainvillea xbuttiana leaves. *Indian J. Exp. Biol.* **2001**, *39*, 600–603.
111. Olivieri, F.; Prasad, V.; Valbonesi, P.; Srivastava, S.; Ghosal-Chowdhury, P.; Barbieri, L.; Bolognesi, A.; Stirpe, F. A systemic antiviral resistance-inducing protein isolated from Clerodendrum inerme Gaertn. is a polynucleotide:adenosine glycosidase (ribosome-inactivating protein). *FEBS Lett.* **1996**, *396*, 132–134. [CrossRef]
112. Prasad, V.; Mishra, S.K.; Srivastava, S.; Srivastava, A. A virus inhibitory protein isolated from Cyamopsis tetragonoloba (L.) Taub. upon induction of systemic antiviral resistance shares partial amino acid sequence homology with a lectin. *Plant Cell Rep.* **2014**, *33*, 1467–1478. [CrossRef]
113. Verma, H.; Srivastava, S.; Kumar, D. Induction of Systemic Resistance in Plants Against Viruses by a Basic Protein from Clerodendrum aculeatum Leaves. *Phytopathology* **1996**, *86*, 485–492. [CrossRef]
114. Verma, H.N.; Tewari, K.K.; Kumar, D.; Tuteja, N. Cloning and characterisation of a gene encoding an antiviral protein from *Clerodendrum aculeatum* L. *Plant Mol. Biol.* **1997**, *33*, 745–751. [CrossRef]
115. Srivastava, A.; Trivedi, S.; Krishna, S.K.; Verma, H.N.; Prasad, V. Suppression of Papaya ringspot virus infection in Carica papaya with CAP-34, a systemic antiviral resistance inducing protein from Clerodendrum aculeatum. *Eur. J. Plant Pathol.* **2008**, *123*, 241–246. [CrossRef]
116. Chen, Y.; Peumans, W.J.; Van Damme, E.J.M. The Sambucus nigratype-2 ribosome-inactivating protein SNA-I′ exhibits in planta antiviral activity in transgenic tobacco. *FEBS Lett.* **2002**, *516*, 27–30. [CrossRef]
117. Iglesias, R.; Pérez, Y.; Citores, L.; Ferreras, J.M.; Méndez, E.; Girbés, T. Elicitor-dependent expression of the ribosome-inactivating protein beetin is developmentally regulated. *J. Exp. Bot.* **2008**, *59*, 1215–1223. [CrossRef]
118. Yang, T.; Zhu, L.-S.; Meng, Y.; Lv, R.; Zhou, Z.; Zhu, L.; Lin, H.-H.; Xi, D. Alpha-momorcharin enhances Tobacco mosaic virus resistance in tobacco NN by manipulating jasmonic acid-salicylic acid crosstalk. *J. Plant Physiol.* **2018**, *223*, 116–126. [CrossRef]
119. Harley, S.M.; Beevers, H. Ricin inhibition of in vitro protein synthesis by plant ribosomes. *Proc. Natl. Acad. Sci. USA* **1982**, *79*, 5935–5938. [CrossRef]
120. Iglesias, R.; Arias, F.; Ángeles Rojo, M.; Escarmis, C.; Ferreras, J.M.; Girbés, T. Molecular action of the type 1 ribosome-inactivating protein saporin 5 on Vicia sativa ribosomes. *FEBS Lett.* **1993**, *325*, 291–294. [CrossRef]
121. Ángeles Rojo, M.; Arias, F.; Ferreras, J.M.; Iglesias, R.; Muñoz, R.; Girbés, T. Development of a cell-free translation system from Cucumis melo: Preparation, optimization and evaluation of sensitivity to some translational inhibitors. *Plant Sci.* **1993**, *90*, 127–134. [CrossRef]
122. Ángeles Rojo, M.; Arias, F.; Iglesias, R.; Ferreras, J.M.; Munoz, R.; Girbés, T. A Cucumis sativus cell-free translation system: Preparation, optimization and sensitivity to some antibiotics and ribosome-inactivating proteins. *Physiol. Plant.* **1993**, *88*, 549–556. [CrossRef]
123. Bonness, M.S.; Ready, M.P.; Irvin, J.D.; Mabry, T.J. Pokeweed antiviral protein inactivates pokeweed ribosomes; implications for the antiviral mechanism. *Plant J.* **1994**, *5*, 173–183. [CrossRef] [PubMed]
124. Taylor, S.; Massiah, A.; Lomonossoff, G.; Roberts, L.M.; Lord, J.M.; Hartley, M. Correlation between the activities of five ribosome-inactivating proteins in depurination of tobacco ribosomes and inhibition of tobacco mosaic virus infection. *Plant J.* **1994**, *5*, 827–835. [CrossRef] [PubMed]
125. Park, S.-W.; Vepachedu, R.; Sharma, N.; Vivanco, J.M. Ribosome-inactivating proteins in plant biology. *Planta* **2004**, *219*, 1093–1096. [CrossRef] [PubMed]
126. Luvisi, A.; Panattoni, A.; Materazzi, A.; Rizzo, D.; De Bellis, L.; Aprile, A.; Sabella, E.; Rinaldelli, E. Early trans-plasma membrane responses to Tobacco mosaic virus infection. *Acta Physiol. Plant.* **2017**, *39*, 225. [CrossRef]
127. Girbés, T.; Ferreras, J.M. Ribosome-inactivating proteins from plants. *Recent Res. Dev. Agric. Biol. Chem.* **1998**, *2*, 1–16.
128. Hudak, K.A.; Wang, P.; Tumer, N.E. A novel mechanism for inhibition of translation by pokeweed antiviral protein: Depurination of the capped RNA template. *RNA* **2000**, *6*, 369–380. [CrossRef]
129. Zoubenko, O.; Hudak, K.; Tumer, N.E. A non-toxic pokeweed antiviral protein mutant inhibits pathogen infection via a novel salicylic acid-independent pathway. *Plant Mol. Biol.* **2000**, *44*, 219–229. [CrossRef]
130. Picard, D.; Kao, C.C.; Hudak, K.A. Pokeweed Antiviral Protein Inhibits Brome Mosaic Virus Replication in Plant Cells. *J. Biol. Chem.* **2005**, *280*, 20069–20075. [CrossRef]
131. Karran, R.A.; Hudak, K.A. Depurination of Brome mosaic virus RNA3 inhibits its packaging into virus particles. *Nucleic Acids Res.* **2011**, *39*, 7209–7222. [CrossRef]
132. Kushwaha, G.S.; Yamini, S.; Kumar, M.; Sinha, M.; Kaur, P.; Sharma, S.; Singh, T.P. First structural evidence of sequestration of mRNA cap structures by type 1 ribosome inactivating protein from Momordica balsamina. *Proteins Struct. Funct. Bioinform.* **2013**, *81*, 896–905. [CrossRef]
133. Zhu, F.; Zhu, P.; Xu, F.; Che, Y.; Ma, Y.; Ji, Z.-L. Alpha-momorcharin enhances Nicotiana benthamiana resistance to tobacco mosaic virus infection through modulation of reactive oxygen species. *Mol. Plant Pathol.* **2020**, *21*, 1212–1226. [CrossRef] [PubMed]

134. Smirnov, S.; Shulaev, V.; Tumer, N.E. Expression of Pokeweed Antiviral Protein in Transgenic Plants Induces Virus Resistance in Grafted Wild-Type Plants Independently of Salicylic Acid Accumulation and Pathogenesis-Related Protein Synthesis. *Plant Physiol.* **1997**, *114*, 1113–1121. [CrossRef] [PubMed]
135. Zoubenko, O.; Uckun, F.; Hur, Y.; Chet, I.; Tumer, N. Plant resistance to fungal infection induced by nontoxic pokeweed antiviral protein mutants. *Nat. Biotechnol.* **1997**, *15*, 992–996. [CrossRef] [PubMed]
136. Vind, A.C.; Genzor, A.V.; Bekker-Jensen, S. Ribosomal stress-surveillance:Three pathways is a magic number. *Nucleic Acids Res.* **2020**, *48*, 10648–10661. [CrossRef] [PubMed]
137. Arunachalam, C.; Doohan, F.M. Trichothecene toxicity in eukaryotes: Cellular and molecular mechanisms in plants and animals. *Toxicol. Lett.* **2013**, *217*, 149–158. [CrossRef] [PubMed]
138. Madan, S.; Ghosh, P.C. Interaction of gelonin with macrophages: Effect of lysosomotropic amines. *Exp. Cell Res.* **1992**, *198*, 52–58. [CrossRef]
139. Chan, W.Y.; Huang, H.; Tam, S.C. Receptor-mediated endocytosis of trichosanthin in choriocarcinoma cells. *Toxicology* **2003**, *186*, 191–203. [CrossRef]
140. Freeman, M.C.; Peek, C.T.; Becker, M.M.; Smith, E.C.; Denison, M.R. Coronaviruses Induce Entry-Independent, Continuous Macropinocytosis. *MBio* **2014**, *5*, e01340-14. [CrossRef]
141. Schelhaas, M. Come in and take your coat off—How host cells provide endocytosis for virus entry. *Cell. Microbiol.* **2010**, *12*, 1378–1388. [CrossRef]
142. Teltow, G.J.; Irvin, J.D.; Aron, G.M. Inhibition of herpes simplex virus DNA synthesis by pokeweed antiviral protein. *Antimicrob. Agents Chemother.* **1983**, *23*, 390–396. [CrossRef]
143. Zarling, J.M.; Moran, P.A.; Haffar, O.; Sias, J.; Richman, D.D.; Spina, C.A.; Myers, D.E.; Kuebelbeck, V.; Ledbetter, J.A.; Uckun, F.M. Inhibition of HIV replication by pokeweed antiviral protein targeted to CD4+ cells by monoclonal antibodies. *Nature* **1990**, *347*, 92–95. [CrossRef] [PubMed]
144. Krivdova, G.; Hudak, K.A. Pokeweed antiviral protein restores levels of cellular APOBEC3G during HIV-1 infection by depurinating Vif mRNA. *Antivir. Res.* **2015**, *122*, 51–54. [CrossRef] [PubMed]
145. Zhabokritsky, A.; Mansouri, S.; Hudak, K.A. Pokeweed antiviral protein alters splicing of HIV-1 RNAs, resulting in reduced virus production. *RNA* **2014**, *20*, 1238–1247. [CrossRef] [PubMed]
146. Wang, Y.-X.; Neamati, N.; Jacob, J.; Palmer, I.; Stahl, S.J.; Kaufman, J.D.; Huang, P.L.; Huang, P.L.; Winslow, H.E.; Pommier, Y.; et al. Solution Structure of Anti-HIV-1 and Anti-Tumor Protein MAP30: Structural insights into its multiple functions. *Cell* **1999**, *99*, 433–442. [CrossRef]
147. Zhao, W.-L.; Feng, D.; Wu, J.; Sui, S.-F. Trichosanthin inhibits integration of human immunodeficiency virus type 1 through depurinating the long-terminal repeats. *Mol. Biol. Rep.* **2009**, *37*, 2093–2098. [CrossRef]
148. Li, M.-X.; Yeung, H.-W.; Pan, L.-P.; Chan, S.I. Trichosanthin, a potent HIV-1 inhibitor, can cleave supercoiled DNA in vitro. *Nucleic Acids Res.* **1991**, *19*, 6309–6312. [CrossRef]
149. Huang, P.L.; Chen, H.C.; Kung, H.F.; Huang, P.; Huang, H.I.; Lee-Huang, S. Anti-HIV plant proteins catalyze topological changes of DNA into inactive forms. *BioFactors* **1992**, *4*, 37–41.
150. Zhao, W.-L.; Zhang, F.; Feng, D.; Wu, J.; Chen, S.; Sui, S.-F. A novel sorting strategy of trichosanthin for hijacking human immunodeficiency virus type 1. *Biochem. Biophys. Res. Commun.* **2009**, *384*, 347–351. [CrossRef]
151. Zhao, W.; Feng, D.; Sun, S.; Han, T.; Sui, S. The anti-viral protein of trichosanthin penetrates into human immunodeficiency virus type 1. *Acta Biochim. Biophys. Sin.* **2009**, *42*, 91–97. [CrossRef]
152. He, N.; Zheng, Y.; Tam, S.-C. The anti-herpetic activity of trichosanthin via the nuclear factor-κB and p53 pathways. *Life Sci.* **2012**, *90*, 673–681. [CrossRef]
153. Ouyang, D.-Y.; Chan, H.; Wang, Y.-Y.; Huang, H.; Tam, S.-C.; Zheng, Y.-T. An inhibitor of c-Jun N-terminal kinases (CEP-11004) counteracts the anti-HIV-1 action of trichosanthin. *Biochem. Biophys. Res. Commun.* **2006**, *339*, 25–29. [CrossRef] [PubMed]
154. He, D.; Yau, K.; He, X.-H.; Shi, H.; Zheng, Y.; Tam, S.-C. Conversion of trichosanthin-induced CD95 (Fas) type I into type II apoptotic signaling during Herpes simplex virus infection. *Mol. Immunol.* **2011**, *48*, 2000–2008. [CrossRef] [PubMed]
155. Huang, H.; Chan, H.; Wang, Y.-Y.; Ouyang, D.-Y.; Zheng, Y.-T.; Tam, S.-C. Trichosanthin suppresses the elevation of p38 MAPK, and Bcl-2 induced by HSV-1 infection in Vero cells. *Life Sci.* **2006**, *79*, 1287–1292. [CrossRef] [PubMed]
156. Wu, C.C.-C.; Peterson, A.; Zinshteyn, B.; Regot, S.; Green, R. Ribosome Collisions Trigger General Stress Responses to Regulate Cell Fate. *Cell* **2020**, *182*, 404–416. [CrossRef]
157. Kahn, J.O.; Gorelick, K.J.; Gatti, G.; Arri, C.J.; Lifson, J.D.; Gambertoglio, J.G.; Boström, A.; Williams, R. Safety, activity, and pharmacokinetics of GLQ223 in patients with AIDS and AIDS-related complex. *Antimicrob. Agents Chemother.* **1994**, *38*, 260–267. [CrossRef]
158. Byers, V.; Levin, A.; Malvino, A.; Waites, L.; Robins, R.; Baldwin, R. A Phase II Study of Effect of Addition of Trichosanthin to Zidovudine in Patients with HIV Disease and Failing Antiretroviral Agents. *AIDS Res. Hum. Retrovir.* **1994**, *10*, 413–420. [CrossRef]
159. Kim, J.-S.; Jun, S.-Y.; Shin, T.-H. Critical Issues in the Development of Immunotoxins for Anticancer Therapy. *J. Pharm. Sci.* **2020**, *109*, 104–115. [CrossRef]
160. Rust, A.; Partridge, L.J.; Davletov, B.; Hautbergue, G.M. The Use of Plant-Derived Ribosome Inactivating Proteins in Immunotoxin Development: Past, Present and Future Generations. *Toxins* **2017**, *9*, 344. [CrossRef]

161. Spiess, K.; Jakobsen, M.H.; Kledal, T.N.; Rosenkilde, M.M. The future of antiviral immunotoxins. *J. Leukoc. Biol.* **2016**, *99*, 911–925. [CrossRef]
162. Berger, E.A.; Pastan, I. Immunotoxin Complementation of HAART to Deplete Persisting HIV-Infected Cell Reservoirs. *PLoS Pathog.* **2010**, *6*, e1000803. [CrossRef]
163. Tazzari, P.L.; De Totero, D.; Bolognesi, A.; Testoni, N.; Pileri, S.; Roncella, S.; Reato, G.; Stein, H.; Gobbi, M.; Stirpe, F. An Epstein-Barr virus-infected lymphoblastoid cell line (D430B) that grows in SCID-mice with the morphologic features of a CD30+ anaplastic large cell lymphoma, and is sensitive to anti-CD30 immunotoxins. *Haematology* **1999**, *84*, 988–995.
164. Tedder, T.F.; Goldmacher, V.S.; Lambert, J.M.; Schlossman, S.F. Epstein Barr virus binding induces internalization of the C3d receptor: A novel immunotoxin delivery system. *J. Immunol.* **1986**, *137*, 1387–1391. [PubMed]
165. Smee, D.F.; Sidwell, R.W.; Barnett, B.B. Combination of antiviral immunotoxin and ganciclovir or cidofovir for the treatment of murine cytomegalovirus infections. *Antivir. Res.* **1996**, *32*, 165–171. [CrossRef]
166. Till, M.A.; Ghetie, V.; Gregory, T.; Patzer, E.J.; Porter, J.P.; Uhr, J.W.; Capon, D.J.; Vitetta, E.S. HIV-infected cells are killed by rCD4-ricin A chain. *Science* **1988**, *242*, 1166–1168. [CrossRef] [PubMed]
167. Hassan, Y.; Ogg, S.; Ge, H. Expression of novel fusion antiviral proteins ricin a chain-pokeweed antiviral proteins (RTA-PAPs) in Escherichia coli and their inhibition of protein synthesis and of hepatitis B virus in vitro. *BMC Biotechnol.* **2018**, *18*, 47. [CrossRef] [PubMed]
168. Matsushita, S.; Koito, A.; Maeda, Y.; Hattori, T.; Takatsuki, K. Selective Killing of HIV-Infected Cells by Anti-gp120 Immunotoxins. *AIDS Res. Hum. Retrovir.* **1990**, *6*, 193–203. [CrossRef]
169. Kim, Y.W.; Fung, M.S.; Sun, N.C.; Sun, C.R.; Chang, N.T.; Chang, T.W. Immunoconjugates that neutralize HIV virions kill T cells infected with diverse strains of HIV-1. *J. Immunol.* **1990**, *144*, 1257–1262.
170. Pincus, S.H.; McClure, J. Soluble CD4 enhances the efficacy of immunotoxins directed against gp41 of the human immunodeficiency virus. *Proc. Natl. Acad. Sci. USA* **1993**, *90*, 332–336. [CrossRef]
171. Sadraeian, M.; Guimarães, F.E.G.; Araujo, A.P.U.; Worthylake, D.K.; Lecour, L.J.; Pincus, S.H. Selective cytotoxicity of a novel immunotoxin based on pulchellin A chain for cells expressing HIV envelope. *Sci. Rep.* **2017**, *7*, 1–12. [CrossRef]
172. Pincus, S.H.; Song, K.; Maresh, G.A.; Hamer, D.H.; Dimitrov, D.S.; Chen, W.; Zhang, M.-Y.; Ghetie, V.F.; Chan-Hui, P.-Y.; Robinson, J.E.; et al. Identification of Human Anti-HIV gp160 Monoclonal Antibodies That Make Effective Immunotoxins. *J. Virol.* **2017**, *91*, e01955-16. [CrossRef]
173. Pincus, S.H.; Cole, R.L.; Hersh, E.M.; Lake, D.; Masuho, Y.; Durda, P.J.; McClure, J. In vitro efficacy of anti-HIV immunotoxins targeted by various antibodies to the envelope protein. *J. Immunol.* **1991**, *146*, 4315–4324. [PubMed]
174. Till, M.A.; Zolla-Pazner, S.; Gorny, M.K.; Patton, J.S.; Uhr, J.W.; Vitetta, E.S. Human immunodeficiency virus-infected T cells and monocytes are killed by monoclonal human anti-gp41 antibodies coupled to ricin A chain. *Proc. Natl. Acad. Sci. USA* **1989**, *86*, 1987–1991. [CrossRef] [PubMed]
175. Pincus, S.H.; Song, K.; Maresh, G.A.; Frank, A.; Worthylake, D.; Chung, H.-K.; Polacino, P.; Hamer, D.H.; Coyne, C.P.; Rosenblum, M.G.; et al. Design and In Vivo Characterization of Immunoconjugates Targeting HIV gp160. *J. Virol.* **2017**, *91*, e01360-16. [CrossRef] [PubMed]
176. McCoig, C.; Van Dyke, G.; Chou, C.-S.; Picker, L.J.; Ramilo, O.; Vitetta, E.S. An anti-CD45RO immunotoxin eliminates T cells latently infected with HIV-1 in vitro. *Proc. Natl. Acad. Sci. USA* **1999**, *96*, 11482–11485. [CrossRef]
177. Erice, A.; Balfour, H.H.; Myers, D.E.; Leske, V.L.; Sannerud, K.J.; Kuebelbeck, V.; Irvin, J.D.; Uckun, F.M. Anti-human immunodeficiency virus type 1 activity of an anti-CD4 immunoconjugate containing pokeweed antiviral protein. *Antimicrob. Agents Chemother.* **1993**, *37*, 835–838. [CrossRef]
178. Bell, K.D.; Ramilo, O.; Vitetta, E.S. Combined use of an immunotoxin and cyclosporine to prevent both activated and quiescent peripheral blood T cells from producing type 1 human immunodeficiency virus. *Proc. Natl. Acad. Sci. USA* **1993**, *90*, 1411–1415. [CrossRef]
179. Barnett, B.B.; Smee, D.F.; Malek, S.M.; Sidwell, R.W. Selective cytotoxicity of ricin A chain immunotoxins towards murine cytomegalovirus-infected cells. *Antimicrob. Agents Chemother.* **1996**, *40*, 470–472. [CrossRef]
180. Caizhen, G.; Yan, G.; Ronron, C.; Lirong, Y.; Panpan, C.; Xuemei, H.; Yuanbiao, Q.; Li, Q.S. Zirconium phosphatidylcholine-based nanocapsules as an in vivo degradable drug delivery system of MAP30, a momordica anti-HIV protein. *Int. J. Pharm.* **2015**, *483*, 188–199. [CrossRef]
181. Wen, J.; Yan, M.; Liu, Y.; Li, J.; Xie, Y.; Lu, Y.; Kamata, M.; Chen, I.S.Y. Specific Elimination of Latently HIV-1 Infected Cells Using HIV-1 Protease-Sensitive Toxin Nanocapsules. *PLoS ONE* **2016**, *11*, e0151572. [CrossRef]
182. Szafraniec-Szczęsny, J.; Janik-Hazuka, M.; Odrobińska, J.; Zapotoczny, S. Polymer Capsules with Hydrophobic Liquid Cores as Functional Nanocarriers. *Polymers* **2020**, *12*, 1999. [CrossRef]
183. Pizzo, E.; Di Maro, A. A new age for biomedical applications of Ribosome Inactivating Proteins (RIPs): From bioconjugate to nanoconstructs. *J. Biomed. Sci.* **2016**, *23*, 1–8. [CrossRef] [PubMed]
184. Giansanti, F.; Flavell, D.J.; Angelucci, F.; Fabbrini, M.S.; Ippoliti, R. Strategies to Improve the Clinical Utility of Saporin-Based Targeted Toxins. *Toxins* **2018**, *10*, 82. [CrossRef] [PubMed]
185. Nicaise, V. Crop immunity against viruses: Outcomes and future challenges. *Front. Plant Sci.* **2014**, *5*, 660. [CrossRef] [PubMed]
186. Reddy, D.; Sudarshana, M.; Fuchs, M.; Rao, N.; Thottappilly, G. Genetically Engineered Virus-Resistant Plants in Developing Countries: Current Status and Future Prospects. *Adv. Virus Res.* **2009**, *75*, 185–220. [CrossRef] [PubMed]

187. Jiang, G.Y.; Jin, D.M.; Weng, M.L.; Guo, B.T.; Wang, B. Transformation and expression of trichosanthin gene in tomato. *Acta Bot. Sin.* **1999**, *41*, 334–336.
188. Tumer, N.E.; Hwang, D.-J.; Bonness, M. C-terminal deletion mutant of pokeweed antiviral protein inhibits viral infection but does not depurinate host ribosomes. *Proc. Natl. Acad. Sci. USA* **1997**, *94*, 3866–3871. [CrossRef]
189. Pauli, S.; Rothnie, H.M.; Chen, G.; He, X.; Hohn, T. The Cauliflower Mosaic Virus 35S Promoter Extends into the Transcribed Region. *J. Virol.* **2004**, *78*, 12120–12128. [CrossRef]

toxins

MDPI

Article

Antiviral Activity of PD-L1 and PD-L4, Type 1 Ribosome Inactivating Proteins from Leaves of *Phytolacca dioica* L. in the Pathosystem *Phaseolus vulgaris*–Tobacco Necrosis Virus (TNV)

Daniela Bulgari [1], Nicola Landi [2], Sara Ragucci [2], Franco Faoro [3] and Antimo Di Maro [2,*]

1 Agri-food and Environmental Microbiology Platform (PiMiAA), Department of Molecular and Translational Medicine, University of Brescia, viale Europa, 11, 25123 Brescia, Italy; daniela.bulgari@unibs.it
2 Department of Environmental, Biological and Pharmaceutical Sciences and Technologies (DiSTABiF), University of Campania 'Luigi Vanvitelli', Via Vivaldi 43, 81100 Caserta, Italy; nicola.landi@unicampania.it (N.L.); sara.ragucci@unicampania.it (S.R.)
3 Department of Agricultural and Environmental Sciences, University of Milan, Via Celoria 2, 20133 Milan, Italy; franco.faoro@unimi.it
* Correspondence: antimo.dimaro@unicampania.it; Tel.: +39 0823 274409

Received: 16 July 2020; Accepted: 13 August 2020; Published: 14 August 2020

Abstract: Using the pathosystem *Phaseolus vulgaris*–tobacco necrosis virus (TNV), we demonstrated that PD-L1 and PD-L4, type-1 ribosome inactivating proteins (RIPs) from leaves of *Phytolacca dioica* L., possess a strong antiviral activity. This activity was exerted both when the RIPs and the virus were inoculated together in the same leaf and when they were inoculated or applied separately in the adaxial and abaxial leaf surfaces. This suggests that virus inhibition would mainly occur inside plant cells at the onset of infection. Histochemical studies showed that both PD-L1 and PD-L4 were not able to induce oxidative burst and cell death in treated leaves, which were instead elicited by inoculation of the virus alone. Furthermore, when RIPs and TNV were inoculated together, no sign of H_2O_2 deposits and cell death were detectable, indicating that the virus could have been inactivated in a very early stage of infection, before the elicitation of a hypersensitivity reaction. In conclusion, the strong antiviral activity is likely exerted inside host cells as soon the virus disassembles to start translation of the viral genome. This activity is likely directed towards both viral and ribosomal RNA, explaining the almost complete abolition of infection when virus and RIP enter together into the cells.

Keywords: antiviral proteins; ribosome inactivating proteins; *Phytolacca dioica* L.; protein purification; tobacco necrosis virus

Key Contribution: This work demonstrates that two type-1 ribosome inactivating proteins, known as PD-L1 and PD-L4, display antiviral activity against tobacco necrosis virus (TNV), reducing or abolishing lesion formation on leaves of *Phaseolus vulgaris*, possibly by damaging both viral and ribosomal RNA.

1. Introduction

Ribosome inactivating proteins (RIPs) are specific rRNA N-glycosylases present in various plants, fungi, and bacteria and potent inhibitors of protein synthesis [1]. Their mode of action is the specific depurination of major rRNA damaging ribosomes. In particular, these enzymes (EC: 3.2.2.22) cleave a specific adenosine (A_{4324}, in the case of rat 28S rRNA) within a universally conserved region known as the Sarcin Ricin Loop (SRL) [2]. The irreversible cleavage of this single adenosine prevents the association between the elongation factors and ribosome, causing the inhibition of protein synthesis [3].

Furthermore, several authors report that RIPs are also able to remove adenine from different substrates such as polynucleotides, tRNAs and DNAs with a different grade of efficiency for which was proposed the name of adenine polynucleotide glycosylases (APGs; [4]) or are able to cleave phosphodiester bonds (DNase activity; [5,6]).

Structurally, RIPs are divided into two groups considering the presence or absence of a quaternary structure. Classically, these enzymes are categorized in monomeric RIPs (type 1) and dimeric RIPs (type 2). Type-1 RIPs consist in a single polypeptide chain with toxic N-glycosylase activity, while type 2 RIPs are constituted by a polypeptide chain exhibiting N-glycosylase activity (A-chain) linked to a lectin chain (B-chain) through a disulphide bond able to recognize carbohydrate (e.g., galactose/N-acetylgalactosamine) moieties of mammalian cell surface [7]. Alternatively, basing on the domain architecture and evolutionary background, type-1 and type-2 RIPs are called type-A and type-AB RIPs, respectively, while other chimeric forms as type-AX are grouped separately, where X indicates a different (unknown) domain found in the genomes of some Poaceae/cereal species [8,9].

Type 2 RIPs are more toxic in cellular systems (IC_{50} 0.0003–1.7 nM on Hela cell lines) with respect to type 1 (IC_{50} 170–3300 nM on Hela cell lines), while in acellular systems the toxicity of the two groups is comparable [10]. In addition, non-canonical RIPs, such as tetrameric RIPs from *Sambucus* [11] or proteolytic activated forms (pro-RIPs; i.e., maize b-32 [12]) were also found.

At the cellular level, the inhibition of protein synthesis by RIPs promotes cell death by apoptosis pathway considering that many studies report the activation of various caspases, caspase-like and serine proteases and poly(ADP-ribose) cleavage [13,14]. Moreover, the relationship between apoptosis and cell death often shows a difference in events succession due to the variation in intracellular routing of RIPs [15,16]. The ability of RIPs to kill target cells with and without specific carriers (e.g., antibodies, hormones, peptides, cytokines and protease inhibitors) is of great biomedical interest for the construction of specific "bullets" against cancer cells and in the treatment of viral or parasitic diseases [16–19].

RIPs have also been used in agriculture to protect crops from diseases caused by viruses and fungi, and from insect pests. Genes encoding for PAP, trichosanthin, and maize RIP are the most commonly used, while the main noted host plants are tobacco, potato, and tomato. Thus, transgenic plants carrying genes of some RIPs have been obtained with different degrees of resistance to viruses, fungi, and insects [20].

Despite the large number of works on RIPs field, the biological role of RIPs in plants has not been completely unveiled; although all researchers agree that RIPs are involved in plants defense or physiology [10,21]. Indeed, these toxins present antiviral properties, antifungal activities, defense role against antagonists or acting in plant processes such as programmed senescence, stress protection and regulation [20].

RIPs are found in a higher number of plants belonging to Caryophyllaceae, Sambucaceae, Cucurbitaceae, Euphorbiaceae, and Poaceae [22] that express various RIP isoforms encoded by multi-gene families [23]. In particular, a rich group of angiosperms expressing many type-1 RIPs is the Phytolaccaceae family [24], where the prototype of type-1 RIPs, pokeweed antiviral protein (PAP) from leaves of *Phytolacca americana* L., was isolated given its pronounced antiviral activity [25]. Since then, many others type-1 RIPs were isolated from Phytolaccaceae species such as *Phytolacca dodecandra* L'Herit [26], *Phytolacca heterotepala* H. Walter [27,28], *Phytolacca insularis* [29] Nakai, and *Phytolacca dioica* L. [30]. In particular, from seeds and from adult and young leaves of *P. dioica* plant, three (PD-S1-3; [31,32]), four (PD-L1-4; [33,34]), and two (dioicin 1 and 2; [35–37]) type-1 RIPs, respectively, were isolated and extensively characterized. On the other hand, biological and antipathogenic activities [6] and a possible source of antimicrobial peptides [38,39] have been recently reported for type-1 RIPs from *P. dioica*.

Despite the numerous type-1 RIPs isolated and studied from leaves of *P. dioica* very few information on their antiviral activity are available with respect to type-1 RIPs from *P. americana*. Therefore, in this work we report the antiviral activity of both PD-L1 and PD-L4, the major isoforms of type-1

RIPs expressed in leaves of *P. dioica* adult plant, by using the pathosystem tobacco necrosis virus (TNV)-*Phaseolus vulgaris* L. TNV is a member of the genus *Necrovirus* in the family *Tombusviridae* with unsegmented and uncapped TNV genome consisting of a single stranded linear positive sense RNA of 3.8 kb that lacks a poly A tail and replicates itself with the aid of its own RNA-dependent RNA polymerase [40]. In bean leaves, as well in many other plant species, TNV infection induces localized necrotic lesions due to the hypersensitive reaction (HR) elicited by the virus coat protein [40]. These lesions occur 3–4 days after infection and are easily quantifiable, thus this pathosystem is widely used to assess the level and mechanisms of induced plant resistance to viruses [41,42].

In this framework, the new information emerged by this work can be useful for deepening the knowledge on type-1 RIPs in *Phytolaccaceae* family.

2. Results and Discussion

2.1. Type-1 RIP Purification

Native PD-L1 and PD-L4 were purified from fully expanded leaves of *P. dioica* as described previously using a general protocol for the preparation of basic proteins [33,35]. From a raw basic proteins extract from leaves, three chromatographic peaks were obtained from the last cation exchange chromatography step (Figure 1a). The first and the last peaks give homogeneous PD-L1 and PD-L4, respectively, while the second peak (peak a Figure 1a) contains simultaneously PD-L2 and PD-L3, glycosylated isoforms of PD-L1 and PD-L4 that were further not purified [34].

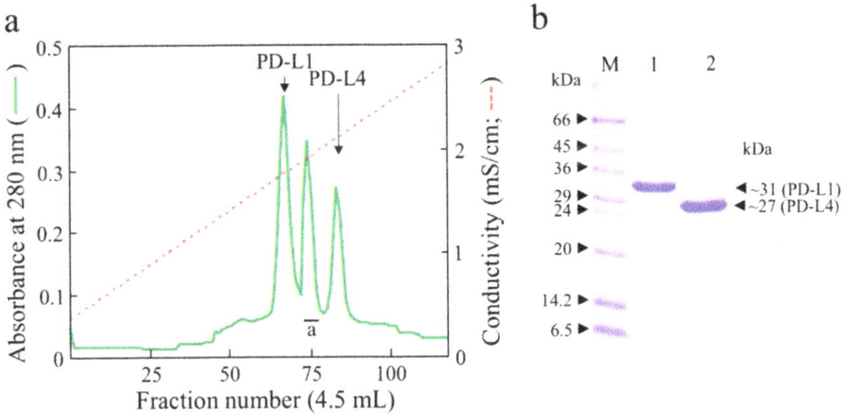

Figure 1. Purification of ribosome inactivating proteins (RIPs) from *Phytolacca dioica* leaves. (**a**) Elution profile from the CM-52 chromatography showing three main peaks. First and last peaks are PD-L1 and PD-L4, respectively. The second peak (peak a) was identified as PD-L2 and PD-L3 minority glycosylated isoforms of PD-L1 and PD-L4, respectively (Di Maro et al., 1999). (**b**) SDS-PAGE on 15% polyacrylamide under reducing conditions of the purified PD-L1 (lane 1) and PD-L4 (lane 2), respectively. M, molecular markers.

Homogenous preparations of both PD-L1 and PD-L4, verified by SDS-PAGE (sodium dodecyl sulfate–polyacrylamide gel electrophoresis; Figure 1b) and capillary electrophoresis (Figure 2), were used for antiviral activity against TNV.

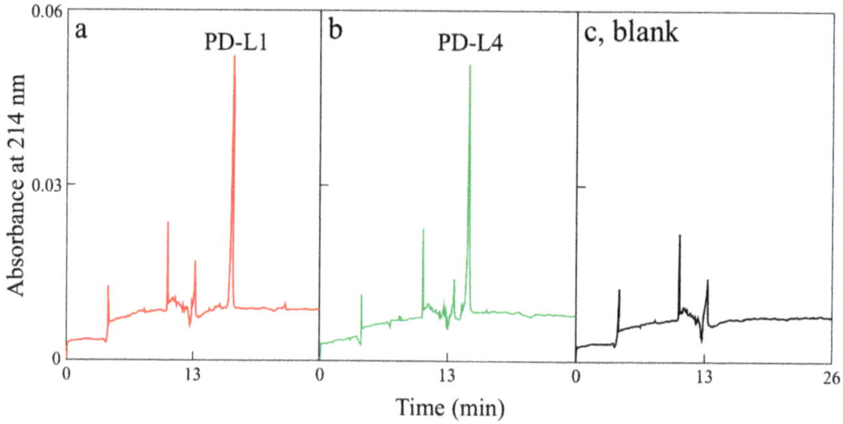

Figure 2. Capillary electrophoresis electropherograms of purified PD-L1 (**a**) and PD-L4 (**b**). In (**c**), reference electropherogram without proteins (blank).

2.2. *Antiviral Activity of PD-L1 and PD-L4*

Both PD-L1 and PD-L4 exerted a strong antiviral activity at the tested concentrations (2 and 10 µg/mL) when applied together with the virus suspension in the same bean leaf surface, as assessed by the reduction of lesion number with respect to bean plants inoculated only with the virus suspension in water (Table 1, Figure 3). However, PD-L4 antiviral activity was greater than PD-L1 and completely inhibited the appearance of visible virus lesions at a concentration of 10 µg/mL (−99.7%).

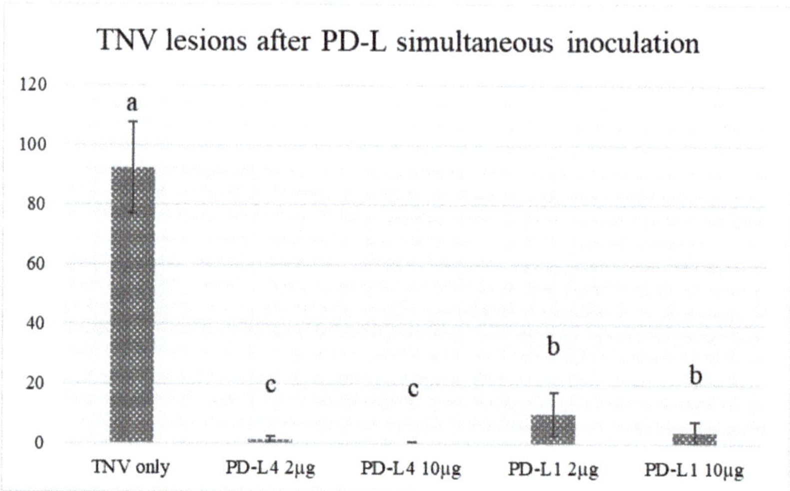

Figure 3. Number of tobacco necrosis virus (TNV) lesions developed after inoculation only with the virus or with the virus mixed to different concentrations of either PD-L1 or PD-L4. Different letters represent significant differences according to Fisher's least significant difference test at $p < 0.05$. The error bars represent standard deviation.

Table 1. Inhibitory effect of PD-L1 and PD-L4, expressed as percent reduction of virus lesions in comparison with control plants inoculated with the virus only. Values are means (± SD) of triplicate analyses ($n = 3$).

Inoculation Method	% Reduction of TNV-Lesions
TNV + PD-L1 2 μg/mL on the same leaf surface	89.1 ± 3
TNV + PD-L1 10 μg/mL on the same leaf surface	95.8 ± 2
TNV + PD-L4 2 μg/mL on the same leaf surface	98.7 ± 4
TNV + PD-L4 10 μg/mL on the same leaf surface	99.7 ± 2
TNV on the adaxial leaf surface and PD-L4 on the abaxial	72.6 ± 3
TNV on the abaxial leaf surface and PD-L4 on the adaxial	78.0 ± 3

To verify if this inhibition was due to a direct effect of the RIPs on viral RNA and not to ribosomes inactivation, TNV was inoculated separately from RIPs. For this purpose, the virus suspension was rubbed either on the adaxial or abaxial leaf surface and PD-L4 on the opposite leaf surface. In these experiments, only PD-L4 was utilized being the most effective at the lowest concentration of 2 μg/mL. Results showed that the inhibitory activity of this RIP against TNV was partially weakened, but still high, with a reduction of lesion number over 70% with respect to controls (Table 1, Figure 4). Though we cannot exclude a direct contact of the virus with PD-L4, i.e., in the extracellular spaces, it is likely that this contact occurs directly in the first damaged cells following virus infection. Here, PD-L4 may exerts its activity both on replicating viral RNA and on ribosomal RNA. Indeed, in a previous paper it was demonstrated that PD-L4 is able to partially degrade tobacco mosaic virus RNA in vitro, thus supporting this hypothesis [6].

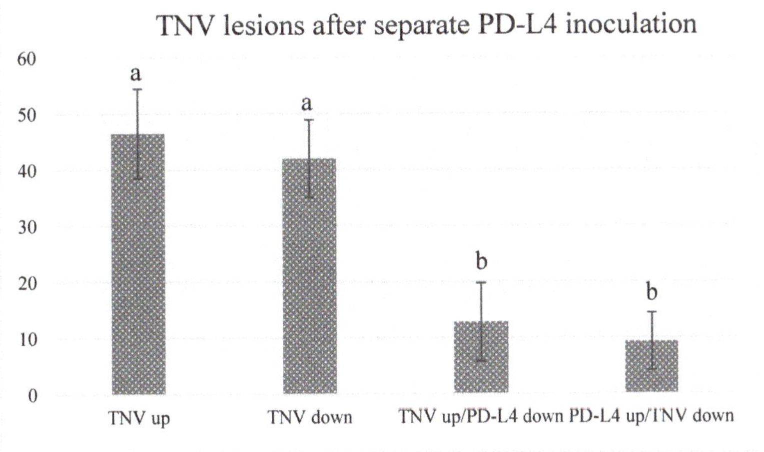

Figure 4. Number of TNV lesions developed when the virus was inoculated alone on the adaxial (up) or abaxial (down) leaf surface, or when it was inoculated separately from PD-L4 in the opposite leaf surface. Different letters represent significant differences according to Fisher's least significant difference test at $p < 0.05$. The error bars represent standard deviation.

2.3. *PD-L1 and PD-L4 Antiviral Activity is not Mediated by Cell Death and Oxidative Burst*

To shed further light on the antiviral activity of these RIPs, bean leaves only treated with PD-L1 or PD-L4 at 2 μg/mL or 10 μg/mL were stained with Evans blue for the detection of cell death at 24, 48, and 72 h after treatment. Dead cells were detected only occasionally after PD-L1 and PD-L4 treatments at both concentrations and at any tested time point (Figure 5a–c) suggesting that these RIPs either are not able to permeate into cells or, once entered, they do not cause enough damages to induce cell death, for at least 72 h (Figure 5a–c). Evans blue staining was also performed on leaves only inoculated with TNV or simultaneously inoculated with PD-L4 (2 μg/mL) and TNV, at 72 h after inoculation on the onset of virus lesion appearance. In leaves inoculated with TNV only the developing lesions were clearly visible as groups of dead cells, surrounded by less damage cells (Figure 5d). These lesions matured in the following two days in typical necrotic lesions clearly visible at naked eyes. Instead, in leaves inoculated simultaneously with a mixture of PD-L4 and TNV (1:1) lesions were very small and formed by a few damaged cells (Figure 5e). Only rare of such damaged cell groups enlarged in the following two days becoming visible necrotic lesions. Finally, in leaves inoculated with TNV (adaxially) and PD-L4 applied abaxially virus lesions were formed by a discrete number of damaged cells, though only few of them appeared dark blue and thus certainly died (Figure 5f). In the following the days, some of these lesions matured in visible necrotic lesions.

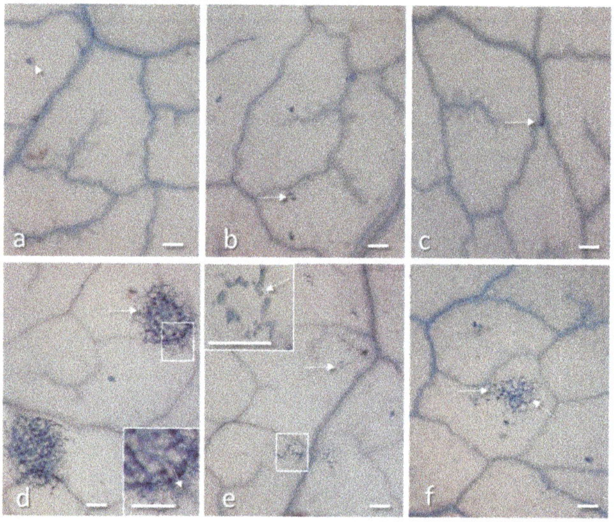

Figure 5. Evans blue staining of bean leaf disks 72 h after treatment with PD-L1 2 μg/mL (**a**) or PD-L4 2 μg/mL (**b**) or water as control (**c**); only very rare dead cells stained in blue (arrows) are present in all the treatments. (**d**) Leaf disk from a non-treated leaf inoculated with 40 μg/mL of TNV, at 72 h after infection: virus lesions formed by numerous dead cells (in dark blue) surrounded by less damaged cells (arrows) are expanding in the tissue; an enlargement of the framed part of a lesion is visible in the inset. (**e**) Leaf inoculated with a mixture of 2 μg/mL PD-L4 and 40 μg/mL TNV (1:1), showing only small groups of damaged cells, enlarged in the inset (arrows); (**f**) Leaf treated with 2 μg/mL PD-L4 in the abaxial surface and inoculated soon after with 40 μg/mL of TNV in the adaxial surface: a TNV lesion is developing at 72 h after infection; the lesion is formed by a few dead cells (in dark blue) and numerous damaged cells around (arrows) and is still invisible at naked eyes. All bars = 200 μm.

3,3′-Diaminobenzidine (DAB) staining of bean leaf disks 48 h after treatment with PD-L1 2 μg/mL or PD-L4 2 μg/mL showed that no H_2O_2 deposits were present in the tissues, except for veins that were stained in brown because they normally contain H_2O_2 for cell wall lignification (Figure 6a–c). This suggest that these RIPs are not able to induce oxidative burst, either because they do not possess this property or are unable to permeate into intact cells. A strong H_2O_2 deposition was instead elicited by TNV infection due to the hypersensitive reaction (HR) caused by the virus on the onset of infection, when lesions were not yet visible at naked eyes (Figure 6d).

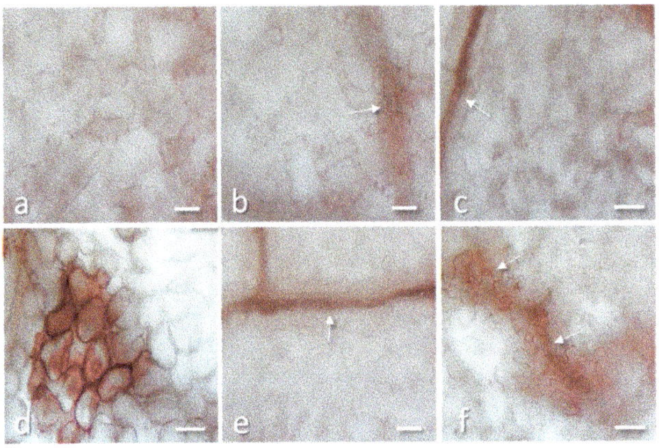

Figure 6. 3,3′-Diaminobenzidine (DAB) staining of bean leaf disks 48 h after treatment with PD-L1 2 μg/mL (**a**) or PD-L4 2 μg/mL (**b**) or water as control (**c**); no staining is present in the tissues, except for veins (arrows) that are stained in brown as they contain H_2O_2 for cell wall lignification. (**d**) Leaf disk from a non-treated leaf inoculated with 40 μg/mL of TNV, at 48 h after infection: a developing virus lesion, densely stained for the presence of H_2O_2 due to the oxidative stress induced by the virus, is shown. (**e**) Leaf inoculated with a mixture of 2 μg/mL PD-L4 and 40 μg/mL TNV (1:1), no staining is present except for the veins (arrow). (**f**) Leaf treated with 2 μg/mL PD-L4 in the abaxial surface and inoculated soon after with 40 μg/mL of TNV in the adaxial surface: some DAB staining is present at 48 h in walls of cells possibly involved in an early formation of a small lesion (arrows), as the one showed in Figure 5f. All bars = 50 μm.

Leaf inoculated simultaneously with a mixture of RIP and TNV showed no DAB staining except for the veins (Figure 6e; arrow), indicating that virus inhibition occurs at a very early stage of infection. Finally, leaf treated with RIP in the abaxial surface and inoculated soon after with TNV in the adaxial surface showed after 48 h some DAB staining in walls of cells possibly involved in an early formation of a small lesion (Figure 6f).

The above microscopic observations confirm that the strongest antiviral activity is exerted when virus and RIP are inoculated together into the leaves, reducing almost completely cell damages. Considering that RIPs should not have access to viral RNA in assembled virus particles [43], it is likely that the antiviral activity is mainly exerted directly into the cells when the virus disassembles, possibly damaging both viral and ribosome RNA, therefore impairing virus replication. In this context, it seems determinant that both virus and RIPs enter the cells at the same time. In fact, by inoculating the virus in the adaxial leaf surface and applying PD-L4 in the abaxial surface, there could be a delay in RIP permeation into the infected cells. This would allow the virus to replicate itself, until it is blocked by the RIP, as suggested by the discrete lesions visible in Figure 5f and by H_2O_2 deposition in Figure 6f.

3. Conclusions

Type-1 RIPs from *P. dioica* are extensively characterized from a structural and enzymatic point of view (for a summary see [30]). Vice versa, few information is known about their biological action that could justify the presence of these enzymes in the various organs of *P. dioica*. On the other hand, recent studies have shown that type-1 RIPs from *P. dioica* display several biological and antipathogenic activities being adept at damaging the tobacco mosaic virus RNA and to inhibit the growth of *Penicillium digitatum* [6]. Thus in this work, we wanted to test the antiviral activity of both PD-L1 and PD-L4, the most expressed type-1 RIPs in the leaves of *P. dioica*, against TNV by using as a host *P. vulgaris*, cv. Borlotto Nano Lingua di Fuoco (BLF).

Data show that both toxins have the ability to reduce the number of lesions when applied on *P. vulgaris* leaves in the same leaf surface compared to leaves inoculated with the virus alone. In particular, the decrease in lesions is particularly evident in the presence of PD-L4 with respect to PD-L1. The different action of the two toxins could justify their different expression, considering that PD-L4 is more expressed in the leaves growing during spring and summer and in minor amount in the autumn and winter [35]. On the other hand, PD-L1 expressed in autumn and winter [35] shows a minor antiviral activity that could be implicated in different physiological roles such as leaf senescence [44].

Moreover, a decrease in antiviral activity for PD-L4 occurred when the virus suspension was inoculated either on the adaxial or abaxial leaf surface and PD-L4 on the opposite leaf surface.

Therefore, it is likely that PD-Ls antiviral activity is fully expressed only when both RIPs and virus are present in the same cells, whether they had entered together or separately. At this regard, the absence of dead cells in leaves only treated with PD-L1 and PD-L4 suggests that these proteins cannot be internalized, at least in a sufficient amount to cause cell death. It is more likely that they enter easily in those cells already damaged by the virus, thus accelerating their death and hampering virus spreading. Whether these RIPs, once present in the virus infected cells, inhibit viral RNA translation and replication by depurination of host ribosomal RNA, or directly depurinate viral RNA, as recently demonstrate for PAP [45], it is not known. Possibly there is a synergistic effect of these activities, also taking into account that in vitro PD-L4 degradation capacity against tobacco mosaic virus (TMV) RNA has previously been shown to be only partially effective [6].

4. Material and Methods

4.1. Materials

Materials for chromatography were described elsewhere [28,46]. All other reagents and chemicals were of analytical grade (Sigma-Aldrich/Merck Life Science S.r.l., Milano, Italy).

4.2. Plant Materials for Type-1 RIP Purifications and for Antiviral Assays

Type-1 RIPs, namely PD-L1 and PD-L4 were from fully expanded leaves, collected from a single female adult tree plant of *Phytolacca dioica*, growing in the Botanical Garden of the University of Naples "Federico II" (Italy). *Phaseolus vulgaris* plants, cv. Borlotto Nano Lingua di Fuoco (BLF) were grown in a greenhouse at 24 ± 2 °C, RH 60 ± 5%, 16 h/8 h light/dark period and used when primary leaves were completely expanded.

4.3. Protein Purification

PD-L1 and PD-L4 were purified according to the procedure previously reported [33]. Briefly, the raw extract was acidified with acetic acid and subjected to consecutive chromatographic steps: Streamline™ SP (GE Healthcare, Milano, Italy) step wise; gel-filtration by Sephadex G-75 Hi-load 26/60 column (GE Healthcare) on an Akta purification system. Finally, a final low-pressure cation exchange chromatography step on a CM-52 column (GE Healthcare Whatman, Chicago, Il, USA) eluted with a NaCl gradient. Fractions corresponding to main peaks (PD-Ls) with activity inhibitory to cell-free protein synthesis were checked by SDS-PAGE analysis, pooled dialyzed against water, freeze-dried and stored at −20 °C until use.

4.4. Biochemical Analytical Procedures

General methodology using for analytical biochemical characterization (SDS-PAGE and protein concentration by bicinchoninic acid (BCA) assay) are reported in detail in previously publisher paper [46]. As molecular markers for SDS-PAGE, SigmaMarker™ (Sigma-Aldrich, St. Louis, Missouri, USA) low range, mol wt 6500–66,000 Da (product code M3913) were used.

4.5. Homogeneity of Protein by Capillary Electrophoresis

Capillary Electrophoresis (CE) in sodium dodecyl sulphate (SDS) was carried out on a Beckman P/ACE System 5550, using the eCAP^{TM} SDS 14–200 kit in a 47 cm capillary, monitoring at 214 nm following manufacturer's instructions (Beckman Coulter SRL, Cassina de'Pecchi (Milano), Italy).

4.6. Tobacco Necrosis Virus (TNV) Purification

Symptomatic *Nicotiana benthamiana* plants, previously inoculated with a TNV-D strain gently supplied by Institute for Sustainable Plant Protection (IPSP), Turin, Italy, were used for virus purification. Frozen material was ground in cold 0.1 M ammonium citrate buffer (1:3 *w*/*v*), filtered, clarified by differential centrifugation as described by [47] and further purified by ultracentrifugation through sucrose density gradient (10–40% in distilled water), at 150,000× *g* for 40 min. A light scattering virus band was recovered from the gradients and concentrated by centrifugation at 150,000× *g* for 4 h. Finally, virus sediment was resuspended in distilled water at the final concentration of 40 µg/mL and stored at −80 °C until utilization.

4.7. PD-Ls Treatments and Virus Inoculation

Plants were inoculated on the first developed leaf by rubbing with 1 mL/leaf mixture of PD-L1 or PD-L4 (2 µg/mL or 10 µg/mL) and purified virus (1:1), using a 600 mesh carborundum as an abrasive. The final virus concentration in the inoculum was of 100 ng/mL. In some other experiments, PD-L4 and virus were inoculated separately on the abaxial and adaxial surface of the same leaf. As controls, some plants were inoculated respectively with: (1) PD-L1; (2) PD-L4; (3) TNV; (4) water.

4.8. Evaluation of Antiviral Activity

Plants were observed for the development of lesions for 4–5 days. When lesions were fully developed, infected leaves were detached and immediately scanned at 300 DPI resolution. Images were analyzed with Global Lab (Data Translation, Marlborough, MA, USA) to determine the number of lesions and PD-Ls inhibitory activity was calculated as percent reduction of this number in respect to controls.

4.9. Histo-Cytochemistry

Five leaf disks, 1 cm in diameter, were randomly punched with a cork-bore from PD-Ls or TNV or PD-Ls + TNV inoculated bean leaves at two days post-inoculation (dpi) and stained with Evans blue, to identify dead cell in the tissues and with 3,3′- Diaminobenzidine (DAB) to localize H_2O_2 deposits, following a previously reported protocol [48]. Samples were examined with an Olympus B×50 light microscope (Olympus, Shinjuku, Tokyo, Japan), equipped with differential interference contrast (DIC) and epi-polarization filters.

4.10. Statistical Analysis

For the antiviral assay on bean plants, results are expressed as mean ± standard deviation of data collected from at least three independent experiments, with 5 bean plants and 10 treated/inoculated leaves for each treatment. Data were subjected to one-way analysis of variance, and comparison among means was determined according to Fisher's least significant difference test. Significant differences were determined at $p < 0.05$.

Author Contributions: D.B. Performed biological experiments in vitro; N.L. and S.R. performed type-1 ribosome inactivating proteins purification; F.F. and A.D.M. conceived the idea, planned experiments, analyzed the data, and co-wrote the manuscript. All authors have read and agreed to the published version of the manuscript.

Funding: This work was supported by the University of Campania "Luigi Vanvitelli," by "DicoVale" project program, Campania region (Italy; P.S.R. 2014–2020; Typology 10.2.1.).

Acknowledgments: The abnegation of all authors has made this study possible without dedicated funds considering the chronic difficulties afflicting the Italian research.

Conflicts of Interest: The authors declare no conflict of interest.

References

1. Stirpe, F. Ribosome-inactivating proteins: From toxins to useful proteins. *Toxicon* **2013**, *67*, 12–16. [CrossRef]
2. Endo, Y.; Tsurugi, K. RNA *N*-glycosidase activity of ricin A-chain. Mechanism of action of the toxic lectin ricin on eukaryotic ribosomes. *J. Biol. Chem.* **1987**, *262*, 8128–8130.
3. Brigotti, M.; Rambelli, F.; Zamboni, M.; Montanaro, L.; Sperti, S. Effect of alpha-sarcin and ribosome-inactivating proteins on the interaction of elongation factors with ribosomes. *Biochem. J.* **1989**, *257*, 723–727. [CrossRef] [PubMed]
4. Bolognesi, A.; Polito, L.; Lubelli, C.; Barbieri, L.; Parente, A.; Stirpe, F. Ribosome-inactivating and adenine polynucleotide glycosylase activities in *Mirabilis jalapa* L. tissues. *J. Biol. Chem.* **2002**, *277*, 13709–13716. [CrossRef] [PubMed]
5. Aceto, S.; Di Maro, A.; Conforto, B.; Siniscalco, G.G.; Parente, A.; Delli Bovi, P.; Gaudio, L. Nicking activity on pBR322 DNA of ribosome inactivating proteins from *Phytolacca dioica* L. leaves. *Biol. Chem.* **2005**, *386*, 307–317. [CrossRef] [PubMed]
6. Iglesias, R.; Citores, L.; Ragucci, S.; Russo, R.; Di Maro, A.; Ferreras, J.M. Biological and antipathogenic activities of ribosome-inactivating proteins from *Phytolacca dioica* L. *Biochim. Biophys. Acta Gen. Subj.* **2016**, *1860*, 1256–1264. [CrossRef] [PubMed]
7. Barbieri, L.; Battelli, M.G.; Stirpe, F. Ribosome-inactivating proteins from plants. *Biochim. Biophys. Acta* **1993**, *1154*, 237–282. [CrossRef]
8. De Zaeytijd, J.; Van Damme, E.J. Extensive evolution of cereal ribosome-inactivating proteins translates into unique structural features, activation mechanisms, and physiological roles. *Toxins* **2017**, *9*, 123. [CrossRef]
9. Peumans, W.J.; Shang, C.; Van Damme, E.J.M. Updated model of the molecular evolution of RIP genes. In *Ribosome-Inactivating Proteins*; Stirpe, F., Lappi, D.A., Eds.; John Wiley & Sons, Inc.: Ames, IA, USA, 2014; pp. 134–150.
10. Stirpe, F.; Gilabert-Oriol, R. Ribosome-inactivating proteins: An overview. In *Plant Toxins*; Carlini, C.R., Ligabue-Braun, R., Eds.; Springer: Dordrecht, The Netherlands, 2017; pp. 153–182.
11. Ferreras, J.M.; Citores, L.; Iglesias, R.; Jiménez, P.; Girbés, T. Sambucus ribosome-inactivating proteins and lectins. In *Toxic Plant Proteins*; Lord, J.M., Hartley, M.R., Eds.; Springer: Berlin/Heidelberg, Germany, 2010; pp. 107–131.
12. Hey, T.D.; Hartley, M.; Walsh, T.A. Maize ribosome-inactivating protein (b-32). Homologs in related species, effects on maize ribosomes, and modulation of activity by pro-peptide deletions. *Plant Physiol.* **1995**, *107*, 1323–1332. [CrossRef]
13. Stirpe, F. Ribosome-inactivating proteins. *Toxicon* **2004**, *44*, 371–383. [CrossRef]
14. Zeng, M.; Zheng, M.; Lu, D.; Wang, J.; Jiang, W.; Sha, O. Anti-tumor activities and apoptotic mechanism of ribosome-inactivating proteins. *Chin. J. Cancer* **2015**, *34*, 325–334. [CrossRef] [PubMed]
15. Vago, R.; Marsden, C.J.; Lord, J.M.; Ippoliti, R.; Flavell, D.J.; Flavell, S.-U.; Ceriotti, A.; Fabbrini, M.S. Saporin and ricin A chain follow different intracellular routes to enter the cytosol of intoxicated cells. *FEBS J.* **2005**, *272*, 4983–4995. [CrossRef] [PubMed]
16. de Virgilio, M.; Lombardi, A.; Caliandro, R.; Fabbrini, M.S. Ribosome-inactivating proteins: From plant defense to tumor attack. *Toxins* **2010**, *2*, 2699–2737. [CrossRef] [PubMed]
17. Pizzo, E.; Di Maro, A. A new age for biomedical applications of Ribosome Inactivating Proteins (RIPs): From bioconjugate to nanoconstructs. *J. Biomed. Sci.* **2016**, *23*, 54. [CrossRef] [PubMed]
18. Rust, A.; Partridge, L.J.; Davletov, B.; Hautbergue, G.M. The use of plant-derived ribosome inactivating proteins in immunotoxin development: Past, present and future generations. *Toxins* **2017**, *9*, 344. [CrossRef] [PubMed]
19. Polito, L.; Djemil, A.; Bortolotti, M. Plant toxin-based immunotoxins for cancer therapy: A short overview. *Biomedicines* **2016**, *4*, 12. [CrossRef]
20. Zhu, F.; Zhou, Y.-K.; Ji, Z.-L.; Chen, X.-R. The plant ribosome-inactivating proteins play important roles in defense against pathogens and insect pest attacks. *Front. Plant Sci.* **2018**, *9*, 146. [CrossRef]

21. Fabbrini, M.S.; Katayama, M.; Nakase, I.; Vago, R. Plant ribosome-inactivating proteins: Progesses, challenges and biotechnological applications (and a few digressions). *Toxins* **2017**, *9*, 314. [CrossRef]
22. Di Maro, A.; Citores, L.; Russo, R.; Iglesias, R.; Ferreras, J.M. Sequence comparison and phylogenetic analysis by the Maximum Likelihood method of ribosome-inactivating proteins from angiosperms. *Plant Mol. Biol.* **2014**, *85*, 575–588. [CrossRef]
23. Lapadula, W.J.; Ayub, M.J. Ribosome inactivating proteins from an evolutionary perspective. *Toxicon* **2017**, *136*, 6–14. [CrossRef]
24. Parente, A.; Chambery, A.; Di Maro, A.; Russo, R.; Severino, V. Ribosome-inactivating proteins from Phytolaccaceae. In *Ribosome-Inactivating Proteins*; Stirpe, F., Lappi, D.A., Eds.; John Wiley & Sons, Inc.: Hoboken, NJ, USA, 2014; pp. 28–43.
25. Domashevskiy, A.V.; Goss, D.J. Pokeweed antiviral protein, a ribosome inactivating protein: Activity, inhibition and prospects. *Toxins* **2015**, *7*, 274–298. [CrossRef] [PubMed]
26. Koch, P.E.; Bonness, M.S.; Lu, H.; Mabry, T.J. Protoplasts from *Phytolacca dodecandra* L'Herit (endod) and *P. americana* L. (pokeweed). *Plant Cell Rep.* **1996**, *15*, 824–828. [CrossRef] [PubMed]
27. Corrado, G.; Bovi, P.D.; Ciliento, R.; Gaudio, L.; Di Maro, A.; Aceto, S.; Lorito, M.; Rao, R. Inducible expression of a *Phytolacca heterotepala* ribosome-inactivating protein leads to enhanced resistance against major fungal pathogens in tobacco. *Phytopathology* **2005**, *95*, 206–215. [CrossRef] [PubMed]
28. Di Maro, A.; Chambery, A.; Daniele, A.; Casoria, P.; Parente, A. Isolation and characterization of heterotepalins, type 1 ribosome-inactivating proteins from *Phytolacca heterotepala* leaves. *Phytochemistry* **2007**, *68*, 767–776. [CrossRef] [PubMed]
29. Moon, Y.H.; Song, S.K.; Choi, K.W.; Lee, J.S. Expression of a cDNA encoding *Phytolacca insularis* antiviral protein confers virus resistance on transgenic potato plants. *Mol. Cells* **1997**, *7*, 807–815. [PubMed]
30. Parente, A.; Berisio, R.; Chambery, A.; Di Maro, A. Type 1 ribosome-Inactivating Proteins from the ombú tree (*Phytolacca dioica* L.). In *Toxic Plant Proteins*; Lord, J.M., Hartley, M.R., Eds.; Springer: Berlin/Heidelberg, Germany, 2010; pp. 79–106.
31. Chambery, A.; Di Maro, A.; Parente, A. Primary structure and glycan moiety characterization of PD-Ss, type 1 ribosome-inactivating proteins from *Phytolacca dioica* L. seeds, by precursor ion discovery on a Q-TOF mass spectrometer. *Phytochemistry* **2008**, *69*, 1973–1982. [CrossRef]
32. Di Maro, A.; Berisio, R.; Ruggiero, A.; Tamburino, R.; Severino, V.; Zacchia, E.; Parente, A. Structural and enzymatic properties of an in vivo proteolytic form of PD-S2, type 1 ribosome-inactivating protein from seeds of *Phytolacca dioica* L. *Biochem. Biophys. Res. Commun.* **2012**, *421*, 514–520. [CrossRef]
33. Di Maro, A.; Valbonesi, P.; Bolognesi, A.; Stirpe, F.; De Luca, P.; Siniscalco Gigliano, G.; Gaudio, L.; Delli Bovi, P.; Ferranti, P.; Malorni, A.; et al. Isolation and characterization of four type-1 ribosome-inactivating proteins, with polynucleotide:adenosine glycosidase activity, from leaves of *Phytolacca dioica* L. *Planta* **1999**, *208*, 125–131. [CrossRef]
34. Di Maro, A.; Chambery, A.; Carafa, V.; Costantini, S.; Colonna, G.; Parente, A. Structural characterization and comparative modeling of PD-Ls 1-3, type 1 ribosome-inactivating proteins from summer leaves of *Phytolacca dioica* L. *Biochimie* **2009**, *91*, 352–363. [CrossRef]
35. Parente, A.; Conforto, B.; Di Maro, A.; Chambery, A.; De Luca, P.; Bolognesi, A.; Iriti, M.; Faoro, F. Type 1 ribosome-inactivating proteins from *Phytolacca dioica* L. leaves: Differential seasonal and age expression, and cellular localization. *Planta* **2008**, *228*, 963–975. [CrossRef]
36. Russo, R.; Chambery, A.; Severino, V.; Parente, A.; Di Maro, A. Structural characterization of dioicin 1 from *Phytolacca dioica* L. gains novel insights into phylogenetic relationships of Phytolaccaceae type 1 RIPs. *Biochem. Biophys. Res. Commun.* **2015**, *463*, 732–738. [CrossRef] [PubMed]
37. Faoro, F.; Conforto, B.; Di Maro, A.; Parente, A.; Iriti, M. Activation of plant defence response contributes to the antiviral activity of Diocin 2 from *Phytolacca dioica*. In Proceedings of the IOBC/WPRS Working Group "Induced Resistance in Plants Against Insects and Diseases", Crete, Greece, 27–29 April 2006; Volume 44, pp. 53–57.
38. Pizzo, E.; Zanfardino, A.; Di Giuseppe, A.M.A.; Bosso, A.; Landi, N.; Ragucci, S.; Varcamonti, M.; Notomista, E.; Di Maro, A. A new active antimicrobial peptide from PD-L4, a type 1 ribosome inactivating protein of *Phytolacca dioica* L.: A new function of RIPs for plant defence? *FEBS Lett.* **2015**, *589*, 2812–2818. [CrossRef] [PubMed]

39. Pizzo, E.; Pane, K.; Bosso, A.; Landi, N.; Ragucci, S.; Russo, R.; Gaglione, R.; Torres, M.D.T.; de la Fuente-Nunez, C.; Arciello, A.; et al. Novel bioactive peptides from PD-L1/2, a type 1 ribosome inactivating protein from *Phytolacca dioica* L. Evaluation of their antimicrobial properties and anti-biofilm activities. *Biochim. Biophys. Acta Biomembr.* **2018**, *1860*, 1425–1435. [CrossRef] [PubMed]
40. Rubino, L.; Matelli, G.P. Necrovirus. In *Desk Encyclopedia of Plant and Fungal Virology*; Mahy, B.W.J., van Regenmortel, M.H.V., Eds.; Academic Press: Cambridge, MA, USA, 2010; pp. 233–235.
41. Iriti, M.; Faoro, F. Abscisic acid is involved in chitosan-induced resistance to tobacco necrosis virus (TNV). *Plant Physiol. Biochem.* **2008**, *46*, 1106–1111. [CrossRef] [PubMed]
42. Iriti, M.; Sironi, M.; Gomarasca, S.; Casazza, A.P.; Soave, C.; Faoro, F. Cell death-mediated antiviral effect of chitosan in tobacco. *Plant Physiol. Biochem.* **2006**, *44*, 893–900. [CrossRef] [PubMed]
43. Battelli, M.G.; Stirpe, F. Ribosome-inactivating poteins from plants. In *Antiviral Proteins in Higher Plants*; Chessin, M., DeBorde, D., Zipf, A., Eds.; CRC Press: Boca Raton, FL, USA, 1995; pp. 39–64.
44. Stirpe, F.; Barbieri, L.; Gorini, P.; Valbonesi, P.; Bolognesi, A.; Polito, L. Activities associated with the presence of ribosome-inactivating proteins increase in senescent and stressed leaves. *FEBS Lett.* **1996**, *382*, 309–312. [CrossRef]
45. Picard, D.; Kao, C.C.; Hudak, K.A. Pokeweed antiviral protein inhibits brome mosaic virus replication in plant cells. *J. Biol. Chem.* **2005**, *280*, 20069–20075. [CrossRef] [PubMed]
46. Landi, N.; Pacifico, S.; Ragucci, S.; Iglesias, R.; Piccolella, S.; Amici, A.; Di Giuseppe, A.M.A.; Di Maro, A. Purification, characterization and cytotoxicity assessment of Ageritin: The first ribotoxin from the basidiomycete mushroom *Agrocybe aegerita*. *Biochim. Biophys. Acta Gen. Subj.* **2017**, *1861*, 1113–1121. [CrossRef]
47. Zhang, L.; French, R.; Langenberg, W.G. Molecular cloning and sequencing of the coat protein gene of a Nebraskan isolate of tobacco necrosis virus: The deduced coat protein sequence has only moderate homology with those of strain A and strain D. *Arch. Virol.* **1993**, *132*, 291–305. [CrossRef]
48. Faoro, F.; Iriti, M. Cell death behind invisible symptoms: Early diagnosis of ozone injury. *Biol. Plant.* **2005**, *49*, 585–592. [CrossRef]

toxins

Article

Kirkiin: A New Toxic Type 2 Ribosome-Inactivating Protein from the Caudex of *Adenia kirkii*

Massimo Bortolotti [1,†], Stefania Maiello [1,†], José M. Ferreras [2], Rosario Iglesias [2], Letizia Polito [1,*] and Andrea Bolognesi [1]

1 Department of Experimental, Diagnostic and Specialty Medicine-DIMES, General Pathology Section, Alma Mater Studiorum—University of Bologna, 40126 Bologna, Italy; massimo.bortolotti2@unibo.it (M.B.); stefaniamaiello@libero.it (S.M.); andrea.bolognesi@unibo.it (A.B.)
2 Department of Biochemistry and Molecular Biology and Physiology, Faculty of Sciences, University of Valladolid, E-47011 Valladolid, Spain; josemiguel.ferreras@uva.es (J.M.F.); riglesia@bio.uva.es (R.I.)
* Correspondence: letizia.polito@unibo.it
† These authors contributed equally to this work.

Citation: Bortolotti, M.; Maiello, S.; Ferreras, J.M.; Iglesias, R.; Polito, L.; Bolognesi, A. Kirkiin: A New Toxic Type 2 Ribosome-Inactivating Protein from the Caudex of *Adenia kirkii*. *Toxins* **2021**, *13*, 81. https://doi.org/10.3390/toxins13020081

Academic Editor: Rodolfo Ippoliti
Received: 29 December 2020
Accepted: 19 January 2021
Published: 22 January 2021

Abstract: Ribosome-inactivating proteins (RIPs) are plant toxins that irreversibly damage ribosomes and other substrates, thus causing cell death. RIPs are classified in type 1 RIPs, single-chain enzymatic proteins, and type 2 RIPs, consisting of active A chains, similar to type 1 RIPs, linked to lectin B chains, which enable the rapid internalization of the toxin into the cell. For this reason, many type 2 RIPs are very cytotoxic, ricin, volkensin and stenodactylin being the most toxic ones. From the caudex of *Adenia kirkii* (Mast.) Engl., a new type 2 RIP, named kirkiin, was purified by affinity chromatography on acid-treated Sepharose CL-6B and gel filtration. The lectin, with molecular weight of about 58 kDa, agglutinated erythrocytes and inhibited protein synthesis in a cell-free system at very low concentrations. Moreover, kirkiin was able to depurinate mammalian and yeast ribosomes, but it showed little or no activity on other nucleotide substrates. In neuroblastoma cells, kirkiin inhibited protein synthesis and induced apoptosis at doses in the pM range. The biological characteristics of kirkiin make this protein a potential candidate for several experimental pharmacological applications both alone for local treatments and as component of immunoconjugates for systemic targeting in neurodegenerative studies and cancer therapy.

Keywords: *Adenia*; apoptosis; kirkiin; lectins; neuroblastoma; ribosome-inactivating proteins; ricin; toxic enzymes

Key Contribution: In this paper, we described for the first time a new type 2 ribosome-inactivating protein, named kirkiin, with high cytotoxicity toward neuronal cell lines. The enzymatic and cytotoxic characteristics of kirkiin make it a promising candidate to be considered for pharmacological purpose.

1. Introduction

Ribosome-inactivating proteins (RIPs) are toxic enzymes widely distributed in the plant kingdom, but also present in some fungal and bacterial species [1–3]. RIP-containing plants are largely used in folk and traditional medicine worldwide, and several derivatives from these plants are still employed for the treatment of numerous pathologies [4,5]. RIPs are classified as rRNA N-glycosylase (EC 3.2.2.22), as they recognize a specific and universally conserved region of 14 nucleotides on 28S rRNA, splitting the N–C glycosidic bond between a specific adenine and its ribose in the sequence GAGA on the rRNA. In the case of rat liver ribosomes, this site is A4324 and is positioned within a single-stranded loop called sarcin-ricin (SRL) [6]. After adenine removal, the apurinic site does not allow the GTPase-dependent binding of elongation factor-1 (EF-1) and elongation factor-2 (EF-2) to the 60S subunit of the ribosome, thus blocking the translation [7].

Further studies showed that RIPs can deadenylate a range of polynucleotides, such as DNA, tRNA, mRNA, and viral RNA and the term polynucleotide:adenosine glycosidases was proposed [8,9]; afterwards, the activity was better defined as adenine polynucleotide glycosylase [10].

RIPs are structurally divided into two main groups: type 1 RIPs, characterized by a single polypeptide chain of about 30 kDa with enzymatic activity, and type 2 RIPs with molecular weight of about 60 kDa, consisting of an enzymatically active A chain, similar to type 1 RIPs, linked through a disulfide bond to a B chain with lectin properties. The B chain has strong affinity for sugar moieties on cell surface and can facilitate the entry of the toxin into the cell, thus conferring to many type 2 RIPs high cytotoxicity [11–13].

Several studies indicate that RIPs have a role in protecting plants from viral and fungal infection or in plant senescence [14]. However, their action in plants defense from pathogens is not still clear. In general, type 2 RIPs seem to be more active on animal ribosomes, whereas type 1 RIPs have a wider specificity. This suggested that adenine polynucleotide glycosylase activity might be responsible for the antiviral action of RIPs [8].

To date, about 80 type 2 RIPs have been purified from a few plant genera [1,2]. In particular, RIPs purified from *Adenia* genus are among the most lethal plant toxins. In addition to modeccin and volkensin, extracted from the roots of *Adenia* (Modecca) *digitata* (Harv.) Engl and *Adenia volkensii* Harms, respectively [15,16], that are known for many years, two other potent toxins from the caudices of *Adenia lanceolata* Engl. (lanceolin) and *Adenia stenodactyla* Harms (stenodactylin) were subsequently described [17]. The high cytotoxicity of *Adenia* RIPs is probably due to their high-affinity cell binding, efficient endocytosis and intracellular routing, resistance to proteolysis, and, regarding stenodactylin, high accumulation into the cell [18]. *Adenia* toxins are retrogradely transported along peripheral nerves and in the central nervous system [19,20]; this property could have different medical and biotechnological applications in neurophysiology and for the experimental treatment of pain [21]. Moreover, because of their high cytotoxicity, RIPs can be used for pharmacological purpose, both native, for local–regional treatments, and as components of immunotoxins, for systemic therapy of cancer and other pathologies [22–24].

It has been reported that in a neuroblastoma cell line, stenodactylin induced multiple cell death pathways, involving apoptosis, necroptosis, and oxidative stress [25]. Moreover, stenodactylin elicited a quick stress response in leukemia cells, producing pro-inflammatory factors and oxidative stress, triggering apoptosis and other cell death pathways [26].

These peculiar characteristics of *Adenia* toxins prompted us to evaluate whether other species belonging to *Adenia* genus, i.e., *Adenia kirkii* (Mast.) Engl. (hereafter referred as *A. kirkii*), contain lectins or toxic RIPs structurally similar to others, already purified from the same genus, and possibly endowed of peculiar biological properties. In this study, a new toxic type 2 RIP, named kirkiin, was purified from the caudex of *A. kirkii*, and its biochemical, enzymatic, and cytotoxic properties were evaluated.

2. Results

2.1. *Purification and Characterization of Adenia Kirkii Lectins*

The extracts from *A. kirkii* caudex were purified by chromatography on an acid treated-Sepharose CL-6B column. The acidic treatment causes the exposure of galactose residues present in the cross-linked agarose matrix. In this manner, the stationary phase become able to bind lectins. This affinity chromatography method allows the one-step purification of lectins present in the crude extract. The lectin was eluted from the stationary phase with 0.2 M galactose. The purified lectin was assayed for the inhibition of protein synthesis in a rabbit reticulocyte lysate system (Table 1). *A. kirkii* lectin showed high enzymatic activity, with concentration inhibiting 50% of protein synthesis (IC_{50}) values of 9.2 and 4.7 μg/mL for the non-reduced and reduced protein, respectively, comparable with those obtained with other RIPs purified from *Adenia* genus. Moreover, *A. kirkii* lectin had high agglutinating activity for human erythrocytes, showing a minimum concentration causing agglutination of 4.0 μg/mL, a value lower than that obtained with other *Adenia* RIPs.

Table 1. Biological activity of *Adenia* toxic lectins purified by chromatography on acid-treated Sepharose CL-6B.

Adenia Species	Cell Free IC_{50} (µg/mL)		Agglutinating Activity [1] (µg/mL)	Ref.
	Non-Red.	Red.		
A. kirkii	9.2	4.7	4.0	
A. stenodactyla	5.6	0.5	49.9	[17]
A. lanceolate	5.2	1.1	230.9	[17]
A. volkensii	7.5	0.7	15.6	[15,17]

[1] Minimum concentration causing hemagglutination.

As the lectin from *A. kirkii* showed strong toxicity, this prompted us to deepen the study of the new toxin. A further purification procedure was undertaken by chromatography on acid-treated Sepharose CL-6B. A single peak of protein material was eluted with 0.2 M galactose (Figure 1a), resulting in 107.9 mg of total proteins with RIP activity obtained from 100 g of fresh tissue. The yield of purification was 13.1% (Table 2). On gel electrophoresis (Figure 1a), the non-reduced proteins from *A. kirkii* gave two bands with relative mobility (Mr) of about 60 and 30 kDa, approximately. After reduction, proteins from *A. kirkii* showed three bands with Mr of about 30 kDa. This suggests that two lectins with different molecular weights are likely present in *A. kirkii* caudex.

Table 2. *Adenia kirkii* lectins purification summary.

Purification Step	Protein (mg/mL)	Total Protein (mg)	Total Protein (%)	IC_{50} (µg/mL) [1]		Agglutinating Activity [2] (µg/mL)	Specific Activity (U/mg) [3]	Total Activity (U)	Yeld (%)
				Non-Red.	Red.				
Crude extract	3.46	1730.0	100.0	55.0	21.2	45.1	18.2	31,454.5	100.0
Sepharose CL-6B Eluate	2.84	107.9	6.2	9.2	4.7	4.0	108.7	4130.4	13.1
Sephacryl S-100 peak 1	1.43	24.2	1.4	7.4	1.0	175.0	135.1	3270.2	10.4
Sephacryl S-100 peak 2	1.49	62.4	3.6	>50	>50	2.9	-	-	-

[1] Concentration of protein that inhibits the 50% of protein synthesis in a cell free system, measured by linear regression. [2] Minimum concentration causing hemagglutination. [3] Units of IC_{50} (non-reducing conditions) in 1 mg of protein.

Subsequently, a chromatography by gel filtration on Sephacryl S-100 was performed in order to separate the two lectins. The chromatography allowed the complete separation of the two lectins. As shown in Figure 1b, the acid-treated Sepharose CL-6B eluted peak was resolved into two well separated peaks, the first one corresponding to the high molecular weight lectin (double-chain lectin) and the second one to the low molecular weight lectin (single-chain lectin). The yield was approximately 14 mg of double-chain lectin and 36 mg of single-chain lectin per gram of tissue (see Table 2). Both lectins agglutinated human erythrocytes; the minimum agglutinating concentration being 175 and 2.9 µg/mL for double-chain and single-chain lectin, respectively.

The fractions corresponding to each peak were collected and analyzed by SDS-PAGE on a 4–15% gradient gel in order to verify their purity (Figure 1c). High molecular weight lectin revealed the presence of a single band with Mr of 58.5 kDa under non-reducing conditions (Figure 1c, lane 1). After reduction with 2-mercaptoethanol, two bands of 27.1 kDa and 35.3 kDa were obtained (Figure 1c, lane 3). The low molecular weight lectin showed only one band of about 32 kDa both in non-reducing and reducing conditions (Figure 1c, lane 2 and lane 4, respectively).

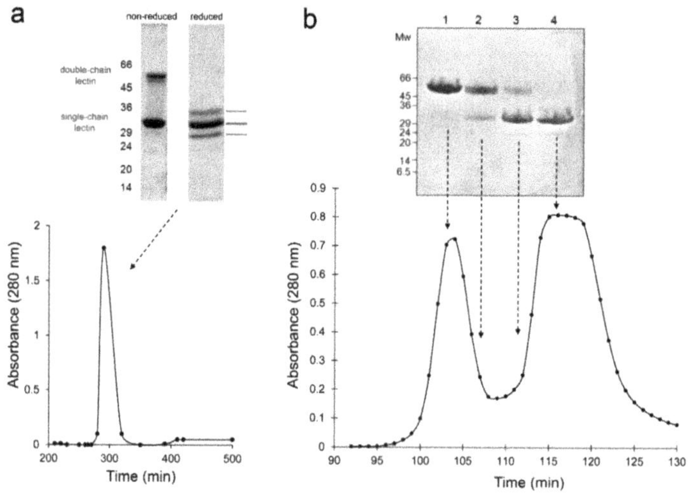

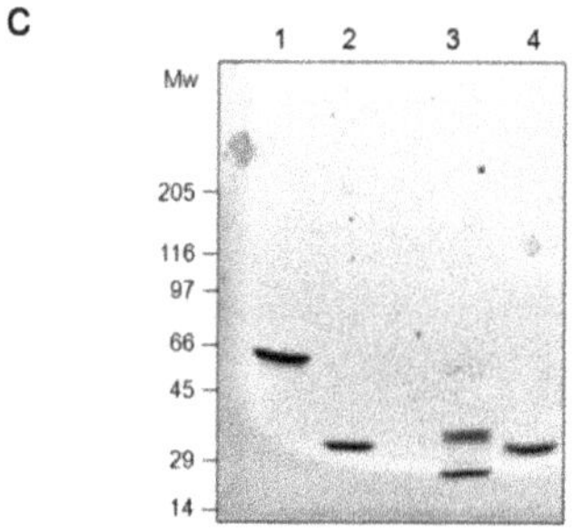

Lectin	Molecular weight (kDa)	
	Non-reduced	*Reduced*
Double-chain	58.5	27.1 35.3
Single-chain	32.9	32.4

Figure 1. (**a**) Chromatography on acid-treated Sepharose CL-6B of *A. kirkii* extracts. Proteins were eluted with 0.2 M galactose in PBS. SDS-PAGE analysis of peak fractions under non-reducing and reducing conditions on 8–25% gradient polyacrylamide gel. (**b**) Chromatography by gel filtration on Sephacryl S-100 of acid-treated Sepharose CL-6B eluate. Proteins were eluted in PBS and peak fractions were analyzed on 8–25% gradient polyacrylamide gel. (**c**) SDS-PAGE of lectins under reducing and non-reducing conditions. Lane 1 and 2 correspond to the non-reduced high and low molecular weight lectins, respectively. Lanes 3 and 4 correspond to the reduced high and low molecular weight lectins, respectively. The electrophoresis was carried out on a 4–15% gradient polyacrylamide gel (staining with Coomassie Blue). Molecular weights of the standard are expressed in kDa. In table, molecular weights of each band, expressed in kDa, are reported after calculation by densitometric analysis of the gel.

2.2. *Enzymatic Properties of Kirkiin*

2.2.1. Effect on Protein Synthesis

The effect of the purified lectins on mammalian ribosomes was evaluated in vitro in a cell-free system consisting of rabbit reticulocyte lysate, by assaying their inhibitory activity on protein synthesis. The two lectins were assayed both in native form and under reducing conditions, thus eliminating the possible steric hindrance given by B chain. As shown in Table 2, double-chain lectin strongly inhibited protein synthesis, with IC_{50} values of 7.4 μg/mL in the native status and of 1 μg/mL after reduction. Instead, single-chain lectin revealed a low inhibition activity with IC_{50} value greater than the highest tested dose (50 μg/mL). For this reason, we chose to continue the research only with the type 2 toxin, hereafter referred as kirkiin.

2.2.2. rRNA N-Glycosylase Activity on Mammalian and Yeast Ribosomes

Kirkiin rRNA N-glycosylase activity was performed through RNA depurination assay of mammalian ribosomes, using rabbit reticulocyte lysate as substrate. The activity was compared to that of the most known type 2 RIP ricin from *Ricinus communis* L. seeds. Both kirkiin and ricin displayed the ability to depurinate mammalian rRNA evidenced by the release of the RNA fragment upon treatment with acid aniline (Endo's fragment), which is diagnostic for RIP action on ribosomes [27]. No RNA fragment was observed in control samples and in samples treated in absence of aniline. These results confirm that inhibition of protein synthesis induced by kirkiin is related to its N-glycosylase activity on mammalian ribosomes (Figure 2a).

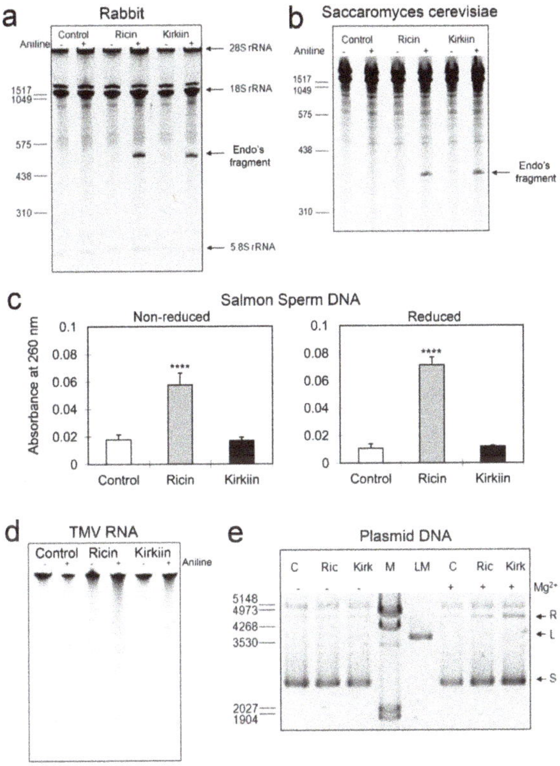

Figure 2. rRNA N-glycosylase activity of kirkiin and ricin on rabbit reticulocyte ribosomes (**a**) and on yeast ribosomes (**b**). Each lane contains 3 μg of RNA. The arrows indicate the 28S, 18S, and the 5.8S rRNAs, and the RNA fragments released as a result of RIP action after aniline acetate treatment at pH 4.5 (+). Numbers indicate the size of the standards in nucleotides. Adenine polynucleotide glycosylase activity of kirkiin and ricin on salmon sperm DNA (**c**) and on Tobacco Mosaic Virus RNA (**d**). In the first case, the amount of released adenine was determined by measuring the absorbance at 260 nm of the supernatant obtained by centrifugation of the samples. The results are the means of two independent experiments, each performed in duplicate. **** $p < 0.0001$, *t*-student test. In the second case, each lane contains 1 μg of RNA. The depurination activity was assayed after aniline acetate treatment at pH 4.5 (+). (**e**) Endonuclease activity of kirkiin (Kirk) and ricin (Ric) on supercoiled plasmid DNA (pCR 2.1) compared to control (C). Each lane contains 100 ng of plasmid DNA. The arrows indicate the supercoiled (S), the linear (L), and the relaxed (R) forms of the plasmid. Numbers indicate the size of the standards (M) in base pairs, and (LM) represents the linear form of the plasmid used as standard.

The effect of kirkiin was also assayed on ribosomes from *Saccharomyces cerevisiae*, which might be homologous to ribosomes from putative plant pathogens. As shown in Figure 2b, kirkiin and ricin displayed rRNA N-glycosylase activity on yeast ribosomes, as indicated by the release of the diagnostic fragment of 360 ± 30 nucleotides upon treatment with aniline acetate, in accordance with that expected for the SRL deglycosylation (368 nucleotides for yeast) [28]. Therefore, kirkiin is able to exert its action also on ribosomes of unicellular eukaryotes.

2.2.3. DNA and RNA Adenine Polynucleotide Glycosylase Activity

Kirkiin adenine polynucleotide glycosylase activity was investigated on salmon sperm DNA (ssDNA) (Figure 2c) and on tobacco mosaic virus RNA (TMVR) (Figure 2d). Kirkiin activity was compared with that of ricin, which possesses a moderate activity. As shown in Figure 2c, kirkiin showed no significant activity on ssDNA both in reduced and non-reduced conditions. On TMVR also, kirkiin exhibited a marginal depurination upon treatment with acid aniline compared to control, resulting slightly less than that shown by ricin (Figure 2d).

2.2.4. Endonuclease Activity on Supercoiled Plasmid DNA

The endonuclease activity of kirkiin was tested on pCR 2.1 plasmid, and it was compared with that of ricin. Both kirkiin and ricin promoted a slight conversion of supercoiled DNA into a relaxed form. This effect was dependent on magnesium ions, and the highest activity was observed at 5 mM of this ion (Figure 2e). Therefore, kirkiin as ricin showed a weak endonuclease activity and acted by cutting only one of the two helices of the plasmid DNA.

2.3. Immunological Properties

The immunological properties of kirkiin were tested with sera against different type 2 RIPs, i.e., the *Adenia* RIPs stenodactylin and volkensin, and ricin. Kirkiin highly cross-reacted with sera against the other two *Adenia* toxins, but no cross-reaction was evidenced with the serum against ricin (Figure 3a). Kirkiin was also tested with sera against some type 1 RIPs, i.e., momordin, pokeweed antiviral protein from seeds (PAP-S), and saporin-S6. Kirkiin showed partial cross-reactivity with anti-momordin and anti-PAP-S sera, while it did not react with the serum against saporin-S6 (Figure 3b).

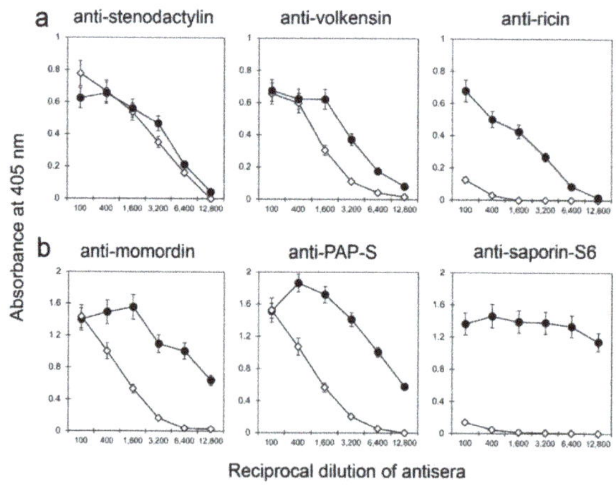

Figure 3. Enzyme-linked immunosorbent assay (ELISA) with (**a**) anti-type 2 and (**b**) anti-type 1 RIP sera. The values of absorbance at 405 nm are expressed in function of the reciprocal of serum dilution. Curves of the RIPs with the respective anti-sera are depicted with black symbols (•), while those related to kirkiin are represented in white symbols (◇). The results are the means of at least three independent experiments.

2.4. *Cytotoxic Effects*

2.4.1. Effect of Kirkiin on NB100 Protein Synthesis and Cell Viability

Cytotoxic effects of kirkiin were compared to ricin and evaluated as protein synthesis inhibition and cell viability reduction in NB100 cells derived from a human neuroblastoma. Cells were treated with scalar concentrations of RIPs, ranging from 1×10^{-15} to 1×10^{-11} M for 72 h, and protein synthesis was assessed by incorporation of ^{3}H-leucine into the new synthesized proteins. Kirkiin and ricin were extremely cytotoxic, showing IC_{50} values of 1.3×10^{-13} and 2.2×10^{-13} M, respectively. Nevertheless, kirkiin showed greater efficacy than ricin in inhibiting protein synthesis completely. At 1×10^{-12} M concentration, kirkiin was able to completely inhibit protein synthesis, whereas the same effect was reached by ricin at 1×10^{-11} M (Figure 4a,c).

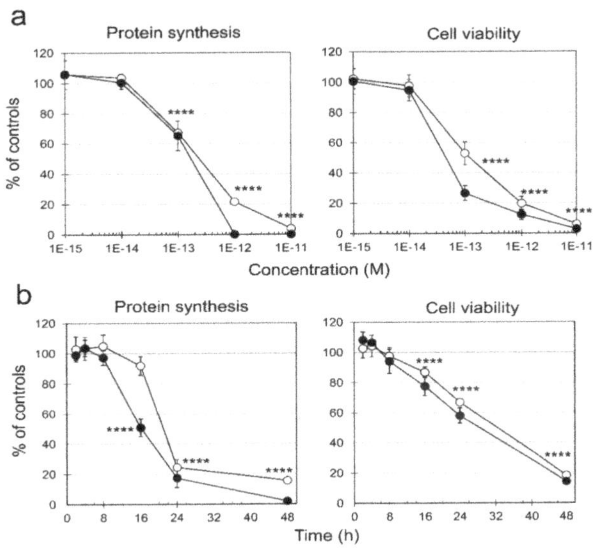

c

	IC_{50}	EC_{50}	IT_{50}	ET_{50}
	(M)		(h)	
Kirkiin	1.3×10^{-13}	4.5×10^{-14}	15.6	25.2
Ricin	2.2×10^{-13}	1.5×10^{-13}	20.6	29.6

Figure 4. (**a**) Concentration–response curves. Comparison of protein synthesis and cell viability in NB100 cells treated with kirkiin (black symbols) or ricin (white symbols) for 72 h. Both parameters are expressed as percentage of controls. (**b**) Time–response curves. Protein synthesis and viability of NB100 cells treated with kirkiin or ricin (1×10^{-11} M) after the indicated times. (**c**) Table reports values of concentration and time that inhibit protein synthesis of 50% (IC_{50} and IT_{50}, respectively), and values of concentration and time that reducing cell viability of 50% (IT_{50} and ET_{50}, respectively). Protein synthesis inhibition was evaluated measuring the ^{3}H-leucine incorporation in the neosynthesized proteins. Viability was evaluated using a colorimetric assay based on MTS reduction. The results are the means of three independent experiments, each performed in triplicate, and are represented as percentage of control values obtained from cultures grown in the absence of RIP. **** $p \leq 0.0001$, ANOVA/Bonferroni, followed by comparison with Dunnett's test.

In cell viability experiments at 72 h, both ricin and kirkiin completely killed all tested cells at 1×10^{-11} M. The concentration that reduces the cell viability of 50% (EC_{50}, effective concentration fifty) was 4.5×10^{-14} and 1.5×10^{-13} M for kirkiin and ricin, respectively (Figure 4a,c). Because both toxins at 1×10^{-11} M resulted in the ability to completely inhibit protein synthesis and reduce cell viability after 72 h incubation, this concentration was chosen for further cytotoxicity experiments carried out in a time range from 2 to 48 h. Time-response curves (Figure 4b) showed that protein synthesis inhibition was faster than cell viability reduction; the time required to inhibit protein synthesis of 50% (IT_{50}) (15.6 and 20.6 h for kirkiin and ricin, respectively) was shorter than that required to reduce cell viability of 50% (ET_{50}, effective time fifty) (25.2 and 29.6 h for kirkiin and ricin, respectively, as shown in Figure 4c).

2.4.2. Evaluation of Apoptosis Induced by Kirkiin in NB100 Cells

In order to evaluate the involvement of apoptosis, we examined the presence of cellular and nuclear morphological changes in NB100 cells treated for 48 h with kirkiin at 1×10^{-11} M concentration, using phase-contrast and fluorescence microscopy, respectively. Morphological characteristics of apoptosis were present in treated cells, such as cell shrinkage, loss of contact with adjacent cells, formation of cytoplasmic protrusions, and apoptotic bodies (Figure 5a). The staining of NB100 cells with DAPI showed that kirkiin intoxication induced a reduction of cell density and an increase of pyknotic and fragmented nuclei (Figure 5b). A typical feature of programmed cell death is the disruption of active mitochondria, which consists in changes in the membrane potential ($\Delta\psi m$) and alterations of the mitochondria redox state. Alterations of $\Delta\psi m$ were detected through fluorescence in cells exposed to kirkiin, after staining with JC-1. Untreated cells showed a strong red fluorescence due to the characteristic J-aggregates in the mitochondria, indicating intact $\Delta\psi m$. In cells treated with kirkiin, JC-1 remained in the monomeric form, yielding green fluorescence and indicating dissipation of the $\Delta\psi m$. These results confirmed that cells undergo programmed cell death after kirkiin intoxication and that mitochondria are involved (Figure 5c).

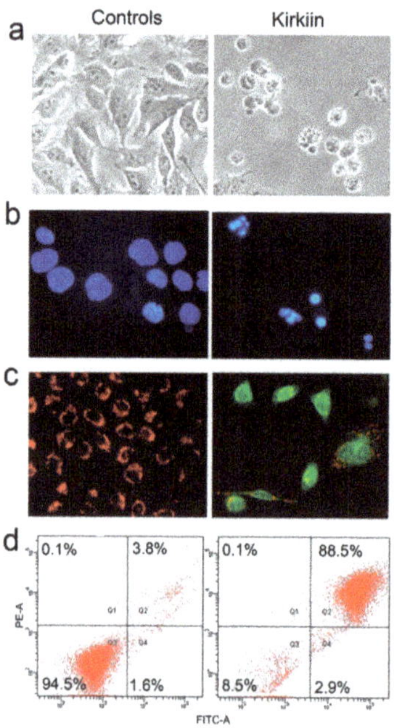

Figure 5. Induction of apoptosis. NB100 untreated (controls) or treated cells with kirkiin 1×10^{-11} M for 48 h were checked for (**a**) cell morphology through phase-contrast microscopy (600× magnification); (**b**) nuclear morphology through fluorescence microscopy after DAPI staining (600× magnification); (**c**) mitochondrial transmembrane potential dissipation through JC-1 staining and an analysis in fluorescence microscopy (600× magnification); (**d**) induction of necrosis/apoptosis by Annexin V-EGFP/PI double staining, followed by flow cytometry analysis. Representative plots of Annexin V (FITC channel)/PI (PE channel).

Apoptosis was also evaluated by double staining with Annexin V-EGFP/Propidium iodide (PI) through flow cytofluorimetric analysis of NB100 cells, in order to quantify the percentage of apoptotic cells and evaluate the eventual involvement of necrosis. PI-positive cells (necrotic cells) are in the upper left quadrant, while apoptotic cells are in the upper (late apoptosis) and lower (early apoptosis) right quadrants. After 48 h of intoxication with the RIP, 88.5% of treated cells were in late-stage apoptosis (Figure 5d). No significative involvement of necrosis was detected after kirkiin intoxication.

To determine the involvement of caspase-dependent apoptosis and to understand the correlation between protein synthesis and apoptosis, caspase 3/7 activation and protein synthesis inhibition were measured in NB100 cells exposed to kirkiin 1×10^{-11} M, in a time range from 4 to 48 h. As shown in Figure 6a, a strong time-dependent activation of caspase 3/7 was observed, which became significant compared to control after 6 h of treatment (143%) and highly significant after 8 h (160%). The level of caspase activity grew exponentially over time, reaching 1093% after 48 h.

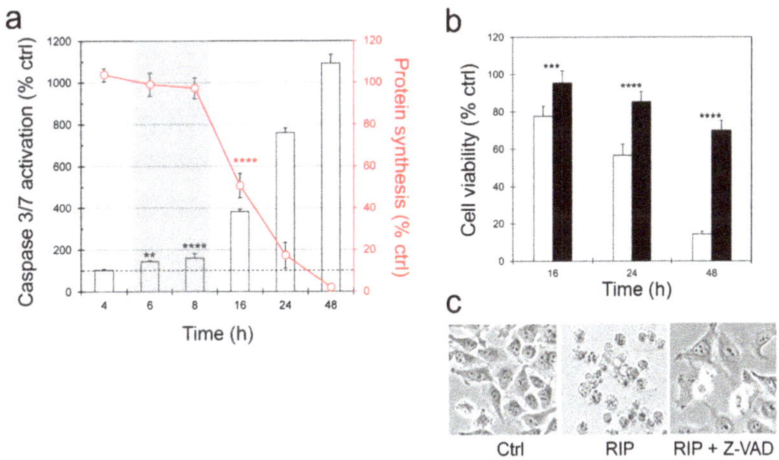

Figure 6. Involvement of caspase-dependent apoptosis. (**a**) Caspase 3/7 activation. Caspase activation (columns) was compared with protein synthesis (red line). Shaded area highlights the time range in which protein synthesis is not inhibited and caspases are significantly activated. Results were expressed as percentage of control values obtained from cultures grown in the absence of RIP. The results are the means of three independent experiments, each performed in triplicate (** $p < 0.01$, **** $p < 0.0001$, ANOVA/Bonferroni test). (**b**) Protection obtained by Z-VAD. NB100 cells were treated with 1×10^{-11} M kirkiin, alone (white columns) or preceded by a 3-h preincubation with 100 μM Z-VAD (black columns). The viability was measured after the indicated times. The statistical analysis was performed using ANOVA/Bonferroni test (confidence range 95%; *** $p \leq 0.001$; **** $p \leq 0.0001$. (**c**) Cell morphology was evaluated at 48h intoxication (400× magnification).

The activation of caspases does not proceed in parallel with the inhibition of protein synthesis. Actually, the inhibition of protein synthesis became significant starting from 16 h of treatment (50% of controls) (Figure 6a, shaded area). These data indicate that protein synthesis and apoptosis are independent events.

To confirm the role of caspase-dependent programmed cell death, the pan-caspase inhibitor Z-VAD was used to selectively inhibit the apoptotic pathway. NB100 cells were pretreated and maintained in 100 μM Z-VAD, and the cell viability was determined after different incubation times with kirkiin (16, 24 and 48 h). As shown in Figure 6b, Z-VAD was able to significantly rescue NB100 cells from death at all the tested times, suggesting the involvement of caspase-dependent cell death. After 48 h-kirkiin intoxication, cells pre-treated with Z-VAD showed a viability of 69.9% versus 14.4% of viability in cells not pre-treated with Z-VAD. These results were confirmed by morphological analysis, showing that after 48 h of intoxication, the most of cells pre-treated with Z-VAD had morphological characteristics similar to those of untreated cells (Figure 6c).

3. Discussion

A. kirkii is a plant spread in Kenya, eastern Tanzania, and Zanzibar with typical glandular-shaped leaves, green flowers, and a caudex as reserve organ, situated at the base of the plant [29].

In the present study, we demonstrate that *A. kirkii* caudex contains a high amount of two lectins that have the characteristics of galactose-specific lectins. The lower molecular weight lectin, in SDS-PAGE gel, showed only one band of 32 kDa both in reducing and non-reducing conditions, and it did not inhibit protein synthesis in a cell free system; this is compatible with a single-chain lectin. The presence of non-toxic lectins in *Adenia* plants has been already described [17]. The higher molecular weight lectin showed one band of about 60 kDa in non-reducing conditions and two bands of 27 and 35 kDa in reducing conditions. Based on the data reported in literature, the A chain of type 2 RIPs weights about 20–30 kDa, whereas the B chain about 30–35 kDa [30]. Therefore, we can assume that the two bands of approximately 27 and 35 kDa represent the A chain and the B chain of a type 2 RIP, respectively.

Both lectins agglutinated erythrocytes; in particular, single-chain lectin showed higher hemagglutination activity than double-chain lectin, probably due to the absence of the steric hindrance of A chain.

Double-chain lectin, named kirkiin, showed a strong inhibition of protein synthesis, displaying IC_{50} values of 7.4 μg/mL in the native status and of 1 μg/mL under reducing conditions. These results were comparable to those obtained for other *Adenia* RIPs already studied that showed IC_{50} values in the range of 2.4–7.5 μg/mL under non-reducing conditions and of 0.4–1.2 μg/mL under reducing conditions [31].

RIPs are commonly known as plant toxins able to recognize and remove a specific adenine from the universally conserved SRL of the 28S rRNA [6]. Kirkiin displayed rRNA N-glycosylase activity against mammalian ribosomes, as indicated by the RIP diagnostic RNA fragment upon treatment with acid aniline [27]. This result confirms that kirkiin ability to inhibit protein synthesis is related to the N-glycosylase activity on mammalian ribosomes. As several evidences suggest that rRNA N-glycosylase activity might play a role in plant defense [32], for example against fungi, the effect of kirkiin was assayed on ribosomes from *Saccharomyces cerevisiae*, which might be homologous to ribosomes from putative plant pathogens. Kirkiin displayed rRNA N-glycosylase activity on yeast ribosomes, as indicated by the release of the diagnostic fragment [33] upon treatment with aniline acetate. Therefore, as kirkiin showed ribosome-inactivating activity on unicellular eukaryotes, it might enter into the fungal cells and inactivate their ribosomes, avoiding the propagation of the pathogen.

Many RIPs are potent inhibitors of animal and/or plant viruses, although the mode of action for the antiviral activity is still not clear [34]. The discovery of a depurinating activity of RIPs on viral RNA allowed hypothesizing a possible use of RIPs as antiviral agents [35]. RIPs have shown a very variable activity on different types of nucleic acids. Adenine polynucleotide glycosylase activity of all toxic type 2 RIPs is significantly lower than type 1 RIPs [8]. No significant activity was detected with kirkiin on viral RNA and eukaryotic DNA with respect to ricin, although the latter has an adenine polynucleotide glycosylase activity substantially lower than type 1 RIPs. Kirkiin did not increase its activity after the reduction of the interchain disulphide bridge, according to what already observed with toxic type 2 RIPs (except for ricin) [36]. Endonuclease activity on plasmid DNA was reported for some RIPs, promoting the conversion of the plasmid from the supercoiled form to the relaxed or linear one [37]. Kirkiin showed a weak activity against supercoiled plasmid DNA. This ability can be important in order to understand the possible biological roles of RIPs, for example in plant defense against pathogenic micro-organisms or viruses. These data indicate that kirkiin, as many other type 2 RIPs, shows high toxicity to mammalian and yeast ribosomes, but slight or no adenine polynucleotide glycosylase activity on other nucleotide substrates.

As RIPs are highly immunogenic, the cross-reactivity between kirkiin and some antibodies-containing sera against various double-chain and single-chain RIPs was evaluated. This analysis may be of interest in order to identify toxins useful for prolonged therapeutic treatment with immunotoxins. In fact, it is possible to reduce the immune response in cancer therapy by varying the type of toxin and prolong the use of RIPs, preserving their therapeutic efficacy. Kirkiin highly cross-reacted with sera against *Adenia* toxins. This strong interaction is not surprising, since all these toxins have been purified from plants belonging to the *Adenia* genus and have a high homology in their amino acid sequences. Kirkiin partially cross-reacted with sera against type 1 RIPs momordin and PAP-S, while it did not react with anti-ricin and anti-saporin-S6 sera. This result represents a remarkable advantage, as ricin and saporin-S6 are the RIPs most used as components of immunoconjugates [38–40]. This is of great interest in prospecting the use of an immunotoxin containing kirkiin A chain in prolonged therapeutic treatments in substitution to immunotoxins containing ricin A chain or saporin-S6.

In order to clarify pathogenetic mechanisms of kirkiin intoxication, inhibition of protein synthesis and cell toxicity were tested on NB100 cells. This cell line was chosen because in previous experiments it resulted very sensitive to *Adenia* RIPs [17,25]. Moreover, NB100 cells could represent a good in vitro model for future neurophysiological studies with kirkiin. Kirkiin resulted very efficient in cell protein synthesis inhibition and cell killing experiments, showing IC_{50} and EC_{50} values comparable to those observed with ricin and other *Adenia* RIPs [17,25]. In time-course experiments, the effects of kirkiin on protein synthesis and viability were examined, showing that the inhibition of protein synthesis precedes the loss of cell viability. Actually, at 24 h, cell viability was 60% of controls, whereas protein synthesis was 20%.

Numerous studies demonstrated that RIPs induce apoptosis as main cell death pathway [41–43]. Kirkiin was able to trigger apoptosis showing cellular and nuclear alterations compatible with an apoptotic pattern, elevated Annexin V positivity, altered mitochondrial transmembrane potential, and strong and fast caspase 3/7 activation. Interestingly, the pan-caspase inhibitor Z-VAD caused a high rescue of cells from death after kirkiin exposure, demonstrating that the apoptotic pathway is the dominant death mechanism. However, the lack of total protection at incubation periods longer than 16 h indicates that the toxin activates other cell death mechanisms, as already described for other RIPs [25,42,44,45].

Caspase activation is an early event with respect to inhibition of protein synthesis. In fact, while caspases are significantly activated starting from 4 h after intoxication, the inhibition of protein synthesis becomes significant only starting from 16 h. These results suggest that caspase activation is independent of inhibition of protein synthesis. This phenomenon has already been described for other RIPs [25,43,46].

4. Conclusions

In this paper, we demonstrated that the new type 2 RIP kirkiin is a galactose-binding lectin able to efficiently inhibit protein synthesis and to agglutinate erythrocytes. In addition, kirkiin showed biochemical, enzymatic, and cytotoxic characteristics typical of type 2 RIPs, possessing N-glycosylase activity on mammalian and yeast ribosomes, but little or no activity on other nucleotide substrates. This toxin is able to completely inhibit cell protein synthesis and to induce cell death by apoptosis at very low doses. The high cytotoxicity of kirkiin, similar to that of other toxins derived from plants belonging to *Adenia* genus, represents an important opportunity for the present and future development of new drugs. Indeed, kirkiin in native form could find application for loco-regional treatments, whereas kirkiin A chain could be used as a component of immunotoxins, for systemic treatments, mainly against hematological tumors [1,47]. Moreover, the assessment of the ability of these toxins to induce apoptosis and to be transported in a retrograde manner in the central nervous system may have very interesting applications in neuroanatomy, neurophysiology, and in the study of degenerative diseases affecting muscle tissue and the nervous system.

5. Materials and Methods

5.1. Materials

A. kirkii caudex was purchased from Mbuyu–Sukkulenten, Bielefeld, Germany. Stenodactylin [17], volkensin [48], ricin [49], and type 1 RIPs [50] were obtained as previously described.

Human neuroblastoma-derived NB100 cell line was from long term culture in our department [25] and was maintained at the logarithmic phase of growth in Roswell Park Memorial Institute medium 1640 (RPMI-1640), supplemented with 10% (*v*/*v*) heat-inactivated fetal bovine serum, 2 mM L-glutamine, 100 U/mL penicillin G, and 100 µg/mL streptomycin (hereafter referred as complete medium) at 37 °C in a humidified atmosphere containing 5% CO_2 in a HeraCell Haraeus incubator (Hanau, Germany). Cells were routinely checked for the absence of mycoplasma infection. Trypan Blue and trypsin/EDTA were obtained from BioWhittaker (Vervies, Belgium). L-[4,5-^{3}H] leucine was purchased by GE Healthcare (Buckingamshire, UK). Flasks and plates were from Falcon (Franklin Lakes, NJ, USA). The pan-caspase inhibitor carbobenzoxy-valyl-alanyl-aspartyl-[O-methyl]-fluoromethylketone (Z-VAD-fmk, hereinafter indicated as Z-VAD) was purchased from Vinci-Biochem (Florence, Italy).

Rabbit sera against ricin, volkensin, type 1 RIPs [51], and stenodactylin [17] were prepared as previously described. The alkaline phosphatase-conjugated antirabbit IgG used for ELISA was purchased from Sigma-Aldrich (St. Louis, MO, USA); the phosphatase substrate (4-nitrophenyl phosphate disodium salt hexahydrate) was purchased from Merck (Darmstadt, Germany).

Caspase activity was evaluated using the luminescent kit Caspase-Glo™3/7 Assay (Promega Corporation, Fitchburg, WI, USA). Morphological membrane changes were detected using Annexin V-EGFP/PI detection kit (Biovision, Mt. View, CA, USA). Viability was measured using the colorimetric CellTiter 96® Aqueous One Solution Cell Proliferation Assay (Promega), which contains the tetrazolium compound [3-(4,5-dimethylthiazol-2-yl)-5-(3-carboxymethoxyphenyl)-2-(4-sulfophenyl)-2H-tetrazolium, MTS] and an electron coupling reagent (1-methoxy phenazine methosulfate, PMS). The mitochondrial potential changes were detected using the Mitochondria Staining Kit (Sigma-Aldrich).

The liquid scintillation cocktail was the Ready-Gel (Beckman Instrument, Fullerton, CA, USA).

Pre-casted gels, molecular weight standards, and buffer strips used for electrophoretic analysis were obtained from GE Healthcare. DAPI-Antifade was from Resnova SRL, Genzano di Roma, Italy. Yeast RNA was purchased from Roche Diagnostics S.L. (Barcelona, Spain). Single-stranded salmon sperm DNA was purchased from Sigma-Aldrich. The water used was prepared with a Milli-Q apparatus (Millipore, Milford, MA, USA). Other reagents used were from Merck (Darmstadt, Germany), Carlo Erba (Milano, Italy), and Sigma. All reagents were of analytical grade, and when possible RNase-free.

5.2. Methods

5.2.1. Adenia Kirkii Lectin Purification

A. kirkii caudex (446 g) was decorticated and homogenized with an Ultra-Turrax (IKA, Staufen, Germany) with 5 mL/g of phosphate-buffered saline (PBS, 0.14 M NaCl containing 5 mM sodium phosphate buffer, pH 7.4). After overnight stirring at 4 °C, the extract was strained through cheesecloth and centrifuged at 18,000× *g* at 4 °C for 30 min. The supernatant (500 mL, corresponding to 1730 mg of proteins) was subjected to affinity chromatography on Sepharose CL-6B matrix (GE Healthcare), pre-treated with 0.2 M HCl for 150 min at 50 °C (acid-treated Sepharose CL-6B), and equilibrated with PBS. The sample was loaded onto the acid-treated Sepharose CL-6B column (7cm h × 5cm Ø) and, after wash with PBS to eliminate the unbound material, the retained protein was eluted stepwise with 0.2 M galactose in PBS (as already described in [17,31]). The determination of the protein content of crude extract and not retained material by acid-treated Sepharose CL-6B was performed by spectrophotometric analysis at 230, 260, and 320 nm, using the Kalb and Bernlohr method [52].

Lectins from acid-treated Sepharose CL-6B were analyzed by 8–25% sodium dodecyl sulfate-polyacrylamide gel electrophoresis (SDS-PAGE) using the PhastSystem (GE-Healthcare) both under reducing and non-reducing conditions.

The volume eluted from acid-treated Sepharose CL-6B (38 mL) was concentrated to 2 mL on YM10 membrane (Merck Millipore, Burlington, MA, USA) under nitrogen pressure and loaded into a Sephacryl S-100 column (94 cm h × 1.5 cm Ø) (GE-Healthcare) in PBS. Peak fractions of the S-100 protein peaks were analyzed on 8–25% PhastGel gradient, following the supplier's protocol. The protein fractions corresponding to the purified lectins were collected and analyzed on a 4–15% PhastGel gradient.

For electrophoretic analysis, proteins were incubated in sample buffer (40 mM Tris/HCl pH 6.8, 2% SDS, 0.005% bromophenol blue) containing 0.5% (*v*/*v*) 2-mercaptoethanol (reducing conditions), or 1 mg/mL iodoacetamide (non-reducing conditions) for 20 min at 37 °C. The gel was stained with 0.1% (*w*/*v*) Coomassie Brilliant Blue G250 in 50% methanol and 10% acetic acid, following the protocol recommended by the manufacturer (GE Healthcare). Densitometric analysis of gels was carried out using ImageJ software, version 1.53a (National Institutes of Health, Bethesda, MD, USA).

5.2.2. Cell Free Protein Synthesis Inhibition

The effect of lectins on protein synthesis was determined through a cell-free system, based on a rabbit reticulocyte lysate. Experiments were carried out both under non-reducing and reducing conditions with the addition of 1% 2-mercaptoethanol for 30 min at 37 °C. Samples were diluted and added to the reaction mixture, as previously described [53]. The radioactivity of L-[^{3}H]leucine incorporated into new synthesized proteins was measured by β-counter (Beckman Instruments). The experiments were conducted in duplicate, and IC_{50} values were calculated by linear regression. Specific activity is expressed as units (U) per mg of protein, where one U is the amount of proteins (in µg) that inhibits 50% protein synthesis in 1 mL of reaction mixture. Total activity was calculated as the specific activity per whole basic-fraction proteins (mg) normalized to the total proteins of the crude extract.

5.2.3. Hemagglutinating Activity

Hemagglutinating activity was determined in 96 wells microtiter plates. Each well contained 50 µL of a 2% suspension of human erythrocytes (group 0, Rh+) and 2-fold serial dilutions of the lectins, in a final volume of 100 µL. The plates were gently shaken and after about 1 h at 25 °C, the presence/absence of agglutination was visually examined.

5.2.4. rRNA Glycosylase Activity

N-glycosylase activity of kirkiin was conducted as previously described [54]. Briefly, rabbit reticulocytes lysate (40 µL) and S-30 lysate from yeast (25 µL) were incubated with 3 µg of kirkiin at 37 °C for 1 h. After treatment, 2 µL of 0.5 M EDTA pH 8 and 500 µL of 50 mM Tris-HCl (pH 7.8) and 0.5% SDS (*w*/*v*) were added, and the samples were vigorously vortexed for 30 s. RNA was extracted by phenolization, treated with 2 M aniline acetate (pH 4.5) on ice for 10 min in the dark, and precipitated with ethanol. The pellet was resuspended in 20 µL of sterile water and the concentration was determined by spectrophotometer at 260 nm. Ribosomal RNA was analyzed using 5% (*w*/*v*) polyacrylamide in denaturating conditions with 7 M urea. RNA samples were incubated in loading buffer containing 150 mg/mL sucrose, 7 M urea, 0.4 µg/mL bromophenol blue, and 1XTBE buffer (45 mM Tris, 45 mM boric acid, 1 mM EDTA pH 8). After boiling the samples for 30 s, the run was performed at 15 mA for 1 h 50 min, approximately, using TBE buffer. The gel was stained with ethidium bromide (20 mg/mL) in TBE buffer for 20–30 min and RNA bands were analyzed by UV-transilluminator (254–312 nm) including in the imaging instrument GelDoc (Biorad).

5.2.5. Adenine Polynucleotide Glycosylase Activity on Salmon Sperm DNA and on Tobacco Mosaic Virus (TMV) RNA

Adenine polynucleotide glycosylase activity was determined by measuring the adenine release from salmon sperm DNA (ssDNA) according to the method reported in [55] with a few modifications. Briefly, 10 µg of ssDNA were incubated with 5 µg of kirkiin, both in reduced and non-reduced conditions, in 300 µL of a reaction mixture containing 1 M KCl and 0.5 M sodium acetate (pH 4.5) at 30 °C for 1 h. After incubation, the DNA was precipitated with ethanol at −80 °C overnight and centrifugated at 13,000 rpm for 15 min at 4 °C. Adenine released from RIP-treated DNA was determined in the supernatants by spectrophotometer at 260 nm.

On TMV, the adenine polynucleotide glycosylase activity of kirkiin was assayed as described in [54]. Briefly, 25 µL samples containing 15 µg of TMV RNA were incubated with 3 µg of kirkiin. After treatment, the RNA was analyzed by extraction, phenolization, treatment with 2 M aniline acetate (pH 4.5), and ethanol precipitation. The RNA was subjected to electrophoresis on 5% (*w*/*v*) polyacrylamide-7 M urea gel at 15 mA for 75 min and stained with ethidium bromide.

5.2.6. Endonuclease Activity on Supercoiled Plasmid DNA

The endonuclease activity of the RIP was assayed on the *E. coli* plasmid pCR 2.1 (Invitrogen). 200 ng of the plasmid were incubated with 3 µg of kirkiin at 37 °C for 1 h in a final volume of 10 µL of 10 mM Tris-HCl (pH 7.8), 50 mM NaCl, and 50 mM KCl in presence/absence of 5 mM $MgCl_2$. The samples were analyzed on 0.8% agarose gel electrophoresis in TAE buffer (0.04 M Tris, 0.04 M acetate, 1 mM EDTA, pH 8.0) and visualized by gel red staining.

5.2.7. Enzyme-Linked Immunosorbent Assay (ELISA)

ELISA assay was performed as described previously [17], using 2 µg per well of kirkiin in 100 µL of 50 mM carbonate buffer pH 9.0 containing 15 mM sodium carbonate and 35 mM sodium bicarbonate. Reciprocal serum dilutions (from 1:100 to 1:12,800) were added. The dilutions were prepared in 50 mM lactose, 50 mM mannose, and 0.05% Tween 20. Rabbit antisera against type 1 and type 2 RIPs were obtained as described in [17,51]. 100 µL of anti-rabbit secondary antibody (1:7000) conjugated to alkaline phosphatase was used and incubated 1 h at 37 °C. 100 µL of 1 mg/mL enzyme substrate (4-nitrophenyl phosphate disodium) dissolved in buffer containing 1 M diethalonamine, 0.5 M $MgCl_2 \times 6H_2O$, and 3 mM NaN_3 were added. The absorption was measured at 405 nm with the Multiskan EX microtiter plate reader (ThermoLabsystem, Helsinki, Finland).

5.2.8. Cell Protein Synthesis Inhibition and Viability Assay

The cytotoxicity of kirkiin was assessed by evaluating both protein synthesis inhibition and viability reduction.

Protein synthesis inhibition was evaluated through L-[^{3}H]leucine incorporation in neosynthesized proteins. NB100 cells (2×10^4/well) were seeded onto 24-well plates in 250 µL of complete medium in the absence (control cultures) or presence of scalar dilutions (from 1×10^{-15} to 1×10^{-11} M) of kirkiin. After 72 h, cell protein synthesis was evaluated as previously described [18]. In addition, time course experiments were conducted on cells exposed to kirkiin (1×10^{-11} M), in a range between 4 and 48 h. The IC_{50} and IT_{50} (kirkiin concentration and time required to inhibit cell protein synthesis by 50%) were calculated using linear regression analysis.

Cell viability was evaluated through the colorimetric cell cytotoxicity assay (CellTiter 96® Aqueous One Solution Cell Proliferation), based on the cellular conversion of a tetrazolium salt into a colored formazan. NB100 cells (2×10^3/100 µL complete medium) were seeded in 96-well microtiter plates. After 24 h, cells were incubated with scalar dilutions of kirkiin (from 1×10^{-15} to 1×10^{-11} M) and left for 72 h. In addition, time course experiments were conducted on cells exposed to kirkiin (1×10^{-11} M), in a range between 4 and 48 h. After the indicated times, the medium was removed and CellTiter 96 Aqueous One Solution Reagent (1: 6 in complete medium). After 1 h of incubation at 37 °C, the absorbance at 492 nm was measured. The EC_{50} and ET_{50} (kirkiin effective concentration and time required to reduce cell viability by 50%) were calculated using linear regression analysis. The results are the means of at least three experiments performed in triplicate.

5.2.9. Evaluation of Apoptosis

The morphological analysis of treated cells (2×10^3/100 µL complete medium) was conducted through phase contrast microscopy, directly in 24-well plate, using an inverted microscope Nikon Eclipse TS100 (Nikon, Melville, NY, USA). For the nuclear analysis, NB100 cells (2×10^4/500 µL complete medium) were seeded directly on a coverslip in 24-well plates 48 h prior to the experiment. After treatment with kirkiin for 48 h, cells were fixed with methanol/acetic acid 1:3 for 20 min. The analysis was conducted under Nikon Eclipse E600W fluorescence microscope with pretreatment of cells with 7 µL DAPI/antifade (4′,6-diamidino-2-phenylindole).

The mitochondrial membrane potential (Δψ m) was measured using the cationic, lipophilic dye JC-1 (5,5′,6,6′-tetrachloro-1,1′,3,3′-tetraethylbenzimidazolocarbocyanineiodide) contained in Mitochondria Staining Kit (Sigma). JC-1 selectively enters the mitochondria and reversibly change color from green to red as the membrane potential increases. Cells (2×10^4/500 µL complete medium) were seeded directly on a coverslip in 24-well plates 48 h prior the experiments. After treatment with kirkiin for 48 h, cells were stained with 500 µL of JC-1 dye (1:100 in complete medium) and incubated at room temperature for 10 min in the dark. The cells were then washed three times with staining buffer purchased from the kit. The coverslips were inverted on glass slide and the cells were observed under Nikon Eclipse E600W fluorescence microscope.

Apoptotic cell death was assessed using a flow cytometry Annexin V-EGFP/PI detection kit and by a luminescent reagent detecting caspase activity. Before flow cytometry, cells (2×10^6/3 mL complete medium) were seeded in 25 cm^2 flasks, and after incubation with kirkiin for 48 h, the cells were pelleted at 400× g for 5 min, washed twice in cold PBS, pelleted, and resuspended in 294 µL of binding buffer provided by the kit. Annexin V-EGFP (3 µL) and PI (3 µL) were added. After 10 min incubation in the dark at room temperature, cells were analyzed by flow cytometry FACSAria (BD) using the FACSDiva software.

The caspase-3/7 activity was assessed by the luminescent Caspase-Glo™3/7 Assay as described in [25]. Briefly, cells (2×10^3/100 µL complete medium) were seeded in 96-well microtiter plates. After incubation with kirkiin for the indicated amounts of time, 50 µL/well of caspase kit reagent (1:2 in complete medium) was added. The plates were shaken at 420 rpm for 1 min and then incubated for 20 min at room temperature in the dark. The luminescence was acquired (integration time 10 s) by a Fluoroskan Ascent FL (Thermo Labsystems) and the values were normalized for cell viability.

5.2.10. Statistical Analyses

Statistical analyses were conducted using XLSTAT-Pro software, version 6.1.9, 2003 (Addinsoft, Inc., Brooklyn, NY, USA). The results are presented as the means ± S.D. of three different experiments. The data were analyzed using ANOVA/Bonferroni test or Student's t-distribution. The Dunnett's test was used in addiction to ANOVA, when necessary.

Author Contributions: Conceptualization: A.B. and L.P.; methodology and validation: A.B., J.M.F., M.B., L.P., R.I. and S.M.; formal analysis and investigation: M.B., R.I. and S.M.; all the authors participated to write, review, and edit the manuscript; funding acquisition: A.B., J.M.F. and L.P. All authors have read and agreed to the published version of the manuscript.

Funding: This work was supported by funds for selected research topics from the Alma Mater Studiorum, University of Bologna and by the Pallotti Legacies for Cancer Research; Fondazione

CARISBO, Project 2019.0539; Grant VA033G19 (Consejería de Educación, Junta de Castilla y León) to the GIR ProtIBio.

Institutional Review Board Statement: Not applicable.

Informed Consent Statement: Not applicable.

Data Availability Statement: Data are available upon request. Please, contact the contributing authors.

Acknowledgments: The present article contains some report on knowledge/insight/data previously included in S.M.'s Ph.D. Dissertation.

Conflicts of Interest: The authors declare no conflict of interest.

References

1. Bolognesi, A.; Bortolotti, M.; Maiello, S.; Battelli, M.G.; Polito, L. Ribosome-Inactivating Proteins from Plants: A Historical Overview. *Molecules* **2016**, *21*, 1627. [CrossRef] [PubMed]
2. Schrot, J.; Weng, A.; Melzig, M.F. Ribosome-inactivating and related proteins. *Toxins* **2015**, *7*, 1556–1615. [CrossRef] [PubMed]
3. Wong, J.H.; Bao, H.; Ng, T.B.; Chan, H.H.L.; Ng, C.C.W.; Man, G.C.W.; Wang, H.; Guan, S.; Zhao, S.; Fang, E.F.; et al. New ribosome-inactivating proteins and other proteins with protein synthesis-inhibiting activities. *Appl. Microbiol. Biotechnol.* **2020**, *104*, 4211–4226. [CrossRef] [PubMed]
4. Polito, L.; Bortolotti, M.; Maiello, S.; Battelli, M.G.; Bolognesi, A. Plants Producing Ribosome-Inactivating Proteins in Traditional Medicine. *Molecules* **2016**, *21*, 1560. [CrossRef]
5. Bortolotti, M.; Mercatelli, D.; Polito, L. *Momordica charantia*, a Nutraceutical Approach for Inflammatory Related Diseases. *Front. Pharmacol.* **2019**, *10*, 486. [CrossRef] [PubMed]
6. Endo, Y.; Mitsui, K.; Motizuki, M.; Tsurugi, K. The mechanism of action of ricin and related toxic lectins on eukaryotic ribosomes. The site and the characteristics of the modification in 28 S ribosomal RNA caused by the toxins. *J. Biol. Chem.* **1987**, *262*, 5908–5912. [CrossRef]
7. Walsh, M.J.; Dodd, J.E.; Hautbergue, G.M. Ribosome-inactivating proteins. *Virulence* **2013**, *4*, 774–784. [CrossRef]
8. Barbieri, L.; Valbonesi, P.; Bonora, E.; Gorini, P.; Bolognesi, A.; Stirpe, F. Polynucleotide:adenosine glycosidase activity of ribosome-inactivating proteins: Effect on DNA, RNA and poly(A). *Nucleic Acids Res.* **1997**, *25*, 518–522. [CrossRef]
9. Battelli, M.G.; Barbieri, L.; Bolognesi, A.; Buonamici, L.; Valbonesi, P.; Polito, L.; Van Damme, E.J.M.; Peumans, W.J.; Stirpe, F. Ribosome-inactivating lectins with polynucleotide:adenosine glycosidase activity. *FEBS Lett.* **1997**, *408*, 355–359. [CrossRef]
10. Bolognesi, A.; Polito, L.; Lubelli, C.; Barbieri, L.; Parente, A.; Stirpe, F. Ribosome-inactivating and adenine polynucleotide glycosylase activities in *Mirabilis jalapa* L. tissues. *J. Biol. Chem.* **2002**, *277*, 13709–13716. [CrossRef]
11. Polito, L.; Bortolotti, M.; Battelli, M.G.; Calafato, G.; Bolognesi, A. Ricin: An Ancient Story for a Timeless Plant Toxin. *Toxins* **2019**, *11*, 324. [CrossRef] [PubMed]
12. Spooner, R.A.; Lord, J.M. Ricin trafficking in cells. *Toxins* **2015**, *7*, 49–65. [CrossRef] [PubMed]
13. Grela, P.; Szajwaj, M.; Horbowicz-Drożdżal, P.; Tchórzewski, M. How Ricin Damages the Ribosome. *Toxins* **2019**, *11*, 241. [CrossRef] [PubMed]
14. De Zaeytijd, J.; Van Damme, E.J. Extensive Evolution of Cereal Ribosome-Inactivating Proteins Translates into Unique Structural Features, Activation Mechanisms, and Physiological Roles. *Toxins* **2017**, *9*, 123. [CrossRef] [PubMed]
15. Stirpe, F.; Barbieri, L.; Abbondanza, A.; Falasca, A.I.; Brown, A.N.; Sandvig, K.; Olsnes, S.; Pihl, A. Properties of volkensin, a toxic lectin from *Adenia volkensii*. *J. Biol. Chem.* **1985**, *260*, 14589–14595. [CrossRef]
16. Gasperi-Campani, A.; Barbieri, L.; Lorenzoni, E.; Montanaro, L.; Sperti, S.; Bonetti, E.; Stirpe, F. Modeccin, the toxin of *Adenia digitata*. Purification, toxicity and inhibition of protein synthesis in vitro. *Biochem. J.* **1978**, *174*, 491–496. [CrossRef]
17. Stirpe, F.; Bolognesi, A.; Bortolotti, M.; Farini, V.; Lubelli, C.; Pelosi, E.; Polito, L.; Dozza, B.; Strocchi, P.; Chambery, A.; et al. Characterization of highly toxic type 2 ribosome-inactivating proteins from *Adenia lanceolata* and *Adenia stenodactyla* (Passifloraceae). *Toxicon* **2007**, *50*, 94–105. [CrossRef]
18. Battelli, M.G.; Scicchitano, V.; Polito, L.; Farini, V.; Barbieri, L.; Bolognesi, A. Binding and intracellular routing of the plant-toxic lectins, lanceolin and stenodactylin. *Biochim. Biophys. Acta* **2010**, *1800*, 1276–1282. [CrossRef]
19. Wiley, R.G.; Kline IV, R.H. Neuronal lesioning with axonally transported toxins. *J. Neurosci. Methods* **2000**, *103*, 73–82. [CrossRef]
20. Monti, B.; D'Alessandro, C.; Farini, V.; Bolognesi, A.; Polazzi, E.; Contestabile, A.; Stirpe, F.; Battelli, M.G. *In vitro* and *in vivo* toxicity of type 2 ribosome-inactivating proteins lanceolin and stenodactylin on glial and neuronal cells. *Neurotoxicology* **2007**, *28*, 637–644. [CrossRef]
21. Pangalos, M.N.; Francis, P.T.; Pearson, R.C.; Middlemiss, D.N.; Bowen, D.M. Destruction of a sub-population of cortical neurones by suicide transport of volkensin, a lectin from Adenia volkensii. *J. Neurosci. Methods* **1991**, *40*, 17–29. [CrossRef]
22. Rust, A.; Partridge, L.J.; Davletov, B.; Hautbergue, G.M. The Use of Plant-Derived Ribosome Inactivating Proteins in Immunotoxin Development: Past, Present and Future Generations. *Toxins* **2017**, *9*, 344. [CrossRef] [PubMed]
23. Polito, L.; Djemil, A.; Bortolotti, M. Plant Toxin-Based Immunotoxins for Cancer Therapy: A Short Overview. *Biomedicines* **2016**, *4*, 12. [CrossRef] [PubMed]

24. Pizzo, E.; Di Maro, A. A new age for biomedical applications of Ribosome Inactivating Proteins (RIPs): From bioconjugate to nanoconstructs. *J. Biomed. Sci.* **2016**, *23*, 54. [CrossRef] [PubMed]
25. Polito, L.; Bortolotti, M.; Pedrazzi, M.; Mercatelli, D.; Battelli, M.G.; Bolognesi, A. Apoptosis and necroptosis induced by stenodactylin in neuroblastoma cells can be completely prevented through caspase inhibition plus catalase or necrostatin-1. *Phytomedicine* **2016**, *23*, 32–41. [CrossRef]
26. Mercatelli, D.; Bortolotti, M.; Andresen, V.; Sulen, A.; Polito, L.; Gjertsen, B.T.; Bolognesi, A. Early Response to the Plant Toxin Stenodactylin in Acute Myeloid Leukemia Cells Involves Inflammatory and Apoptotic Signaling. *Front. Pharmacol.* **2020**, *11*, 630. [CrossRef]
27. Iglesias, R.; Pérez, Y.; de Torre, C.; Ferreras, J.M.; Antolín, P.; Jiménez, P.; Rojo, M.A.; Méndez, E.; Girbés, T. Molecular characterization and systemic induction of single-chain ribosome-inactivating proteins (RIPs) in sugar beet (*Beta vulgaris*) leaves. *J. Exp. Bot.* **2005**, *56*, 1675–1684. [CrossRef]
28. Iglesias, R.; Citores, L.; Di Maro, A.; Ferreras, J.M. Biological activities of the antiviral protein BE27 from sugar beet (*Beta vulgaris* L.). *Planta* **2015**, *241*, 421–433. [CrossRef]
29. Eggli, U.; Newton, L.E. *Etymological Dictionary of Succulent Plant. Names*, 1st ed.; Springer: Berlin/Heidelberg, Germany, 2013; Volume 3, p. 127. [CrossRef]
30. Fabbrini, M.S.; Katayama, M.; Nakase, I.; Vago, R. Plant Ribosome-Inactivating Proteins: Progesses, Challenges and Biotechnological Applications (and a Few Digressions). *Toxins* **2017**, *9*, 314. [CrossRef]
31. Pelosi, E.; Lubelli, C.; Polito, L.; Barbieri, L.; Bolognesi, A.; Stirpe, F. Ribosome-inactivating proteins and other lectins from *Adenia* (Passifloraceae). *Toxicon* **2005**, *46*, 658–663. [CrossRef]
32. Barbier, J.; Gillet, D. Ribosome Inactivating Proteins: From Plant Defense to Treatments against Human Misuse or Diseases. *Toxins* **2018**, *10*, 160. [CrossRef] [PubMed]
33. Endo, Y.; Tsurugi, K. The RNA N-glycosidase activity of ricin A-chain. The characteristics of the enzymatic activity of ricin A-chain with ribosomes and with rRNA. *J. Biol. Chem.* **1988**, *263*, 8735–8739. [CrossRef]
34. Zhu, F.; Zhou, Y.K.; Ji, Z.L.; Chen, X.R. The Plant Ribosome-Inactivating Proteins Play Important Roles in Defense against Pathogens and Insect Pest Attacks. *Front. Plant. Sci.* **2018**, *9*, 146. [CrossRef] [PubMed]
35. Parikh, B.A.; Tumer, N.E. Antiviral activity of ribosome inactivating proteins in medicine. *Mini Rev. Med. Chem.* **2004**, *4*, 523–543. [CrossRef] [PubMed]
36. Barbieri, L.; Ciani, M.; Girbés, T.; Liu, W.Y.; Van Damme, E.J.; Peumans, W.J.; Stirpe, F. Enzymatic activity of toxic and non-toxic type 2 ribosome-inactivating proteins. *FEBS Lett.* **2004**, *563*, 219–222. [CrossRef]
37. Barbieri, L.; Valbonesi, P.; Righi, F.; Zuccheri, G.; Monti, F.; Gorini, P.; Samorí, B.; Stirpe, F. Polynucleotide:Adenosine glycosidase is the sole activity of ribosome-inactivating proteins on DNA. *J. Biochem.* **2000**, *128*, 883–889. [CrossRef]
38. Gilabert-Oriol, R.; Weng, A.; von Mallinckrodt, B.; Melzig, M.F.; Fuchs, H.; Thakur, M. Immunotoxins constructed with ribosome-inactivating proteins and their enhancers: A lethal cocktail with tumor specific efficacy. *Curr. Pharm. Des.* **2014**, *20*, 6584–6643. [CrossRef]
39. Polito, L.; Bortolotti, M.; Mercatelli, D.; Battelli, M.G.; Bolognesi, A. Saporin-S6: A useful tool in cancer therapy. *Toxins* **2013**, *5*, 1698–1722. [CrossRef]
40. Lu, J.Q.; Zhu, Z.N.; Zheng, Y.T.; Shaw, P.C. Engineering of Ribosome-inactivating Proteins for Improving Pharmacological Properties. *Toxins* **2020**, *12*, 167. [CrossRef]
41. Narayanan, S.; Surendranath, K.; Bora, N.; Surolia, A.; Karande, A.A. Ribosome inactivating proteins and apoptosis. *FEBS Lett.* **2005**, *579*, 1324–1331. [CrossRef]
42. Polito, L.; Bortolotti, M.; Farini, V.; Battelli, M.G.; Barbieri, L.; Bolognesi, A. Saporin induces multiple death pathways in lymphoma cells with different intensity and timing as compared to ricin. *Int. J. Biochem. Cell Biol.* **2009**, *41*, 1055–1061. [CrossRef] [PubMed]
43. Sikriwal, D.; Ghosh, P.; Batra, J.K. Ribosome inactivating protein saporin induces apoptosis through mitochondrial cascade, independent of translation inhibition. *Int. J. Biochem. Cell Biol.* **2008**, *40*, 2880–2888. [CrossRef] [PubMed]
44. Hodges, A.L.; Kempen, C.G.; McCaig, W.D.; Parker, C.A.; Mantis, N.J.; LaRocca, T.J. TNF Family Cytokines Induce Distinct Cell Death Modalities in the A549 Human Lung Epithelial Cell Line when Administered in Combination with Ricin Toxin. *Toxins* **2019**, *11*, 450. [CrossRef] [PubMed]
45. Polito, L.; Mercatelli, D.; Bortolotti, M.; Maiello, S.; Djemil, A.; Battelli, M.G.; Bolognesi, A. Two Saporin-Containing Immunotoxins Specific for CD20 and CD22 Show Different Behavior in Killing Lymphoma Cells. *Toxins* **2017**, *9*, 182. [CrossRef] [PubMed]
46. Sowa-Rogozińska, N.; Sominka, H.; Nowakowska-Gołacka, J.; Sandvig, K.; Słomińska-Wojewódzka, M. Intracellular Transport and Cytotoxicity of the Protein Toxin Ricin. *Toxins* **2019**, *11*, 350. [CrossRef]
47. Puri, M.; Kaur, I.; Perugini, M.A.; Gupta, R.C. Ribosome-inactivating proteins: Current status and biomedical applications. *Drug Discov. Today* **2012**, *17*, 774–783. [CrossRef]
48. Barbieri, L.; Falasca, A.I.; Stirpe, F. Volkensin, the toxin of *Adenia volkensii* (kilyambiti plant). *FEBS Lett.* **1984**, *171*, 277–279. [CrossRef]
49. Nicolson, G.L.; Blaustein, J.; Etzler, M.E. Characterization of two plant lectins from *Ricinus communis* and their quantitative interaction with a murine lymphoma. *Biochemistry* **1974**, *13*, 196–204. [CrossRef]

50. Barbieri, L.; Stoppa, C.; Bolognesi, A. Large scale chromatographic purification of ribosome-inactivating proteins. *J. Chromatogr.* **1987**, *408*, 235–243. [CrossRef]
51. Strocchi, P.; Barbieri, L.; Stirpe, F. Immunological properties of ribosome-inactivating proteins and of a saporin-IgG conjugate. *J. Immunol. Methods* **1992**, *155*, 57–63. [CrossRef]
52. Kalb, V.F., Jr.; Bernlohr, R.W. A new spectrophotometric assay for protein in cell extracts. *Anal. Biochem.* **1977**, *82*, 362–371. [CrossRef]
53. Polito, L.; Bortolotti, M.; Mercatelli, D.; Mancuso, R.; Baruzzi, G.; Faedi, W.; Bolognesi, A. Protein synthesis inhibition activity by strawberry tissue protein extracts during plant life cycle and under biotic and abiotic stresses. *Int. J. Mol. Sci.* **2013**, *14*, 15532–15545. [CrossRef] [PubMed]
54. Iglesias, R.; Citores, L.; Ragucci, S.; Russo, R.; Di Maro, A.; Ferreras, J.M. Biological and antipathogenic activities of ribosome-inactivating proteins from *Phytolacca dioica* L. *Biochim. Biophys. Acta* **2016**, *1860*, 1256–1264. [CrossRef] [PubMed]
55. Di Maro, A.; Chambery, A.; Daniele, A.; Casoria, P.; Parente, A. Isolation and characterization of heterotepalins, type 1 ribosome-inactivating proteins from *Phytolacca heterotepala* leaves. *Phytochemistry* **2007**, *68*, 767–776. [CrossRef]

toxins

MDPI

Article

Ebulin l Is Internalized in Cells by Both Clathrin-Dependent and -Independent Mechanisms and Does Not Require Clathrin or Dynamin for Intoxication

Rosario Iglesias [1,†], José M. Ferreras [1,†], Alicia Llorente [2,3] and Lucía Citores [1,*]

1 Department of Biochemistry and Molecular Biology and Physiology, Faculty of Sciences, University of Valladolid, E-47011 Valladolid, Spain; riglesia@bio.uva.es (R.I.); josemiguel.ferreras@uva.es (J.M.F.)
2 Department of Molecular Cell Biology, Institute for Cancer Research, Oslo University Hospital, 0379 Oslo, Norway; alillo@rr-research.no
3 Department of Mechanical, Electronics and Chemical Engineering Art and Design, Oslo Metropolitan University, 0130 Oslo, Norway
* Correspondence: luciac@bio.uva.es
† Authors contributed equally to this work.

Abstract: Ebulin l is an A-B toxin, and despite the presence of a B chain, this toxin displays much less toxicity to cells than the potent A-B toxin ricin. Here, we studied the binding, mechanisms of endocytosis, and intracellular pathway followed by ebulin l and compared it with ricin. COS-1 cells and HeLa cells with inducible synthesis of a mutant dynamin (K44A) were used in this study. The transport of these toxins was measured using radioactively or fluorescently labeled toxins. The data show that ebulin l binds to cells to a lesser extent than ricin. Moreover, the expression of mutant dynamin does not affect the endocytosis, degradation, or toxicity of ebulin l. However, the inhibition of clathrin-coated pit formation by acidification of the cytosol reduced ebulin l endocytosis but not toxicity. Remarkably, unlike ricin, ebulin l is not transported through the Golgi apparatus to intoxicate the cells and ebulin l induces apoptosis as the predominant cell death mechanism. Therefore, after binding to cells, ebulin l is taken up by clathrin-dependent and -independent endocytosis into the endosomal/lysosomal system, but there is no apparent role for clathrin and dynamin in productive intracellular routing leading to intoxication.

Keywords: apoptosis; clathrin; dynamin; ebulin; endocytosis; intracellular transport; lectin; rRNA *N*-glycosylase; ribosome-inactivating protein; ricin

Key Contribution: This work contributes to elucidating the endocytosis, intracellular transport, and toxicity mechanisms of the plant toxin ebulin l. This knowledge is important for potential medical and biotechnological use of this toxin.

Citation: Iglesias, R.; Ferreras, J.M.; Llorente, A.; Citores, L. Ebulin l Is Internalized in Cells by Both Clathrin-Dependent and -Independent Mechanisms and Does Not Require Clathrin or Dynamin for Intoxication. *Toxins* **2021**, *13*, 102. https://doi.org/10.3390/toxins13020102

Received: 28 December 2020
Accepted: 27 January 2021
Published: 30 January 2021

Publisher's Note: MDPI stays neutral with regard to jurisdictional claims in published maps and institutional affiliations.

1. Introduction

Ribosome inactivating proteins (RIPs) are a family of well-characterized toxins that specifically and irreversibly inhibit protein synthesis. RIPs belong to a class of enzymes (EC 3.2.2.22) that exhibits rRNA N-glycosylase activity. This activity prevents protein synthesis by causing the release of a specific adenine residue in the sarcin-ricin loop (SRL) of the large rRNA that is crucial for interaction of the elongation factor with the ribosome [1].

Most RIPs are produced by plants, where they may play a role in defense mechanism against predators, fungi, and viruses [2,3]. RIPs also show toxicity towards animal cells, targeting the host protein synthesis machinery. In addition to rRNA damage, RIPs can induce apoptosis [4,5].

There are important differences in toxicity among RIPs depending on their ability to reach the ribosomes in the cytosol of target cells. Since RIPs are unable to cross the plasma membrane directly, they use existing cellular mechanisms designed for uptake of macromolecules. Following initial internalization, RIPs are transported within the cell to the particular membrane where toxin translocation to the cytosol occurs. Type 1 RIPs consisting of a single enzymatic active (A) chain often display lower toxicity than type 2 RIPs which consist of a binding (B) chain with lectin activity linked by a disulfide bond to the enzymatic A chain. The carbohydrate-binding domains of the B chain recognize glycosylated receptors on the cell surface, facilitating the entry of the A chain into the cell [5]. However, the presence of the B chain is not sufficient to confer a high level of cytotoxicity on all type 2 RIPs. Based on their toxicity to mammals, type 2 RIPs are divided into two groups: the toxic and nontoxic type 2 RIPs [6]. The former group includes ricin, abrin, viscumin, volkesin, and stenodactylin, which are among the most potent plant toxins. In contrast, ebulin l, nigrin b, *Ricinus* agglutinin (RCA), *Iris* agglutinin (IRA) b/r, and cinnamomin belonging to the latter group show little or no toxicity in higher animals. The reason for the different toxicities among type 2 RIPs is not clear. It could rather be attributed to differences between the B chains, which are responsible for the interaction with cellular membranes, than to the enzymatic A chains, which inactivate naked ribosomes with apparently similar efficiency.

Ricin, a toxin isolated from *Ricinus communis* L., is the archetype of the toxic type 2 RIP family. The structure, biochemistry, and cytotoxicity of this 64-kD A-B toxin have been extensively examined and reviewed [7–10]. In order to enter and intoxicate cells, ricin first has to bind to cell surface receptors. Ricin binds to both glycoproteins and glycolipids with terminal galactose and then is internalized by different endocytic mechanisms. After being endocytosed, most of the ricin molecules are either recycled or transported to lysosomes for degradation. However, a small proportion (5%) of ricin is transported to the Golgi apparatus and then retrogradely to the endoplasmic reticulum (ER). After a reduction of the internal disulfide bond that connects the A and B chain, the A chain enters the cytosol using the quality control pathway that leads to ER-associated protein degradation (ERAD). Once in the cytosol, a small fraction of the toxin is able to escape ubiquitination and degradation by the proteasome and binds to its ribosomal target [7,9].

In recent years, an extensive study for the presence of RIPs in several species of the genus *Sambucus* has allowed the isolation of more than 20 toxins. All of the type 2 RIPs found in the genus *Sambucus* are considered nontoxic type 2 RIPs since, despite being as toxic as ricin at the ribosomal level, they display much less toxicity to cells and animals. Nontoxic type 2 RIPs specific for galactose [11,12], tetrameric type 2 RIPs specific for sialic acid [13,14], nontoxic type 2 RIPs lacking sugar binding activity [15,16], and nontoxic type 2 RIPs with affinity for N-acetyl-glucosamine oligomers [17] have been described for the first time in the genus *Sambucus*.

Ebulin l, a 56 kD A-B toxin obtained from the leaves of *Sambucus ebulus* L., was one of the first nontoxic type 2 RIPs isolated [11]. The structure of ebulin l has been resolved by X-ray diffraction analysis, and the tertiary structure closely resembles that of ricin [18]. In the A chain, ebulin l has roughly the same positioning of key active site residues as ricin. This is consistent with the fact that both proteins have a similar inhibitory activity of protein synthesis in cell-free systems. The overall fold of the ebulin and ricin B chains is very similar. However, ebulin l has a lower affinity for galactose than ricin due to a change in the structure of the 2-γ subdomain of the ebulin B chain. In fact, it was found that ebulin l has different binding properties to D-galactose-containing matrixes than ricin [16,18]. This reduced affinity for galactosides could alter the ability of the B chain to bind cells and could affect the uptake and the intracellular fate of the toxin. In contrast to the high enzymatic activity on ribosomes, the toxicity of ebulin l on animal cells was found to be about 10^4–10^6 times lower than the toxicity of ricin [11,16]. In mice, the LD_{50} of ebulin l administered by intraperitoneal injection is 2 mg/kg body weight, while for ricin, it is in the range of a few micrograms per kilogram [11].

RIPs are potent inhibitors of protein synthesis that have been used for the construction of conjugates and immunotoxins [5,19]. Linked to a targeting portion such as an antibody or a protein that specifically binds to a receptor, toxins have been used to specifically kill tumor cells. Ebulin l has been used in different conjugates and immunotoxins targeting tumor cells with high selectivity [20]. The main advantage of ebulin l over ricin and its derivatives is its reduced cytotoxicity. Antibodies or ligands led the internalization and promoted the productive translocation of ebulin l to the cytosol, thus allowing for its anti-ribosomal activity. To improve the efficiency of selective targeting of ebulin l to malignant cells, a better understanding of endocytosis and the cellular transport and toxicity mechanisms of ebulin l is essential. However, very little is known about the receptors that mediate the cellular uptake of ebulin l or its intracellular transport. Therefore, in this work, the binding, the mechanism of endocytosis, and the intracellular pathway followed by the nontoxic type 2 RIP ebulin l were investigated. Moreover, the transport of ebulin l and the toxic type 2 RIP ricin were compared. To investigate the mechanism of ebulin l internalization, we used HeLa cells with inducible expression of a mutant dynamin (K44A) that blocks clathrin-dependent and some clathrin-independent pathways (caveolae, RhoA, fast endophilin-mediated endocytosis (FEME), and others). In addition, cytosol acidification was also used to inhibit endocytosis from coated vesicles in COS and HeLa dynK44A cells. Our results show that, after binding to the cells, ebulin l is taken up by clathrin-dependent and -independent endocytosis into the endosomal/lysosomal system, but there is no apparent role for clathrin and dynamin in productive intracellular routing leading to intoxication.

2. Results and Discussion

2.1. Binding, Endocytosis, Recycling, and Degradation of Ebulin l and Ricin in COS Cells

2.1.1. Binding

Ebulin l and ricin are A-B toxins consisting of an enzymatic A chain with rRNA N-glycosylase activity linked by a disulfide bond to a binding B chain with lectin activity. Both RIPs are galactose-binding lectins [16,18] and therefore bind to different molecules with terminal galactose on the cell surface. To test for specific binding to COS cells, toxins labeled with radioactive iodine (Figure 1a) as well as fluorescent labeled ebulin l were used.

Crosslinking experiments after preincubation of the cells for 1 h at 4 °C with ^{125}I-labeled ebulin l demonstrated that there are receptors for ebulin l at the surface of these cells. In addition to labeled toxin, bands migrating higher than 100 kD were observed (Figure 1b). As expected, the addition of 0.1 M lactose to parallel cultures incubated in the same conditions prevented crosslinking of ^{125}I-ebulin l to the receptors (Figure 1b). Moreover, when the cells were treated with CY3-ebulin l at 4 °C, the toxin was bound to the cell surface in a homogenous manner (Figure 1c). To better quantify the number of receptors for ebulin l and ricin, COS cells were incubated with increasing concentrations of the labeled toxins at 4 °C. The binding experiments showed that ^{125}I-ebulin l was bound in a saturable way to COS cells with a K_d of 1.5×10^{-7} M and 2.8×10^{6} binding sites per cell (Figure 1d), whereas for ^{125}I-ricin, a K_d value of 4.6×10^{-7} M and 5.6×10^{7} binding sites per cell were measured (Figure 1e). The binding affinity determined by the K_d values was low and comparable for the two toxins. It is believed that 10^{6}–10^{8} ricin molecules can be bound to the cell surface [9]. Thus, we found that the number of cell receptors for ebulin l was 20 and 150 times lower than that for ricin in COS and HeLa dynK44A cells (see Section 2.2), respectively. It has been shown that ebulin l has a lower affinity for galactose than ricin due to a change in the structural disposition of the 2γ-subdomain of the ebulin B chain, which limits its ability to bind galactosides on cell surfaces [18]. This may explain the differences in the number of cell surface receptors observed. For ricin, it is likely that the binding to many different receptors results in multiple intracellular transport pathways that could deliver ricin to the appropriate compartment for membrane translocation to the cytosol. It is then possible that the differential affinity of ebulin l for galactosides determines its intracellular fate and possibly its cytotoxicity. However, differences in binding (20–150 times higher for ricin) cannot explain the lower cytotoxicity of ebulin l compared to ricin (10^{5} times higher for ricin). Accordingly, it has been shown that HeLa cells have a similar number of receptors for the nontoxic RIP nigrin b as that for modeccin and volkensin, which are more toxic than ricin, and two-log lower receptor numbers than for ricin [21]. It has also been shown that cinnamomin, a nontoxic type 2 RIP, and ricin share similar binding sites on BA/F3β cells with different affinity and that the lower cytotoxicity of cinnamomin is due to its B-chain [22]. However, the type 2 RIP articulatin-D, which lacks lectin activity, has been shown to display a cytotoxicity comparable to that of highly toxic type 2 RIPs, indicating that, for uptake and subsequent toxicity of all type 2 RIPs, recognition by the B chain of glycosylated receptors on the cell surface may not be essential [23].

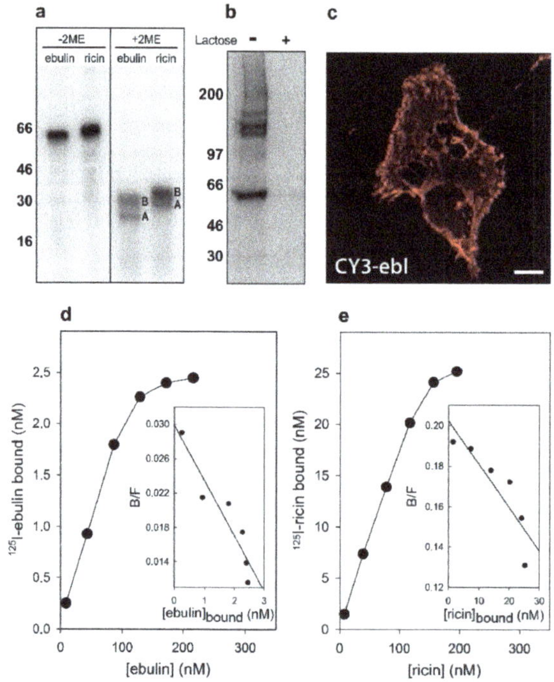

Figure 1. Binding of ebulin l and ricin to COS cells: (**a**) the ^{125}I-labeled ebulin l and ^{125}I-labeled ricin were analyzed by sodium dodecyl sulphate–polyacrylamide gel electrophoresis (SDS-PAGE) either in the absence or the presence of 2-mercaptoethanol (2ME) followed by autoradiography. A and B indicate the corresponding A and B chains of the toxins. Molecular weight standards (kD) are indicated on the left of the gels. (**b**) Crosslinking of bound ^{125}I-labeled ebulin l to COS cells: ^{125}I-ebulin l was added to the cells for 1 h at 4 °C in the presence or absence of 0.1 M lactose. Then, the cells were treated for 20 min at 4 °C with 0.3 mM disuccinimidyl suberate to induce crosslinking and were analyzed by SDS-PAGE and autoradiography. (**c**) Binding of fluorescent ebulin l to COS cells: The cells were incubated at 4 °C for 1 h with CY3-ebulin l (CY3-ebl) to allow binding and then fixed immediately. Bar, 50 µm. (**d**,**e**) Toxin binding and Scatchard plots: the binding of ^{125}I-ebulin l and ^{125}I-ricin to cells was measured by adding increasing concentrations of the labeled toxin to cells at 4 °C. After 1 h, any unbound toxin was removed by washing and the amount of radioactivity associated with the cells was measured. The insets are Scatchard plots of the binding data. These experiments were repeated twice with similar results. (B/F) Bound/Free.

2.1.2. Endocytosis, Recycling, and Degradation

Ebulin l has been shown to be internalized upon binding to glycoproteins and glycolipids containing terminal galactose [16]. To study this process further, COS cells were incubated with ^{125}I-labeled ebulin l or ricin at 37 °C for different periods of time. Considering that lactose removes surface-bound toxins but not internalized toxins, toxin internalization was measured as the ratio of endocytosed to surface-bound ^{125}I-labeled toxin at different time points. As shown in Figure 2a, the amount of internalized ebulin l and ricin levelled off after 30 min, indicating that toxin uptake and intracellular processing approached equilibrium at this time. Approximately 28% of total cell-associated ^{125}I-ebulin l and 23% of total cell-associated ^{125}I-ricin were internalized into the COS cells during a 30 min incubation period at 37 °C.

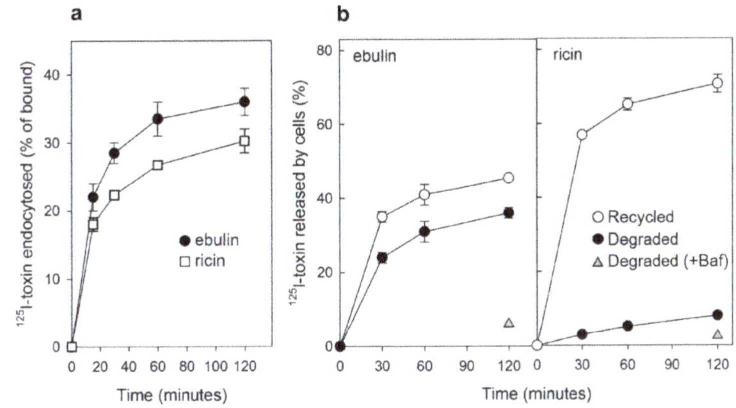

Figure 2. (**a**) Kinetics of ebulin l (circles) or ricin (squares) internalization in COS cells: the cells were incubated with ^{125}I-ebulin or ^{125}I-ricin at 37 °C for 0 to 120 min and the amount of bound and endocytosed toxins were quantified as described in Section 4.2.2. The data are expressed as the internalized radioactivity in percentage of the total radioactivity associated with the cells. (**b**) Recycling and degradation of ebulin l and ricin in COS cells: the cells were incubated with ^{125}I-ebulin l or ^{125}I-ricin for 20 min at 37 °C. Surface-bound toxins were removed by 0.1 M lactose and the incubation continued for the times indicated. Recycling (open circles) was measured as the amount of trichloroacetic acid (TCA)-precipitable toxin in the medium and at the cell surface. Degradation (closed circles) was measured as the amount of radioactivity that could not be precipitated by TCA. Degradation in the presence of bafilomycin A1 (Baf) was measured after 120 min (triangles). In both cases, the data are expressed as a percent of the total radioactivity. The data represent the mean ± SD of two experiments.

We next studied the ability of COS cells to recycle internalized toxins back to the surface and to the culture medium. As shown in Figure 2b, recycling of toxins, measured as trichloroacetic acid (TCA)-insoluble radioactivity in the medium, occurred as a rapid phase that lasted for less than 30 min, followed by a slower phase. Approximately 45% of ebulin l and 70% of ricin were found in the TCA-insoluble fraction after 2 h of incubation at 37 °C. We also measured the degradation of toxins internalized by COS cells. As shown in Figure 2b, only 8% of the total internalized ^{125}I-ricin was found to be degraded in COS cells after 2 h incubation at 37 °C. However, a higher percentage of ebulin l (36%) was found in the TCA-soluble fraction. Bafilomycin A1, an inhibitor of the vacuolar H^+-ATPase [24], inhibited the degradation of ebulin l and ricin, indicating that this process takes place in an acidic compartment. According to this, nigrin b, a nontoxic type 2 RIP, was shown to enter HeLa cells in a similar way to ricin; however, it was much faster and widely degraded. Moreover, the nigrin b-derived material released by cells was completely inactive [21,25].

2.1.3. Mechanism of Endocytosis

Several endocytic mechanisms have been documented, including macropinocytosis, clathrin-dependent endocytosis, caveolae-dependent endocytosis, and clathrin- and caveolae-independent endocytosis [26]. It has been demonstrated that ricin is able to employ different endocytic mechanisms, probably because it can recognize and bind to a great variety of cell surface components [9]. Clathrin-independent endocytosis was first described by studying the uptake of ricin, which continued after the inhibition of clathrin-dependent endocytosis [27]. To determine if ebulin l and ricin are internalized from clathrin-coated pits in COS cells, the cytosol was acidified to inhibit internalization from clathrin-coated pits [27]. When the cytosolic pH falls below 6.5, clathrin-coated pits at the cell surface can no longer pinch off and form clathrin-coated vesicles. In this study, the cytosol was acidified by preloading the cells with increasing concentrations of NH_4Cl followed by its removal [27]. The data in Figure 3a show that, in COS cells, the uptake of ebulin l and ricin were reduced by about 50% after acidification of the cytosol. Control experiments showed that the uptake of transferrin, which is endocytosed by clathrin-dependent endocytosis [28], was reduced by more than 95% under the same conditions. When the cytosol of COS cells was acidified by incubation with acetic acid, similar results to that with NH_4Cl pre-pulsing were obtained. Thus, while the endocytosis of transferrin was strongly reduced, there was only an approximately 50% and 55% reduction in the uptake of ^{125}I-ebulin and ^{125}I-ricin, respectively (Figure 3b). This indicates that ebulin l and ricin uptake is mediated by both clathrin-dependent and clathrin-independent endocytosis in COS cells.

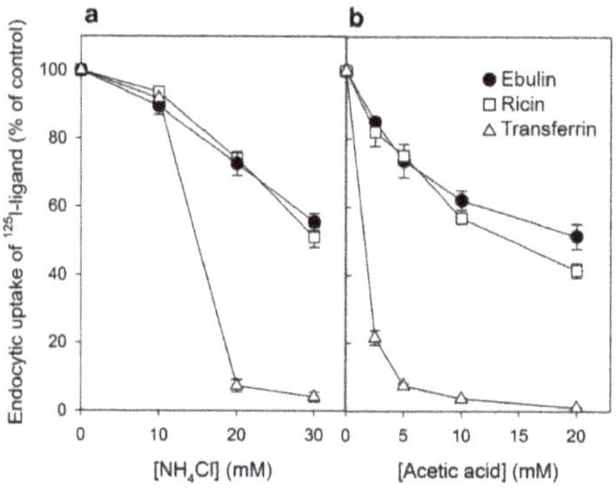

Figure 3. Effect of acidification of the cytosol on the ability of COS cells to internalize ^{125}I-ebulin l, ^{125}I-ricin, and ^{125}I-transferrin: (**a**) COS cells were incubated for 30 min at 37 °C in 4-(2-hydroxyethyl)-1-piperazineethanesulfonic acid (HEPES) medium pH 7 with the indicated concentrations of NH_4Cl. The medium was removed, and a solution containing 0.14 M KCl, 2 mM $CaCl_2$, 1 mM amiloride, 1 mM $MgCl_2$, and 20 mM HEPES, pH 7.0, was added. After 5 min of incubation at 37 °C, ^{125}I-ebulin l, ^{125}I-ricin, or ^{125}I-transferrin were added, and cell bound and endocytosed proteins were measured after 20 min of incubation for ebulin l and ricin and after 5 min for transferrin. Symbols: ●, ebulin l; □, ricin; and △, transferrin. (**b**) The cells were incubated for 5 min at 37 °C in HEPES medium, pH 5.5, with increasing concentrations of acetic acid. ^{125}I-ebulin l, ^{125}I-ricin, or ^{125}I-transferrin were then added, and after 20 min of incubation for ebulin l and ricin and after 5 min for transferrin, the amount of endocytosed proteins was measured as described above. Symbols: ●, ebulin l; □, ricin; and △, transferrin. The data represent the mean ± SD of two experiments.

2.1.4. Intracellular Transport of Ebulin l in COS Cells

After endocytosis, RIPs move within the endosomal system until they reach the appropriate compartment for entry into the cytosol, where they inactivate ribosomes. To determine the intracellular transport of ebulin l in COS cells, ebulin l was labeled with the fluorophore CY3 and incubated with COS cells. The cellular distribution of ebulin l was studied after incubation at 37 °C for various periods of time. The data in Figure 1c demonstrate that, when the cells were treated with ebulin l at 4 °C, the toxin is evenly bound all over the cell surface. When the cells were subsequently incubated at 37 °C, the amount of ebulin l at the surface was reduced and the fluorescent ebulin l appeared as intracellular dots, suggesting uptake in vesicles (Figure 4). We next performed double-labeling experiments with EEA1 (early endosome antigen 1), a protein that is associated with early endosomes [29]. As shown in Figure 4A, incubation for 10 min at 37 °C resulted in good colocalization of EEA1 and the ebulin l-labeled intracellular structures (Pearson's correlation coefficient (PCC) = 0.42). After 60 min at 37 °C, ebulin l remained in vesicles but did not colocalize to a large extent with EEA1 (data not shown). Instead, some of the structures stained positive for CD63, a late endosome/lysosome marker, colocalized with ebulin l (PCC = 0.24), indicating that the toxin is transported to lysosomes for degradation (Figure 4B). Moreover, ebulin l did not colocalize with mannose-6-phosphate receptor (M6PR), a marker for late endosomes and the trans-Golgi network [30] (PCC = −0.05) (Figure 4C). There is a possibility that only a very small fraction of ebulin l is transported to the Golgi apparatus and that this was not detectable in our assay. However, these results together with further experiments (see below) suggest binding and entry of ebulin l into the endosomal/lysosomal compartment but not to the Golgi apparatus.

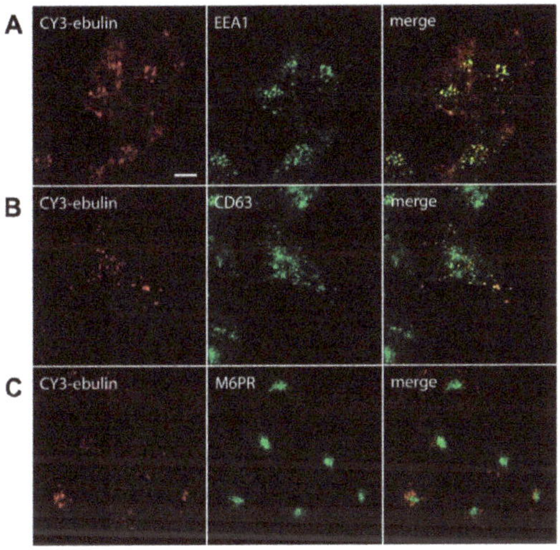

Figure 4. Transport of fluorescent ebulin l in COS cells and colocalization with markers for different intracellular organelles: the cells were incubated at 4 °C for 1 h with CY3-ebulin l to allow binding of the protein and then incubated at 37 °C for 10 min (**A**) and 60 min (**B** and **C**) before fixation. Then, the cells were stained with anti-EEA1 (early endosomes) (**A**), anti-CD63 (late endosomes/lysosomes) (**B**), and anti M6PR (Golgi) (**C**) antibodies followed by fluorescein isothiocyanate (FITC)-conjugated secondary antibody. Colocalizations between CY3-ebulin and the different markers were quantified using Pearson's correlation coefficient (PCC). PCC between CY3-ebulin and EEA1 = 0.42 ± 0.08; PCC between CY3-ebulin and CD63 = 0.24 ± 0.07; and PCC between CY3-ebulin and M6PR = −0.05 ± 0.03. The PCC values represent the mean ± SD of 5 images analyzed for each marker. Bar, 50 µm.

Some bacterial toxins, such as diphtheria toxin and anthrax toxin, are translocated to the cytosol from acidic endosomes [31,32]. By contrast, ricin and other bacterial toxins such as cholera toxin and Shiga toxin follow the retrograde pathway from endosomes to the Golgi complex and further to the endoplasmic reticulum before the A chain is translocated to the cytosol [33]. To better understand how ebulin l enters the cytosol, we studied the effects of agents interfering with intracellular routing on the cytotoxic process. Lysosomotrophic amines such as NH_4Cl and chloroquine raise the pH within acidic intracellular vesicles. Preincubation of the COS cells with chloroquine and NH_4Cl enhanced the cytotoxicity of ebulin l as well as that of ricin (Figure 5). This indicates that ebulin l, similar to ricin, does not require a low pH for translocation to the cytosol. Moreover, the lysosomotrophic amines may stimulate cytotoxicity by preventing toxin degradation by inactivating the lysosomal enzymes, possibly due to an increase in the intralysosomal pH. In addition, we investigated the effect of the fungal inhibitor brefeldin A, which causes Golgi complex disassembly and has been shown to inhibit ricin toxicity. As shown in Figure 5, preincubation of COS cells with brefeldin A markedly reduced as expected the cytotoxicity of ricin but had no significant effect on the cytotoxicity of ebulin l (Figure 5). This indicates that ebulin l follows a Golgi-independent pathway to the cytosol.

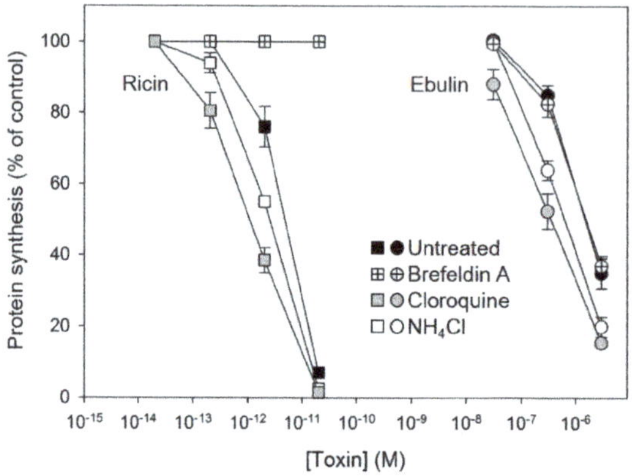

Figure 5. Effect of brefeldin A, chloroquine, and NH_4Cl on protein synthesis in COS cells treated with ebulin l (circles) and ricin (squares): the cells were left untreated (black symbols) or preincubated with 5 µg/mL brefeldin A (crossed symbols), 25 µM cloroquine (grey symbols), and 20 mM NH_4Cl (open symbols) for 1 h and then incubated with different concentrations of ebulin l and ricin for 18 h. Protein synthesis was finally measured as indicated in Section 4.2.4. The data represent the mean ± SD of two experiments performed in duplicate.

2.2. Binding, Endocytosis, and Degradation of Ebulin l and Ricin in HeLa Cells Overexpressing dynK44A

As shown in Figure 3, in COS cells, ebulin l, similar to ricin, was still endocytosed when the formation of clathrin-coated vesicles was inhibited by acidification of the cytosol. This suggests that clathrin-independent endocytosis is responsible for approximately 50% of the ebulin l uptake in those cells. In order to investigate further the mechanism of ebulin l internalization, we used HeLa dynK44A cells, which are HeLa cells with inducible synthesis of a mutant dynamin (K44A) [34]. It has been reported that the GTPase dynamin mediates the scission of clathrin-coated pits and that it is also involved in the budding of caveolae. It has been shown that, in those cells, ricin is internalized by clathrin- and caveolae-independent endocytosis [35]. HeLa dynK44A cells express, under tetracycline regulation, the dominant negative dynamin K44A mutant unable to bind and hydrolyze GTP. When the mutant dynamin is induced by the removal of tetracycline for two days, the cells are defective in clathrin-mediated endocytosis as well as in endocytosis from caveolae [34,36,37]. It has been shown that the prolonged inhibition of clathrin-dependent endocytosis in HeLa dynK44A allows the induction of compensatory mechanisms activating clathrin-independent endocytosis [35,38].

First, we studied the binding of ^{125}I-ebulin l and ^{125}I-ricin and found that HeLa dynK44A cells bound approximately 150 times more ricin than ebulin l. Scatchard analysis indicated that the number of binding sites for ebulin l was approximately 2.5×10^5 and 3.7×10^7 for ricin (Figure S1). Cells bound approximately the same amount of ^{125}I-ebulin l independently of mutant expression (data not shown).

We next studied the internalization of ebulin l and ricin both in cells where the mutant dynamin was overexpressed and in cells where its expression was repressed by the presence of tetracycline. Control experiments showed that the endocytosis of ^{125}I-transferrin, which occurs from coated pits [28], was inhibited by more than 90% by overexpression of the mutant dynamin (Figure 6b). By contrast, ^{125}I-ebulin l and ^{125}I-ricin uptake were unchanged in cells expressing the mutant dynamin (Figure 6a). Approximately 25% of total cell-associated ^{125}I-ebulin l and 15% of total cell-associated ^{125}I-ricin were internalized into HeLa dynK44A cells, with and without the induction of mutant dynamin, during 30 min of incubation at 37 °C. To study whether the transport of ebulin l to lysosomes was affected by the expression of mutant dynamin, toxin degradation was measured. As shown in Figure 6c, essentially the same degradation rates were obtained whether the mutant dynamin was expressed by the removal of tetracycline or not. The percentage of ebulin l released by cells in the TCA-soluble fraction was higher (34%) than that of ricin (9%) (Figure 6c). Bafilomycin A1 inhibited toxin degradation, indicating that the process took place in a low-pH compartment (Figure 6c). Moreover, recycling of ebulin l and ricin was not affected by the overexpression of dyn K44A (data not shown) and the values obtained were comparable to those observed in COS cells (Figure 2b). Therefore, the data indicate that the endocytic uptake of ebulin l and ricin, at least in HeLa cells with dynK44A overexpression, does not occur by clathrin-coated pits or caveolae. According to this, it has been shown earlier that ricin endocytosis continued to the same level after dynK44A expression [35]. The cells therefore seem then to be able to upregulate clathrin-independent endocytosis under conditions where a prolonged inhibition of clathrin-dependent endocytosis takes place [35,38].

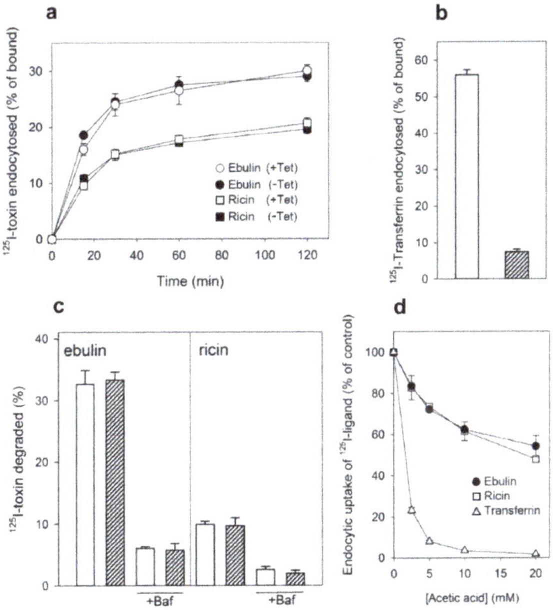

Figure 6. (**a**) Rate of internalization of ebulin l (circles) or ricin (squares) in HeLa dynK44A cells, with (closed symbols) and without (open symbols) the induction of mutant dynamin: HeLa dynK44A cells were grown in the presence or the absence of tetracycline (Tet) for 2 days. The cells were then washed, and ^{125}I-toxins were added. The cells were incubated at 37 °C for 0 to 120 min, and bound and endocytosed toxins were quantified as described in Section 4.2.2. (**b**) Endocytosis of ^{125}I-labeled transferrin in HeLa dynK44A cells with (filled bar) and without (open bar) the induction of mutant dynamin: endocytosed transferrin was quantified after 5 min of internalization. (**c**) Degradation of ebulin l and ricin in HeLa dynK44A cells: the cells were grown in the presence (open bar) or the absence (filled bar) of tetracycline for 2 days. The cells were then transferred to a HEPES-containing medium and preincubated without or with (+Baf) bafilomycin A1 (1 mM) for 30 min at 37 °C. ^{125}I-toxins were then added, and 20 min later, the surface-bound toxins were removed with a 0.1 M lactose solution at 37 °C. The incubation was continued in the presence or in the absence of bafilomycin A1, and after 2 h, further incubation toxin degradation was measured as described in Section 4.2.2. The data represent the mean ± SD of two experiments. (**d**) Uptake of ^{125}I-transferrin (△),^{125}I-ebulin l (●), and ^{125}I-ricin (□) by HeLa dynK44A cells. The cells grown in the presence of tetracycline were incubated for 5 min at 37 °C in HEPES medium, pH 5.5, with increasing concentrations of acetic acid. ^{125}I-transferrin or ^{125}I-toxins were then added as described above, and after 5 and 20 min of incubation, the amount of endocytosed proteins was measured as in (a). The data represent the mean ± SD of two experiments performed in duplicate.

Our results indicated that, after a prolonged inhibition of clathrin-dependent endocytosis in HeLa dyn K44A expressing the mutant dynamin, ebulin l endocytosis continues (Figure 6a), while in COS cells, there was a 50% reduction of the endocytic uptake of ebulin l when endocytosis from coated pits was acutely inhibited by acidification of the cytosol (Figure 3). To test if this is also the case in HeLa dynK44A cells, the uptake of ^{125}I-labeled ebulin l and ricin was measured in cells grown with tetracycline that had been acidified by incubation with acetic acid. Figure 6d shows that, in these cells, the uptake of ebulin l was reduced by about 45% after acidification of the cytosol. The uptake of transferrin was reduced by more than 95% under the same conditions. These results therefore indicate that there is no difference between COS cells and HeLa cells in the way they endocytose ebulin l and ricin. Thus, prolonged inhibition of clathrin-dependent uptake in cells expressing the mutant dynamin can induce an increase in clathrin-independent endocytosis [35,38] while acute inhibition of clathrin-dependent endocytosis by acidifying the cytosol cannot. Endocytosis by mechanisms not involving clathrin-coated pits has been shown for several bacterial toxins such as tetanus toxin, cholera toxin, and plant RIPs such as ricin, lanceolin, and stenodactylin [35,39,40].

2.3. *Effect of DynK44A Overexpression and Cytosol Acidification on Ebulin l and Ricin Cytotoxicity*

After endocytosis and transport to the Golgi apparatus and the endoplasmic reticulum, a small number of ricin molecules reach the cytosol, inhibiting protein synthesis. It has been shown earlier that cells overexpressing dynK44A were more resistant to ricin than dynK44A cells expressing endogenous dynamin [35]. The expression of mutant dynamin inhibits transport of endocytosed ricin to the Golgi apparatus, and this transport is important for ricin intoxication [35]. To investigate whether the overexpression of dynK44A changes the ability of ebulin l to intoxicate cells, we measured protein synthesis 18 h after the addition of increasing concentrations of ebulin l or ricin to dynK44A cells grown with and without tetracycline. As shown in Figure 7a, the toxicity of ebulin l was not affected by the expression of mutant dynamin while, as expected, it protected the cells against ricin. Under these conditions, uninduced HeLa dynK44A cells were about 4.8 times more sensitive to ricin that dynK44A-induced cells. In addition, we also studied whether ebulin l and ricin internalized by the clathrin- and caveolae-independent pathway in these cells must be transported through the Golgi apparatus to inhibit protein synthesis. In these experiments, HeLa cells with mutant dynamin were pretreated with brefeldin A, which disrupts the Golgi apparatus, and protein synthesis was measured 18 h later. As shown in Figure 7a and consistent with our previous observations (Figure 5), brefeldin A did not protect cells from the inhibition of protein synthesis by ebulin l but the cells were completely protected against ricin. These data clearly indicate that transport through the Golgi is not required for ebulin l intoxication.

Since the endocytic uptake of ebulin l and ricin in COS cells was reduced by 50% by acidification of the cytosol (Figure 3), we decided to study if toxin internalized under such conditions (by clathrin-independent endocytosis) can intoxicate cells. In these experiments, we used the NH_4Cl pre-pulse method to acidify the cytosol. Ebulin l or ricin were added to COS cells, and endocytosis was allowed to proceed for 20 min. Then, the cell surface-bound toxins were removed with a medium containing 0.1 M lactose and the cells were incubated 18 h in normal medium to allow the internalized toxin to intoxicate the cells. After that, the ability of the cells to incorporate [^{3}H] leucine was measured. The data in Figure 7b show that protein synthesis in cells treated in this way was inhibited by ebulin l and ricin to the same extent as in control cells, where endocytic uptake of the RIPs occurred at normal internal pH. The data suggest that, in COS cells, the endocytosis of ebulin l that leads to intoxication of cells takes place predominantly from clathrin-independent mechanisms, and thus, there is no apparent role for clathrin in productive intracellular transport.

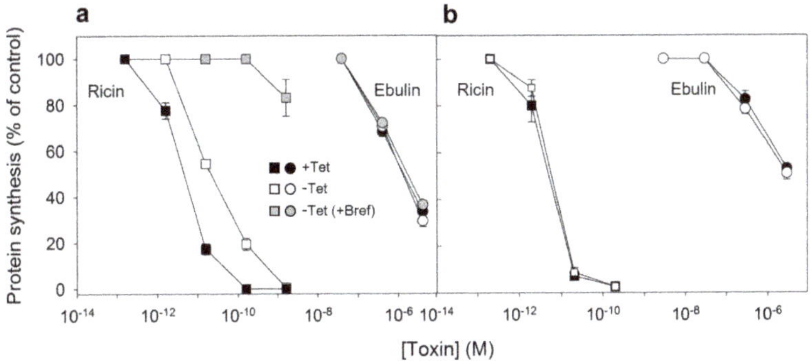

Figure 7. Ability of ebulin l and ricin to inhibit protein synthesis in HeLa dynK44A (**a**) and COS cells (**b**): dynK44A cells were grown with (closed symbols) and without (open symbols) tetracycline (Tet) for 2 days. Then, the cells were incubated with different concentrations of ebulin l (circles) and ricin (squares) for 18 h, and protein synthesis was measured as indicated in Section 4.2.4. To investigate the effect of brefeldin A (grey symbols) on protein synthesis of K44A cells grown without tetracycline, the cells were preincubated for 1 h with brefeldin A and then incubated with different concentrations of ebulin l and ricin for 18 h, and protein synthesis was measured. The data represent the mean ± SD of two experiments performed in duplicate. (**b**) The toxic effect of ebulin l (circles) or ricin (squares) endocytosed at normal (open symbols) and acidic internal pH (closed symbols): COS cells were incubated for 30 min at 37 °C in HEPES medium, pH 7, with and without 25 mM of NH_4Cl. The medium was removed, and a solution containing 0.14 M KCl, 2 mM $CaCl_2$, 1 mM amiloride, 1 mM $MgCl_2$, and 20 mM HEPES, pH 7.0, was added. After 5 min of incubation at 37 °C, increasing concentrations of ebulin or ricin were added and, after 20 min of further incubation, a growth medium containing 0.1 M lactose was added. The cells were then incubated for 18 h, and protein synthesis was measured. The data represent the mean ± SD of two experiments performed in duplicate.

2.4. Ebulin l Induces Apoptosis in COS Cells

In addition to rRNA damage, RIPs are capable of inducing cell death by apoptosis [4]. We therefore decided to investigate the death pathways involved in the cytotoxicity of ebulin l in COS cells. Cells treated with ebulin l exhibited the morphological features characteristic of apoptosis such as cell rounding and blebbing (data not shown). To demonstrate the involvement of caspase-dependent apoptosis, caspase-3/7 activation was measured in COS cells exposed to 10^{-6} and 10^{-7} M ebulin l for 24, 48, and 72 h. As shown in Figure 8a, a time- and dose-dependent activation of effector caspases was observed in COS cells. Caspase activity seems to be significantly induced after treating the cells with 10^{-6} M ebulin l for 24 h, and at that concentration, protein synthesis was inhibited by 50% after 18 h (Figure 5). Thus, protein synthesis inhibition seems to be an earlier event than apoptosis in these cells and suggests that apoptosis might be a consequence of the ribotoxic stress induced by the A chain after entry into the cytosol. However, we cannot rule out the possibility that both processes run in parallel. To evaluate the role of the different cytotoxic mechanisms induced by ebulin l, COS cells were pretreated with two inhibitors, the pan-caspase inhibitor Z-VAD, which irreversibly binds to the catalytic site of caspases and was used to selectively inhibit the apoptotic pathway, and the inhibitor of necroptosis, Necrostatin. COS cells were pretreated and maintained in 100 µM Z-VAD, and the cell viability was determined for different ebulin l concentrations. As shown in Figure 8b, caspase inhibition by Z-VAD largely prevented the cytotoxicity of ebulin l after 48 h. At a concentration of 10^{-6} M, viability increased from 20% to 72% in the presence of Z-VAD. In contrast, the necroptosis inhibitor Necrostatin did not rescue ebulin l-induced cell death. (Figure 8b). Therefore, these data indicate that apoptosis is the predominant pattern of cell death induced by ebulin l.

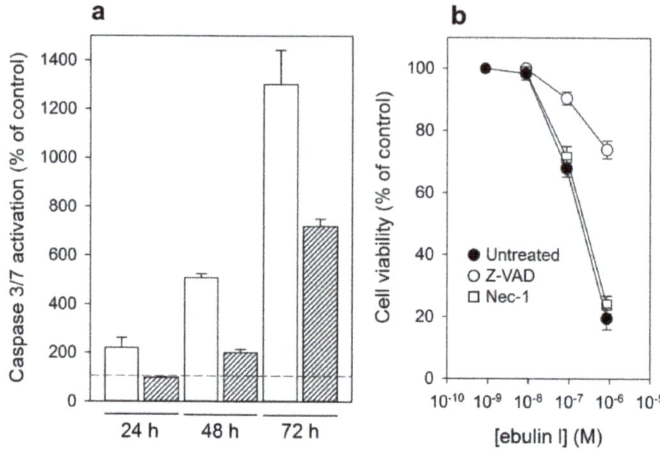

Figure 8. (**a**) Caspase-3/7 activation in COS cells treated with 10^{-6} M (empty bar) or 10^{-7} M (filled bar) ebulin l for 24, 48, and 72 h: activity is expressed as the percentage of control values obtained from cells grown in the absence of ebulin l (dashed line). The data represent the mean ± SD of two experiments performed in duplicate. (**b**) Effect of Z-VAD and Necrostatin (Nec-1) on cytotoxicity of ebulin l on COS cells: the cells were preincubated with Z-VAD or Necrostatin for 3 h or left untreated, and then, different concentrations of ebulin l were added and the cells were incubated for 48 h. Cell viability was assessed by a colorimetric assay as indicated in Section 4.2.4. The data represent the mean ± SD of two experiments performed in triplicate. Symbols: ●, untreated; ○, +Z-VAD; and □, +Nec-1.

3. Conclusions

This work contributes to elucidating the mechanisms involved in the endocytosis and intracellular transport of the plant toxin ebulin l. Our results demonstrate that ebulin l has a lower number of binding receptors than ricin in COS and HeLa dynK44A cells. This may be due to a change in the structural disposition of the 2γ-subdomain of the ebulin B chain, which limits its ability to bind galactosides. Following binding, ebulin l is internalized by both clathrin-dependent and -independent mechanisms. A short time after internalization, ebulin l is localized to early endosomes and later to lysosomes but apparently not to the Golgi apparatus. The ebulin l molecules that lead to intoxication of cells are internalized via clathrin-independent mechanisms. Moreover, the cytotoxic effect of ebulin l occurs independently of low endosomal pH and does not require transport of the toxin through the Golgi apparatus. Moreover, ebulin l induces a caspase-dependent apoptosis as the predominant cell death mechanism. Importantly, toxins have a potential as therapeutic agents if the toxicity can be targeted to malignant cells. The low unspecific toxicity of ebulin l together with its strong anti-ribosomal activity and induction of apoptosis make it an excellent candidate as a toxic moiety in the construction of immunotoxins and conjugates directed against specific targets. Knowledge of the mechanisms of transport and action of the toxin is essential in achieving this goal.

4. Materials and Methods

4.1. Reagents and Cells

The sources of the chemicals and cells were described previously [16,29,34,35,41]. Particular details are given in the Supplementary File S1.

4.2. Methods

Particular experimental details are given in the Supplementary File S1.

4.2.1. Binding of ^{125}I-Labeled Toxins to Cells and Crosslinking of ^{125}I-ebulin l to Membrane Receptors

Confluent cells were washed twice with ice-cold HEPES medium and incubated at 4 °C for 15 min before increasing concentrations of ebulin l or ricin were added. The cells were incubated for 1 h with the toxins and then washed five times with ice-cold phosphate buffered saline (PBS). The radioactivity was measured after dissolving the cells in 0.1 M KOH. Nonspecific binding was estimated by the incubation of cells in the presence of 0.1 M lactose. Receptor dissociation constants and the number of binding sites were estimated by the Scatchard method [42].

Crosslinking of bound ^{125}I-ebulin l to cells was carried out with disuccinimidyl suberate [43].

4.2.2. Measurements of Endocytosis, Recycling, and Degradation

Endocytosis of transferrin was measured after incubation for 5 min with transferrin (100 ng/mL, labeled with ^{125}I) [35]. Internalized toxin (400 ng/mL) was measured as the amount of ^{125}I-labeled toxin that was not removed after incubating the cells with a 0.1 M lactose solution for 5 min at 37 °C [44]. Recycling and degradation of toxins were measured as previously described [44].

Endocytosis in cells with acidified cytosol by NH_4Cl pre-pulsing or by incubation with acetic acid was assessed as previously described [27]. Cell-bound and endocytosed proteins were measured after 20 min of incubation for ^{125}I-ebulin and ^{125}I-ricin and 5 min for ^{125}I-transferrin.

4.2.3. Immunofluorescence Microscopy

Binding of CY3-ebulin l and double-staining experiments were carried out as previously described [43]. The coverslips were examined with a Zeiss LSM 510 META confocal microscope (Carl Zeiss, Jena, Germany). Colocalization analysis were performed using Coloc2 (version 3.0.0) in Fiji (ImageJ 1.53c) (http://fiji.sc/wiki/index.php/Fiji).

4.2.4. Other Measurements

Protein synthesis was measured as previously described [35]. Cell viability was determined with a colorimetric assay based on cleavage of the tetrazolium salt WST-1 to formazan and the caspase-3/7 activity was assessed by the luminescent assay Caspase-GloTM 3/7 [16].

Supplementary Materials: The following are available online at https://www.mdpi.com/2072-6651/13//102/s1, Figure S1: Toxin binding and Scatchard plots, File S1: Description of the materials and methods.

Author Contributions: Conceptualization, L.C., R.I., and J.M.F.; methodology and investigation, R.I., J.M.F., A.L., and L.C.; writing—review and editing, L.C., R.I., J.M.F., and A.L.; funding acquisition, L.C. and A.L. All authors have read and agreed to the published version of the manuscript.

Funding: This work was funded by grants BIO39/VA39/14 and BIO/VA17/15 (Consejería de Sanidad, Junta de Castilla y León) to L.C. and grant VA033G19 (Consejería de Educación, Junta de Castilla y León) to the GIR ProtIBio. A.L. acknowledges the South-Eastern Norway Regional Health Authority, The Norwegian Cancer Society, and The Norwegian Research Council.

Institutional Review Board Statement: Not applicable.

Informed Consent Statement: Not applicable.

Data Availability Statement: Data are available upon request; please contact the contributing authors.

Acknowledgments: We thank Sjur Olsnes for encouragement and unconditional support and Anne Engen for her excellent assistance in cell culture.

Conflicts of Interest: The authors declare no conflict of interest.

References

1. Endo, Y.; Tsurugi, K. The RNA N-glycosidase activity of ricin A-chain. The characteristics of the enzymatic activity of ricinA-chain with ribosomes and with rRNA. *J. Biol. Chem.* **1988**, *263*, 8735–8739. [CrossRef]
2. Schrot, J.; Weng, A.; Melzig, M.F. Ribosome-inactivating and related proteins. *Toxins* **2015**, *7*, 1556–1615. [CrossRef] [PubMed]
3. Di Maro, A.; Citores, L.; Russo, R.; Iglesias, R.; Ferreras, J.M. Sequence comparison and phylogenetic analysis by the Maximum Likelihood method of ribosome-inactivating proteins from angiosperms. *Plant Mol. Biol.* **2014**, *85*, 575–588. [CrossRef] [PubMed]
4. Narayanan, S.; Surendranath, K.; Bora, N.; Surolia, A.; Karande, A.A. Ribosome inactivating proteins and apoptosis. *FEBS Lett.* **2005**, *579*, 1324–1331. [CrossRef] [PubMed]
5. Bolognesi, A.; Bortolotti, M.; Maiello, S.; Battelli, M.; Polito, L. Ribosome-Inactivating Proteins from Plants: A Historical Overview. *Molecules* **2016**, *21*, 1627. [CrossRef]
6. Citores, L.; Iglesias, R.; Ferreras, J.M. Ribosome inactivating proteins from plants: Biological properties and their use in experimental therapy. In *Antitumor Potential and Other Emerging Medicinal Properties of Natural Compounds*; Fang, E.F., Ng, T.B., Eds.; Springer: Dordrecht, The Netherlands, 2013; pp. 127–143.
7. Spooner, R.A.; Lord, J.M. Ricin trafficking in cells. *Toxins* **2015**, *7*, 49–65. [CrossRef]
8. Polito, L.; Bortolotti, M.; Battelli, M.G.; Calafato, G.; Bolognesi, A. Ricin: An Ancient Story for a Timeless Plant Toxin. *Toxins* **2019**, *11*, 324. [CrossRef]
9. Sowa-Rogozińska, N.; Sominka, H.; Nowakowska-Gołacka, J.; Sandvig, K.; Słomińska-Wojewódzka, M. Intracellular Transport and Cytotoxicity of the Protein Toxin Ricin. *Toxins* **2019**, *11*, 350. [CrossRef]
10. Robertus, J. The structure and action of ricin, a cytotoxic N-glycosidase. *Semin. Cell Biol.* **1991**, *2*, 23–30.
11. Girbés, T.; Citores, L.; Iglesias, R.; Ferreras, J.M.; Muñoz, R.; Rojo, M.A.; Arias, F.J.; García, J.R.; Méndez, E.; Calonge, M. Ebulin 1, a nontoxic novel type 2 ribosome-inactivating protein from Sambucus ebulus L. leaves. *J. Biol. Chem.* **1993**, *268*, 18195–18199. [CrossRef]
12. Girbés, T.; Citores, L.; Ferreras, J.M.; Rojo, M.A.; Iglesias, R.; Muñoz, R.; Arias, F.J.; Calonge, M.; García, J.R.; Méndez, E. Isolation and partial characterization of nigrin b, a non-toxic novel type 2 ribosome-inactivating protein from the bark of Sambucus nigra L. *Plant Mol. Biol.* **1993**, *22*, 1181–1186. [CrossRef] [PubMed]
13. Van Damme, E.J.M.; Barre, A.; Rougé, P.; Van Leuven, F.; Peumans, W.J. The NeuAc(α-2,6)-Gal/GalNAc-Binding Lectin from Elderberry (Sambucus Nigra) Bark, a type-2 Ribosome-Inactivating Protein with an Unusual Specificity and Structure. *Eur. J. Biochem.* **1996**, *235*, 128–137. [CrossRef] [PubMed]
14. Iglesias, R.; Citores, L.; Ferreras, J.M.; Pérez, Y.; Jiménez, P.; Gayoso, M.J.; Olsnes, S.; Tamburino, R.; Di Maro, A.; Parente, A.; et al. Sialic acid-binding dwarf elder four-chain lectin displays nucleic acid N-glycosidase activity. *Biochimie* **2010**, *92*, 71–80. [CrossRef] [PubMed]
15. De Benito, F.M.; Citores, L.A.; Iglesias, R.; Ferreras, J.M.; Camafeita, E.; Méndez, E.; Girbés, T. Isolation and partial characterization of a novel and uncommon two-chain 64-kDa ribosome-inactivating protein from the bark of elder (*Sambucus nigra* L.). *FEBS Lett.* **1997**, *413*, 85–91. [CrossRef]
16. Iglesias, R.; Ferreras, J.M.; Di Maro, A.; Citores, L. Ebulin-RP, a novel member of the Ebulin gene family with low cytotoxicity as a result of deficient sugar binding domains. *Biochim. Biophys. Acta Gen. Subj.* **2018**, *1862*, 460–473. [CrossRef]
17. Van Damme, E.J.M.; Barre, A.; Rougé, P.; Van Leuven, F.; Peumans, W.J. Isolation and Molecular Cloning of a Novel Type 2 Ribosome-inactivating Protein with an Inactive B Chain from Elderberry (*Sambucus nigra*) Bark. *J. Biol. Chem.* **1997**, *272*, 8353–8360. [CrossRef]
18. Pascal, J.M.; Day, P.J.; Monzingo, A.F.; Ernst, S.R.; Robertus, J.D.; Iglesias, R.; Pérez, Y.; Férreras, J.M.; Citores, L.; Girbés, T. 2.8-Å crystal structure of a nontoxic type-II ribosome-inactivating protein, ebulin l. *Proteins* **2001**, *43*, 319–326. [CrossRef]
19. Polito, L.; Djemil, A.; Bortolotti, M. Plant Toxin-Based Immunotoxins for Cancer Therapy: A Short Overview. *Biomedicines* **2016**, *4*, 12. [CrossRef]
20. Ferreras, J.M.; Citores, L.; Iglesias, R.; Jimenez, P.; Girbes, T. Use of ribosome-inactivating proteins from Sambucus for the construction of immunotoxins and conjugates for cancer therapy. *Toxins* **2011**, *3*, 420–441. [CrossRef]
21. Battelli, M.G.; Musiani, S.; Buonamici, L.; Santi, S.; Riccio, M.; Maraldi, N.M.; Girbés, T.; Stirpe, F. Interaction of volkensin with HeLa cells: Binding, uptake, intracellular localization, degradation and exocytosis. *Cell. Mol. Life Sci.* **2004**, *61*, 1975–1984. [CrossRef]
22. Wang, B.-Z.; Zou, W.-G.; Liu, W.-Y.; Liu, X.-Y. The lower cytotoxicity of cinnamomin (a type II RIP) is due to its B-chain. *Arch. Biochem. Biophys.* **2006**, *451*, 91–96. [CrossRef] [PubMed]
23. Das, M.K.; Sharma, R.S.; Mishra, V. A cytotoxic type-2 ribosome inactivating protein (from leafless mistletoe) lacking sugar binding activity. *Int. J. Biol. Macromol.* **2011**, *49*, 1096–1103. [CrossRef] [PubMed]
24. Yoshimori, T.; Yamamoto, A.; Moriyama, Y.; Futai, M.; Tashiro, Y. Bafilomycin A1, a specific inhibitor of vacuolar-type H(+)-ATPase, inhibits acidification and protein degradation in lysosomes of cultured cells. *J. Biol. Chem.* **1991**, *266*, 17707–17712. [CrossRef]
25. Battelli, M.G.; Citores, L.; Buonamici, L.; Ferreras, J.M.; de Benito, F.M.; Stirpe, F.; Girbés, T. Toxicity and cytotoxicity of nigrin b, a two-chain ribosome-inactivating protein from *Sambucus nigra*: Comparison with ricin. *Arch. Toxicol.* **1997**, *71*, 360–364. [CrossRef] [PubMed]

26. Sandvig, K.; Kavaliauskiene, S.; Skotland, T. Clathrin-independent endocytosis: An increasing degree of complexity. *Histochem. Cell Biol.* **2018**, *150*, 107–118. [CrossRef] [PubMed]
27. Sandvig, K.; Olsnes, S.; Petersen, O.W.; van Deurs, B. Acidification of the cytosol inhibits endocytosis from coated pits. *J. Cell Biol.* **1987**, *105*, 679–689. [CrossRef]
28. Hopkins, C.R. Intracellular routing of transferrin and transferrin receptors in epidermoid carcinoma A431 cells. *Cell* **1983**, *35*, 321–330. [CrossRef]
29. Mu, F.T.; Callaghan, J.M.; Steele-Mortimer, O.; Stenmark, H.; Parton, R.G.; Campbell, P.L.; McCluskey, J.; Yeo, J.P.; Tock, E.P.; Toh, B.H. EEA1, an early endosome-associated protein. EEA1 is a conserved alpha-helical peripheral membrane protein flanked by cysteine "fingers" and contains a calmodulin-binding IQ motif. *J. Biol. Chem.* **1995**, *270*, 13503–13511. [CrossRef]
30. Goda, Y.; Pfeffer, S.R. Selective recycling of the mannose 6-phosphate/IGF-II receptor to the trans Golgi network in vitro. *Cell* **1988**, *55*, 309–320. [CrossRef]
31. Sandvig, K.; Olsnes, S. Rapid entry of nicked diphtheria toxin into cells at low pH. Characterization of the entry process and effects of low pH on the toxin molecule. *J. Biol. Chem.* **1981**, *256*, 9068–9076. [CrossRef]
32. Bradley, K.A.; Mogridge, J.; Mourez, M.; Collier, R.J.; Young, J.A. Identification of the cellular receptor for anthrax toxin. *Nature* **2001**, *414*, 225–229. [CrossRef] [PubMed]
33. Sandvig, K.; van Deurs, B. Transport of protein toxins into cells: Pathways used by ricin, cholera toxin and Shiga toxin. *FEBS Lett.* **2002**, *529*, 49–53. [CrossRef]
34. Damke, H.; Baba, T.; Warnock, D.E.; Schmid, S.L. Induction of mutant dynamin specifically blocks endocytic coated vesicle formation. *J. Cell Biol.* **1994**, *127*, 915–934. [CrossRef] [PubMed]
35. Llorente, A.; Rapak, A.; Schmid, S.; Deurs, B.; Sandvig, K. Expression of Mutant Dynamin Inhibits Toxicity and Transport of Endocytosed Ricin to the Golgi Apparatus. *J. Cell Biol.* **1998**, *140*, 553–563. [CrossRef] [PubMed]
36. Oh, P.; McIntosh, D.P.; Schnitzer, J.E. Dynamin at the neck of caveolae mediates their budding to form transport vesicles by GTP-driven fission from the plasma membrane of endothelium. *J. Cell Biol.* **1998**, *141*, 101–114. [CrossRef] [PubMed]
37. Henley, J.R.; McNiven, M.A. Association of a dynamin-like protein with the Golgi apparatus in mammalian cells. *J. Cell Biol.* **1996**, *133*, 761–775. [CrossRef] [PubMed]
38. Damke, H.; Baba, T.; van der Bliek, A.M.; Schmid, S.L. Clathrin-independent pinocytosis is induced in cells overexpressing a temperature-sensitive mutant of dynamin. *J. Cell Biol.* **1995**, *131*, 69–80. [CrossRef]
39. Battelli, M.G.; Scicchitano, V.; Polito, L.; Farini, V.; Barbieri, L.; Bolognesi, A. Binding and intracellular routing of the plant-toxic lectins, lanceolin and stenodactylin. *Biochim. Biophys. Acta Gen. Subj.* **2010**, *1800*, 1276–1282. [CrossRef]
40. Torgersen, M.L.; Skretting, G.; van Deurs, B.; Sandvig, K. Internalization of cholera toxin by different endocytic mechanisms. *J. Cell Sci.* **2001**, *114*, 3737–3747.
41. Fraker, P.J.; Speck, J.C., Jr. Protein and cell membrane iodinations with a sparingly soluble chloroamide, 1,3,4,6-tetrachloro-3a,6a-diphrenylglycoluril. *Biochem. Biophys. Res. Commun.* **1978**, *80*, 849–857. [CrossRef]
42. Scatchard, G. The Attractions of Proteins for Small Molecules and Ions. *Ann. N. Y. Acad. Sci.* **1949**, *51*, 660. [CrossRef]
43. Citores, L.; Wesche, J.; Kolpakova, E.; Olsnes, S. Uptake and intracellular transport of acidic fibroblast growth factor: Evidence for free and cytoskeleton-anchored fibroblast growth factor receptors. *Mol. Biol. Cell* **1999**, *10*, 3835–3848. [CrossRef] [PubMed]
44. Sandvig, K.; Olsnes, S. Effect of temperature on the uptake, excretion and degradation of abrin and ricin by HeLa cells. *Exp. Cell Res.* **1979**, *121*, 15–25. [CrossRef]

toxins

MDPI

Article

Sapovaccarin-S1 and -S2, Two Type I RIP Isoforms from the Seeds of *Saponaria vaccaria* L.

Louisa Schlaak [1], Christoph Weise [2], Benno Kuropka [2] and Alexander Weng [1,*]

[1] Institute of Pharmacy, Freie Universität Berlin, Königin-Luise-Str. 2+4, 14195 Berlin, Germany; l.schlaak@fu-berlin.de
[2] Institute of Chemistry and Biochemistry, Freie Universität Berlin, Thielallee 63, 14195 Berlin, Germany; chris.weise@biochemie.fu-berlin.de (C.W.); kuropka@zedat.fu-berlin.de (B.K.)
* Correspondence: alexander.weng@fu-berlin.de; Tel.: +49-30-838-51265

Abstract: Type I ribosome-inactivating proteins (RIPs) are plant toxins that inhibit protein synthesis by exerting rRNA *N*-glycosylase activity (EC 3.2.2.22). Due to the lack of a cell-binding domain, type I RIPs are not target cell-specific. However once linked to antibodies, so called immunotoxins, they are promising candidates for targeted anti-cancer therapy. In this study, sapovaccarin-S1 and -S2, two newly identified type I RIP isoforms differing in only one amino acid, were isolated from the seeds of *Saponaria vaccaria* L. Sapovaccarin-S1 and -S2 were purified using ammonium sulfate precipitation and subsequent cation exchange chromatography. The determined molecular masses of 28,763 Da and 28,793 Da are in the mass range typical for type I RIPs and the identified amino acid sequences are homologous to known type I RIPs such as dianthin 30 and saporin-S6 (79% sequence identity each). Sapovaccarin-S1 and -S2 showed adenine-releasing activity and induced cell death in Huh-7 cells. In comparison to other type I RIPs, sapovaccarin-S1 and -S2 exhibited a higher thermostability as shown by nano-differential scanning calorimetry. These results suggest that sapovaccarin-S1 and -S2 would be optimal candidates for targeted anti-cancer therapy.

Keywords: plant toxin; ribosome-inactivating protein (RIP); type I RIP; rRNA glycosylase activity (EC 3.2.2.22); protein isolation; protein sequencing; mass spectrometry

Key Contribution: In this work, the complete amino acid sequence of two RIP isoforms from *Saponaria vaccaria* L. was determined and their enzyme properties were characterized.

Citation: Schlaak, L.; Weise, C.; Kuropka, B.; Weng, A. Sapovaccarin-S1 and -S2, Two Type I RIP Isoforms from the Seeds of *Saponaria vaccaria* L. *Toxins* **2022**, *14*, 449. https://doi.org/10.3390/toxins14070449

Received: 7 June 2022
Accepted: 27 June 2022
Published: 30 June 2022

1. Introduction

Saponaria vaccaria L., also known as cow cockle or prairie carnation, is an annual herbaceous plant belonging to the carnation family (Caryophyllaceae). The flowering plant is a single species in its genus *Vaccaria* and was originally widespread in Eurasia, but nowadays is also found in North and South America, South Africa, and Australia [1]. Many different synonyms for *Saponaria vaccaria* L. exist in the literature, for example *Vaccaria pyramidata* Medik., *Gypsophila vaccaria* (L.) Sm, and *Vaccaria hispanica* (Mill.) Rauschert. For more than 1000 years, Wang-Bu-Liu-Xing (*Vaccariae semen* in Chinese) have been used in traditional Chinese medicine to treat dysmenorrhea, amenorrhea, and lactation failure [1]. The seeds of *Saponaria vaccaria* L. contain triterpenoid saponins including gypsogenin bisdesmosides, cyclic peptides, flavonoids, and polysaccharides [1–4]. In 1995, Bolognesi et al. reported on a 28 kDa type I ribosome-inactivating protein (RIP) from the seeds of *Vaccaria pyramidata* Medik. for the first time [5].

RIPs are distributed in over more than 100 species [6]. However, they have been discovered primarily in the families of Caryophyllaceae, Cucurbitaceae, Euphorbiaceae, Fabaceae, Phytolaccaceae, and Poaceae [6]. RIPs exhibit rRNA *N*-glycosylase activity (EC 3.2.2.22). It was first shown on rat ribosomes that RIPs cleave off a specific adenine residue (A_{4324}) from a conserved GAGA motif in the alpha-sarcin loop of the 28S rRNA [7].

As a consequence, the integrity of the 60S ribosomal subunit is compromised, the elongation factor 2 cannot interact with the ribosome any longer and the translation process at the ribosome is stalled [8]. As a result of the cleavage protein synthesis is inhibited, which in the end leads to cell death [8].

A distinction is made between type I and type II RIPs. Type II RIPs such as ricin from *Ricinus communis* L. are composed of an enzymatically active A-chain and a B-chain with lectin-like properties which promotes the binding to galactose molecules on the cell surface [9,10]. That allows them to enter the cell by receptor-mediated endocytosis, using the retrograde transport to the endoplasmic reticulum and reach the cytosol by the endoplasmic-reticulum-associated protein degradation (ERAD) pathway to exert their cytotoxicity [11,12]. Type I RIPs are enzymatically active single-chain proteins [13]. Lacking the B-chain, they enter the cell by receptor-independent endocytosis and accumulate in the late endosomes and lysosomes, where degradation takes place [14]. Only a small quantity escapes from the lysosomes which results in low cytotoxicity [14]. Very prominent and well-characterized type I RIPs are dianthin 30 from *Dianthus caryophyllus* L. and saporin-S6 from *Saponaria officinalis* L. [15,16].

The primary function of RIPs in the plant is not yet completely clear [17]. Different hypotheses on the role of RIPs in plants are under discussion. RIPs possess antiviral, antifungal, antibacterial and insecticidal activities, which may contribute in the protection against plant pests and predators [18,19]. In addition, RIP activity is increased in senescent and stressed leaves, suggesting that RIPs play an important role in regulating the death of plant cells. Also, roles as regulators of protein synthesis and protein storage have been proposed [20].

RIPs have increasingly become a focus of research due to their very efficient protein synthesis inhibitory activity and the resulting high cytotoxicity. Currently, research is being conducted primarily with regard to their promising application in targeted anti-tumor therapy as conjugates with target-specific antibodies, so-called immunotoxins. More than 450 immunotoxins constructed with RIPs are published in the literature [21]. Most of the described immunotoxins have been tested in vitro, some of them also in vivo. Only a few RIP-immunotoxins have been used for clinical studies [21]. Because of the high relevance of this topic, there is a great interest in the scientific community to explore and characterize new RIPs with the aim of identifying additional suitable candidates for immunotoxins. Beyond their potential in anti-cancer therapy, a second area of application for plant extracts containing saponins and RIPs arose in the last few years: the use of RIPs for crop protection. The new application was derived from the function that RIPs have in the plants, in which they are synthesized. Antiviral properties of RIPs have been demonstrated for a wide range of plant viruses [22]. The RIP plant extract can either be directly applied to the plant surface or, alternatively, a virus-resistant transgenic plant may be constructed [23,24]. In terms of crop protection, the antiviral activity of RIPs is complemented by insecticidal effects [19,25].

In 1995, the group of Fiorenzo Stirpe succeeded in isolating a type I RIP from *Saponaria vaccaria* L. (*Vaccaria pyramidata* Medik.) with a size of approximately 28 kDa and an isoelectric point (pI) of 9.5 and could identify its 30 N-terminal amino acids [5]. In 1997, the isolated RIP was used for immunoconjugate construction and successful inhibition of tumor growth in mice by these immunotoxins was demonstrated [26]. The full amino acid sequence, however, has never been published. The authors of the same study reported increased cytotoxicity of the isolated RIP from *Saponaria vaccaria* L. when compared to other RIPs on different cell lines [5]. Due to the increased cytotoxicity, RIPs from *Saponaria vaccaria* L. appear to be promising candidates for use in targeted tumor therapy and crop protection. We now aimed to isolate the type I RIP from *Saponaria vaccaria* L., to determine its precise molecular mass and its complete amino acid sequence, and to characterize its thermal stability and enzymatic activity. Additionally, we attempted to investigate the distribution of type I RIPs across differently processed seed material, as well as identify the main location of type I RIPs within the seed. The main location of type I RIPs within the seed has never been characterized before.

2. Results

2.1. Protein Extraction from the Seeds of Saponaria vaccaria L.

Whole dried seeds of *Saponaria vaccaria* L. were available for protein purification. A crude extract with complex protein composition was obtained by aqueous extraction. In SDS-PAGE the crude extract showed promising protein bands in the mass range of 20–35 kDa typical for RIPs (Figure 1A, lane III) [27]. In order to further separate and concentrate these proteins, the aqueous extraction was followed by an ammonium sulfate precipitation at three different concentrations (30%, 60% and 90%). Similar to the crude extract, the three ammonium sulfate fractions also revealed protein bands in the expected mass range for RIPs (Figure 1A, lane IV–VI). The 90%-ammonium sulfate fraction displayed the strongest bands in the mass range of interest (Figure 1A, lane VI). Similar results were obtained in qualitative analysis of the *N*-glycosylase activity. While all three fractions exhibited *N*-glycosylase activity, the 90%-ammonium sulfate fractions possessed by far the highest enzyme activities and therefore were chosen for further protein isolation [28]. These were further subjected to cation exchange chromatography (see Supplementary Figures S1 and S2). At a concentration of 0.3 M NaCl (in 50 mM HEPES, pH 7.0), an apparent single protein eluted with a molecular mass of approximately 25 kDa as estimated from SDS-PAGE (Figure 1A, lane VII). This band had already been among the most intense bands in the crude extract as well as in the 90%-ammonium sulfate fraction. The final yield of protein isolated from the 90% fraction was 8.56 mg per 100 g of whole seeds. We were also able to demonstrate enzyme activity for the isolated protein fraction. High-resolution mass spectrometry of the isolated protein yielded a mono-isotopic mass of 28,763.24 Da and an additional mass peak of 43% relative abundance at 28,793.24 Da (mass difference $\Delta m = 30$ Da, Figure 1B), suggesting that the isolated protein fraction contains two RIPs that in the following will be referred to as sapovaccarin-S1 and -S2.

2.2. Protein Sequencing of Sapovaccarin-S1 and -S2

At the beginning of this study, only the 30 N-terminal amino acids sequenced by Bolognesi et al. had been identified [5]. Here, the determination of the complete amino acid sequence of sapovaccarin-S1 and -S2 was achieved by combining MS-based peptide analysis with PCR experiments. The isolated protein was in-gel digested with trypsin and the peptide map recorded by MALDI-TOF-MS. The resulting peptide mass fingerprint was compared to a protein sequence database using the software Mascot. Since some peptide masses could be matched with known type I RIP sequences, it was confirmed that the isolated protein is a type I RIP. However, many dominant peptide signals from the spectra could not be assigned to any known RIP sequence indicating that the isolated type I RIP has a different sequence. Additionally, the sequences of six tryptic peptides with 64 amino acid residues were determined de novo by MS/MS analysis. Based on the identified peptide sequences, two PCR experiments were performed (see Supplementary Figure S3). DNA of gypsophilin-S, a type I RIP from *Gypsophila elegans* M.Bieb., saporin-S6 and dianthin 30 served as templates for primer design [29–31]. Both experiments combined resulted in a 728-bp DNA sequence that was translated into the 242 amino acid sequence shown in Figure 2.

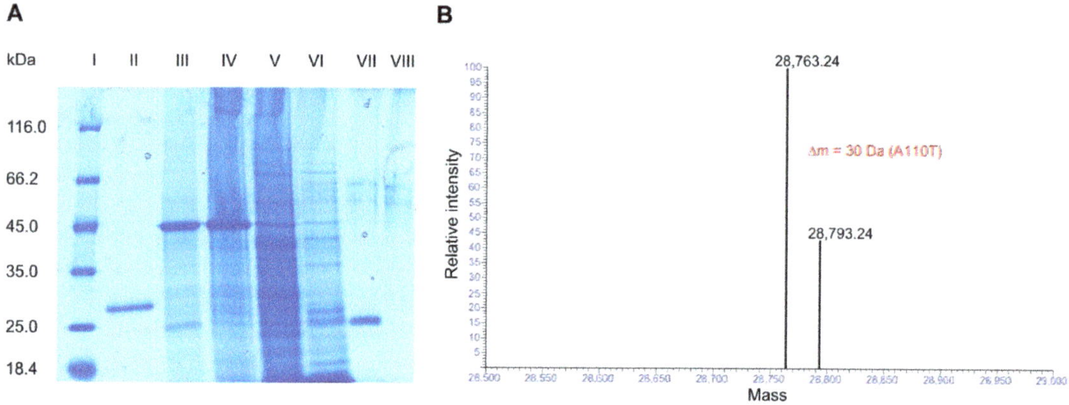

Figure 1. Isolation of sapovaccarin-S1 and -S2 from the seeds of *Saponaria vaccaria* L. and determination of its protein mass. (**A**) SDS-PAGE (12.5%) of the protein purification of sapovaccarin-S1 and -S2 using the whole-seed fraction, Coomassie Brilliant Blue stain. I: Protein marker (in kDa); II: N-terminally His-tagged dianthin (His-dianthin; 0.66 μg), a type I RIP from *Dianthus caryophyllus* L., as a reference; III: Crude extract from *Saponaria vaccaria* L. (4.83 μg); IV: 30%-ammonium sulfate fraction (40.03 μg); V: 60%-ammonium sulfate fraction (45.76 μg); VI: 90%-ammonium sulfate fraction (28.06 μg); VII: The 90%-ammonium sulfate fraction was subjected to cation exchange chromatography. At 0.3 M NaCl sapovaccarin-S1 and -S2 eluted (0.55 μg); VIII: Blank, phosphate buffered saline (PBS). (**B**) High-resolution mass spectrometric analysis of the isolated protein revealed two mono-isotopic masses of 28,763.24 Da and 28,793.24 Da. The y-axis represents the relative signal intensity and the x-axis shows the deconvoluted mass in Da.

```
Sapovaccarin-S1    VVTITLNLANPSKGQYSSFVDRIRNNVRDPKLKYGGTDIAVIGAPPTREKYLRINLQGPR   60
Sapovaccarin-S2    VVTITLNLANPSKGQYSSFVDRIRNNVRDPKLKYGGTDIAVIGAPPTREKYLRINLQGPR   60
V. pyramidata      VVTITLNLANPSKGQYSSFVDRIRNNVRDP------------------------------   30

Sapovaccarin-S1    GTVSLGLRRENLYVVAYLAMDNTNTNKAYYFRNQITSAELRTVFPEATAANQIVIQYGED   120
Sapovaccarin-S2    GTVSLGLRRENLYVVAYLAMDNTNTNKAYYFRNQITSAELRTVFPEATATNQIVIQYGED   120
                                                                    110

Sapovaccarin-S1    YQSIERNAQITQGSQSRKELGLGIDLLVTSIDGVNRKARVVRNEARFLLIAIQMTAEAAR   180
Sapovaccarin-S2    YQSIERNAQITQGSQSRKELGLGIDLLVTSIDGVNRKARVVRNEARFLLIAIQMTAEAAR   180

Sapovaccarin-S1    FRYIQNLVTFNFPKKFDSDNKVIQFEVSWGKISRAIYGDCKNGVFNKDYDFGFGKVRQAK   240
Sapovaccarin-S2    FRYIQNLVTFNFPKKFDSDNKVIQFEVSWGKISRAIYGDCKNGVFNKDYDFGFGKVRQAK   240

Sapovaccarin-S1    QLQMGLLMYLGRPG   254
Sapovaccarin-S2    QLQMGLLMYLGRPG   254
```

Figure 2. Protein sequence determination of sapovaccarin-S1 and -S2. Sequence alignment of sapovaccarin-S1 and -S2 and the N-terminus of *Vaccaria pyramidata* Medik. identified by Bolognesi et al. [5]. The final sequences of sapovaccarin-S1 and -S2 are the result of combining MS-based peptide analysis and PCR experiments. Amino acids shown in black were identified by PCR experiments and amino acids (N-terminal and C-terminal region) highlighted in green were not covered by PCR analysis. Peptide sequences identified by MALDI-MS-based peptide sequencing are highlighted by red boxes. Amino acids which could be confirmed by ISD data are underlined in blue. In addition, a triangle is highlighting the amino acid at position 110, indicating the position where sapovaccarin-S1 (alanine) and -S2 (threonine) differ.

This sequence was named sapovaccarin-S1. In the N- and C-terminal region, five and seven amino acids, respectively, could not be identified by the described PCR method (Figure 2). MALDI in-source decay (ISD) was conducted in order to obtain additional information on the N- and C-terminal regions. C-terminal ISD measurements revealed an ion series, enabling to complete the C-terminus (see Supplementary Figure S4). The N-terminus published by Bolognesi et al. matched the sequence identified by PCR and ISD data, finally allowed to confirm the five missing N-terminal residues (Figure 2). The complete protein sequence of mature sapovaccarin-S1 consisting of 254 amino acids is shown in Figure 2.

In addition, in-gel digestions by LysC, chymotrypsin and AspN were performed and peptides were analyzed by LC-ESI-MS. Combining these data sets with the trypsin digest and allowing for unspecific cleavage in the database search, we obtained a sequence coverage of 100% for the sapovaccarin-S1 sequence, thus confirming its correctness. In the trypsin digest, for the peptide with a mass of 2842.4 (pos. 102–126, TVFPEATA**A**NQIVIQYGEDYQSIER) we consistently found a second form with lower signal intensity (roughly one third) with a mass of 2872.4 (Δm = 30 Da) (see Supplementary Figure S5). In the fragment spectrum of this peptide, the b ions beyond b8 and the y ions beyond y16 were shifted by exactly 30 Da, indicative of a substitution A > T at position 110 (see Supplementary Figures S6 and S7). The protein carrying this substitution is referred to as sapovaccarin-S2 (Figure 2). The sapovaccarin-S2 isoform could be detected not only at the peptide level, but also at the DNA level: the codon for A110 revealed two peaks at position 328—a major guanine and a minor adenine peak (see Supplementary Figure S8). The substitution of GCG (sapovaccarin-S1) to ACG (sapovaccarin-S2) implies the amino acid substitution from alanine to threonine at position 110.

As shown in Figure 1B, Orbitrap-based intact protein measurement yielded a monoisotopic mass of 28,763.24 Da, which is almost identical to the theoretical mass calculated from the obtained sapovaccarin-S1 sequence (28,763.12 Da). Additionally, a minor peak was observed (43% relative abundance compared with the main peak) at a monoisotopic mass of 28,793.24 Da which represents sapovaccarin-S2.

The sequence of sapovaccarin-S1 determined here is highly similar to other well-characterized type I RIP sequences with 83% sequence identity to gypsophilin-S and 79% sequence identity to each saporin-S6 and dianthin 30, as shown in Figure 3A [29,31,32]. The protein sequence data reported in this paper will appear in the UniProt Knowledgebase under the accession numbers Q7M1L6 and C0HM39.

To predict the protein structure of sapovaccarin-S1 a homology model was built (Figure 3B). The high-resolution crystal structure of dianthin 30 (1.4 Å, PDB ID 1RL0) was chosen as template [33]. The template sequence showed 79% identity and 95% similarity with the target sequence of sapovaccarin-S1—an excellent starting point for homology modelling. The root-mean-square deviation (rmsd) between the template structure and the homology model was 0.52 Å. The low rmsd value indicated a correct global fold of the final model. The protein geometry did not show any Phi-Psi outliers nor atom clashes. The three-dimensional structure of sapovaccarin-S1 corresponded to the common fold of type I RIPs consisting of two domains: the N-terminal domain rich in β-strands and the C-terminal domain rich in α-helices (Figure 3B). The active site was located at the cleft between both domains (Figure 3C). In type I and II RIPs the active-site key residues Tyr73, Tyr121, Glu177 and Arg180 are highly conserved (Figure 3A) [33]. The same applied to sapovaccarin-S1—all key residues were preserved at the same position. The protein surface of sapovaccarin-S1 with its hydrophilic binding pocket is shown in Figure 3C,D. The same structural elements were known from dianthin 30 and saporin-S6 (PDB ID 1QI7) [33,34].

A

```
Sapovaccarin-S1    VVTITLNLANPSKGQYSSFVDRIRNNVRDPKLKYGGTDIAVIGAPPTREKYLRINLQGPR   60
Saporin-S6         VTSITLDLVNPTAGQYSSFVDKIRNNVKDPNLKYGGTDIAVIG-PPSKEKFLRINFQSSR   59
Dianthin 30        ATAYTLNLANPSASQYSSFLDQIRNNVRDTSLIYGGTDVAVIGAPSTTDKFLRLNFQGPR   60
Gypsophilin-S      -TTITLNLATPTAGQYSSFLDRIRNNVRDTKLKYGGTDIAVIG-PPSRDKFLRINFQGPR   58
                    .: **:*..*: .*****:*:*****:* .* *****:**** * : :*:**:*:*. *

Sapovaccarin-S1    GTVSLGLRRENLYVVAYLAMDNTNTNKAYYFRNQITSAELRTVFPEATAANQIVIQYGED   120
Saporin-S6         GTVSLGLKRDNLYVVAYLAMDNTNVNRAYYFKSEITSAELTALFPEATTANQKALEYTED   119
Dianthin 30        GTVSLGLRRENLYVVAYLAMDNANVNRAYYFKNQITSAELTALFPEVVVANQKQLEYGED   120
Gypsophilin-S      GTVSLGLRRENLYVVSYLAMDNTGVNKAYYFKNQITTVELTTLFPEATVANHKLLEYAED   118
                   *******:*:*****:******:..*:****:.:**:.** ::***...**:  ::* **

Sapovaccarin-S1    YQSIERNAQITQGSQSRKELGLGIDLLVTSIDGVNRKARVVRNEARFLLIAIQMTAEAAR   180
Saporin-S6         YQSIEKNAQITQGDKSRKELGLGIDLLLTFMEAVNKKARVVKNEARFLLIAIQMTAEVAR   179
Dianthin 30        YQAIEKNAKITTGDQSRKELGLGINLLITMIDGVNKKVRVVKDEARFLLIAIQMTAEAAR   180
Gypsophilin-S      YQSIEKNAKITQGDKSRKELGLGIDLLVTSIDGVNRKVRVVRDEARFLLIGIQMTAEAAR   178
                   **:**:**:** *.:*********:**:* ::.**:*.***::*******.******.**

Sapovaccarin-S1    FRYIQNLVTFNFPKKFDSDNKVIQFEVSWGKISRAIYGDCKNGVFNKDYDFGFGKVRQAK   240
Saporin-S6         FRYIQNLVTKNFPNKFDSDNKVIQFEVSWRKISTAIYGDAKNGVFNKDYDFGFGKVRQVK   239
Dianthin 30        FRYIQNLVTKNFPNKFDSENKVIQFQVSWSKISTAIFGDCKNGVFNKDYDFGFGKVRQAK   240
Gypsophilin-S      FRYIQNLVTKNFPKKFESENKVIQFQISWSKISKAIYGDCKKGVFNKDYDFGFGKVRQAK   238
                   ********* ***:**:*:******::** *** **:**.*:****************.*

Sapovaccarin-S1    QLQMGLLMYLGRPG   254
Saporin-S6         DLQMGLLMYLGKPK   253
Dianthin 30        DLQMGLLKYLGRPK   254
Gypsophilin-S      QLQMGLLMYLGRPK   252
                   :****** ***:**
```

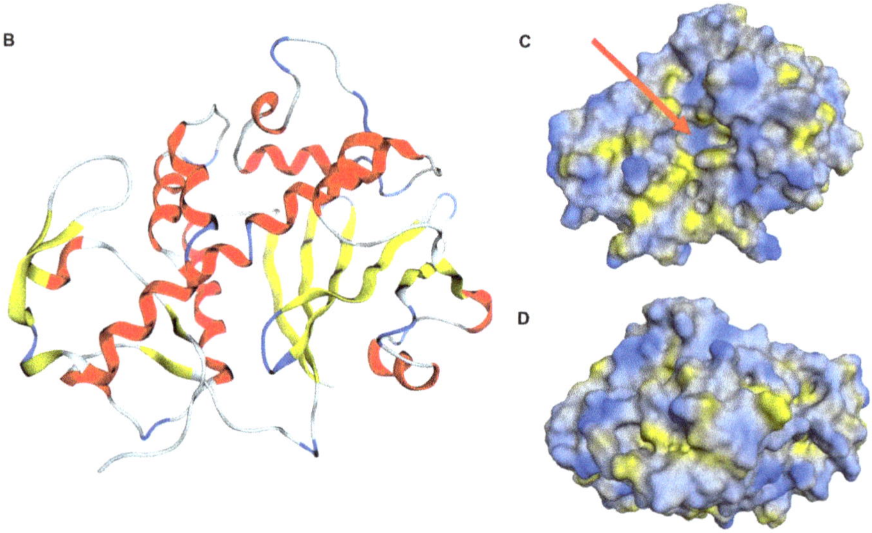

Figure 3. Sequence comparison of sapovaccarin-S1 with well-characterized type I RIPs; tertiary structure model of sapovaccarin-S1. (**A**) Sequence alignment of sapovaccarin-S1, saporin-S6, dianthin 30 and gypsophilin-S. Highly conserved active-site resiudes are highlighted in yellow [33]. The alanine at position 110, substitued in sapovaccarin-S2, is highlighted in red. Aligned amino acids labeled with a star symbol (*) are fully conserved in all four sequences; those with a dot symbol (.) are identical in three out of four sequences and those with colon symbol (:) show moderate identity. Alignment was performed using the Clustal Omega multiple sequence alignment tool (https://www.ebi.ac.uk/Tools/msa/clustalo/, accessed on 11 January 2022). (**B**) Homology model of sapovaccarin-S1 showing its tertiary structure. (**C**) Front view on the protein surface of sapovaccarin-S1. Hydrophilic regions of the protein are highlighted in blue and lipophilic areas in yellow. The substrate binding pocket (red arrow) is located in the cavity in the middle of the protein. (**D**) Back view on the protein surface.The panels (**B–D**) have been produced using MOE (version 2020.0901, Chemical Computing Group, Montreal, Canada).

2.3. N-Glycosylase Activity

Due to their cytotoxic effects type I and type II RIPs have now been studied for over 40 years for their promising use in anti-cancer therapy [35]. The cytotoxic effects of the plant toxins are dependent, on the one hand, on the extent of the endosomal release and on the other hand on their characteristic adenine releasing activity from DNA, RNA, and other polynucleotides [14,36,37]. The group of Fiorenzo Stirpe reported *N*-glycosylase activity on rabbit-reticulocyte lysate and purified rat liver ribosomes for the RIP isolated from *V. pyramidata* [5]. The *N*-glycosylase activity of isolated RIPs from *S. vaccaria* can also be extended to an A30-oligonucleotide substrate (A30). Weng developed an assay, which allows to quantify the enzyme activity of RIPs by measuring the adenine release from an A30-oligonucleotide substrate and the subsequent detection at 260 nm on a thin-layer chromatography (TLC) plate [28]. *N*-glycosylase activity of the isolated sapovaccarin-S1 and -S2 and—for comparative purposes—the isolated gypsophilin-S and recombinant His-dianthin were analyzed with the adenine-releasing assay (Figure 4A). The adenine release of sapovaccarin-S1 and -S2 was higher that of the other investigated RIPs. However, His-dianthin which was recombinantly expressed, released significantly less adenine than the RIPs isolated from the seeds (Figure 4A).

2.4. Cytotoxicity of Sapovaccarin-S1 and -S2

In order to confirm the previously proven in vitro *N*-glycosylase activity of sapovaccarin-S1 and -S2 in cells, we used label-free live-cell microscopy to investigate the cytotoxicity of sapovaccarin-S1 and -S2 in Huh-7 cells. The confluence analysis of the raw data was performed by the analysis algorithms of the software package CytoSMART. No effect on cell viability could be observed for the different concentrations shown in Figure 4B 24 h after the addition of sapovaccarin-S1 and -S2. After 35 h, 1000 nM sapovaccarin-S1 and -S2 incubation the confluence began to be reduced significantly compared to the control (*t*-test, $p \leq 0.05$). For 100 nM, 10 nM, and 1 nM the significant reduction in confluence occurred after 46 h, 45 h, and 47 h, respectively (*t*-test, $p \leq 0.05$). At the end of the sapovaccarin-S1 and -S2 incubation time, a concentration-dependent cytotoxicity of sapovaccarin-S1 and -S2 was observable. A concentration of 1000 nM had the strongest cytotoxic effect compared to the others (*t*-test, $p \leq 0.05$).

2.5. Thermal Stability

Protein thermal stability plays a key role in the development of new anti-cancer drugs, for both science and pharmaceutical industrial processes. The thermostability of sapovaccarin-S1 and -S2 was analyzed by nano-differential scanning calorimetry (DSC). In addition, DSC profiles of gypsophilin-S, which has been isolated recently by our group, and recombinantly expressed His-dianthin were recorded as a reference. The DSC profiles and the transition midpoint temperatures (Tm) are presented in Figure 5. Compared to gypsophilin-S (Tm 64.7 °C) and His-dianthin (Tm 65.6 °C), sapovaccarin-S1 and -S2 had the highest Tm of 68.9 °C. Tm values for proteins range typically between 40 and 80 °C. Thus, all three RIPs possessed a moderately high thermal stability.

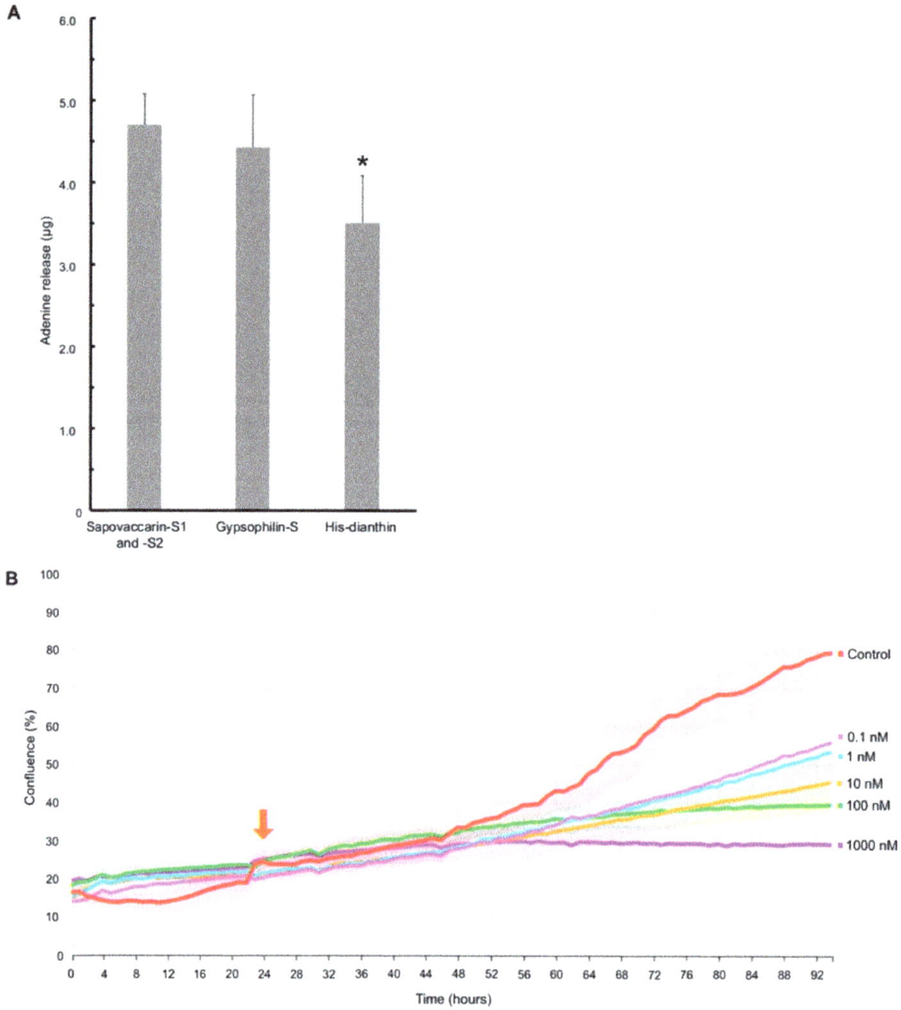

Figure 4. Characterization of the enzymatic activity of sapovaccarin-S1 and -S2. (**A**) Quantitative analysis of *N*-glycosylase activity on an A30-oligonucleotide substrate by TLC-densitometry. Sapovaccarin-S1 and -S2 exhibited *N*-glycosylase activity. Gypsophilin-S and His-dianthin were used as positive controls. Sapovaccarin-S1 and -S2 and gypsophilin-S (each 0.01 nM), that were isolated from the seeds, exhibited significantly higher enzymatic activity than His-dianthin (0.01 nM) that was recombinatly expressed. * significant to sapovaccarin-S1 and -S2 and to gypsophilin-S, t-test, $p \leq 0.05$. (**B**) Live-cell imaging of Huh-7 cells. After an incubation of 24 h (orange arrow) sapovaccarin-S1 and -S2 was added at different concentrations (0.1–1000 nM). Cells were continuously monitored for 94 h. Sapovaccarin-S1 and -S2 exhibited a concentration-dependent effect on the cell viability of Huh-7 cells.

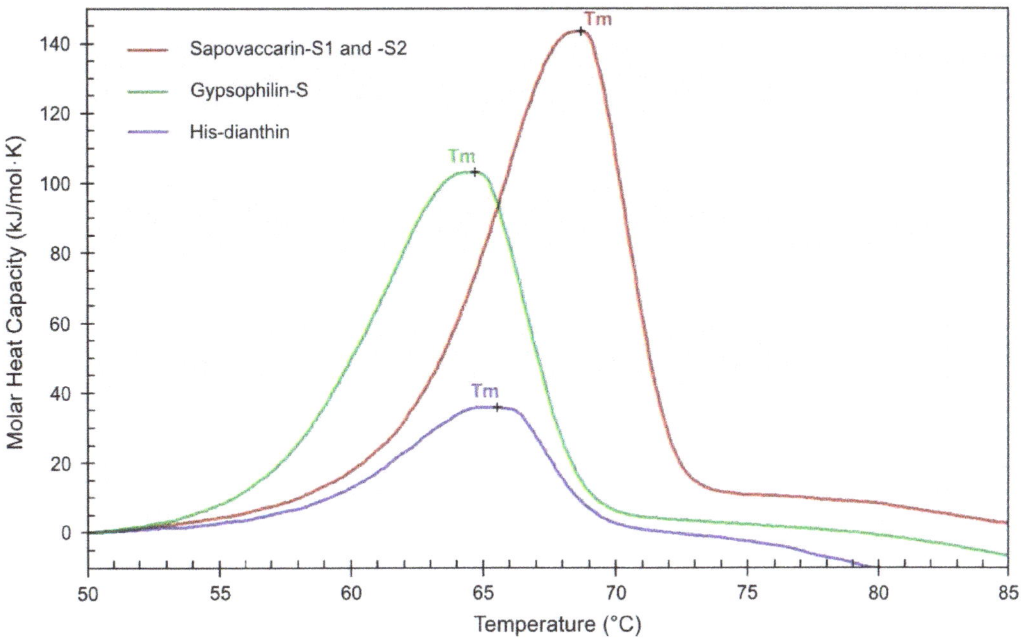

Figure 5. DSC profiles of isolated sapovaccarin-S1 and -S2, isolated gypsophilin-S and recombinantly expressed His-dianthin in PBS, pH 7.4. Each DSC profile was recorded at a protein concentration of 0.4 mg/mL. The transition midpoint temperatures (Tm) were recorded for sapovaccarin-S1 and -S2 at 68.9 °C, for gypsophilin-S at 64.7 °C and for His-dianthin at 65.6 °C. The figure has been produced using NanoAnalyze software (version 3.11.0, TA instruments, New Castle, DE, USA).

2.6. Distribution of Sapovaccarin-S1 and -S2 in Differently Processed Seed Material from Saponaria vaccaria L.

Beside the whole seeds, the Canadian Carnation Biocompany provided differently processed seed material, which allowed us to study the exact distribution of sapovaccarin-S1 and -S2 in the processed seed material (see Figure 6). The three ammonium sulfate fractions (30%, 60%, and 90% saturation) of each seed fraction were tested for *N*-glycosylase activity. The whole seeds and all eight seed fractions exerted *N*-glycosylase activity for all three ammonium sulfate fractions. With increasing ammonium sulfate concentration, enzyme activity increased—the highest enzyme activities were consistently found in the 90%-ammonium sulfate fractions. The next aim was to quantify the adenine release activity of the differently processed seed material. For this purpose, the 90%-ammonium sulfate fractions of the whole seeds, the fractionated seed material, the embryo-enriched and the perisperm-enriched seed fraction were investigated. The adenine release correlated with the total protein amount. According to the results shown in Table 1, sapovaccarin-S1 and -S2 was most abundant in the perisperm of the seeds of *Saponaria vaccaria* L.

Figure 6. Differently processed seed material from *Saponaria vaccaria* L. (**A**) Defatted and grinded seed material. Fractionated seed material was obtained by sieve analysis: the numbers indicate the mesh sizes of the residues of the fractions or if <200 the passage. The embryo-enriched (EEF) and the perisperm-enriched seed fractions (PSF) were obtained by separating the embryo from the rest of the seed using an impact and a roller mill and by sieving. (**B**) Whole seeds in corresponding scale. (**C**) PSF in corresponding scale. (**D**) EEF in corresponding scale.

Table 1. Adenine released by the 90%-ammonium sulfate fractions of the differently processed seed material.

Seed Fraction	Adenine Release (µg/mg Total Protein)
Whole seeds	720
EEF	99
PSF	851
Mesh size < 200	487
Mesh size 200	639
Mesh size 100	614
Mesh size 80	556
Mesh size 60	285
Mesh size 40	381

3. Discussion

Here we report the isolation and characterization of sapovaccarin-S1 and -S2, two protein isoforms from *Saponaria vaccaria* L. They were classified as type I RIPs in the course of this study, including their full amino acid sequence. Furthermore, we report for the first time that RIPs in *Saponaria vaccaria* L. are mainly located in the perisperm of the seeds. It should be noted here that the localization of RIPs was determined by evaluating the

N-glycosylase activity of extracts from perisperm- and endosperm-enriched seed fractions. The localization in the perisperm might also indicate that sapovaccarin-S1 and -S2 could serve for nitrogen storage in *Saponaria vaccaria* L. This observation might lend some support to the hypothesis that RIPs in addition to having a defense function may also function as storage proteins in some plants. Both RIPs were isolated from the seeds by aqueous extraction, ammonium sulfate precipitation, and cation exchange chromatography. The exact amino acid sequence as well the molecular masses of sapovaccarin-S1 and -S2 were determined by PCR and mass spectrometry. Both isoforms differ only in the substitution of one amino acid at position 110 (Ala110 in sapovaccarin-S1 substituted by Thr110 in sapovaccarin-S2). Their intact masses lie within the characteristic mass range of type I RIPs [27]. The ratio of abundance of sapovaccarin-S1 to sapovaccarin-S2 was consistent in the MS spectrum of intact protein mass, the MS/MS spectrum of tryptic peptides and the DNA chromatogram, indicating that sapovaccarin-S1 is the more abundant isoform.

The occurrence of RIP isoforms within the carnation family was already well described in the literature [15,16,38]. The isoforms described herein differ in only one amino acid: the non-polar alanine in position 110 in sapovaccarin-S1 is substituted by a polar threonine residue in sapovaccarin-S2 provoking a minimal increase in mass (Δm = 30 Da) but no changes in pI (calculated pI 9.87). Hence, separation of both isoforms could be achieved neither by cation exchange chromatography nor SDS-PAGE, resulting in a mixture of the two isoforms. Due to the difference in just one amino acid and the resulting very small mass and non-existent pI differences, the separation of both isoforms by other protein purification methods, such as gel filtration or hydrophobic interaction chromatography, probably could not be achieved. We therefore decided to further investigate the mixture of two isoforms. The decision was reinforced by the fact that the single amino acid substitution was located neither within the active site nor in its immediate vicinity. Although in light of the predicted structures it seems highly unlikely that the substitution A110T will have a major impact on *N*-glycosylase activity, in the future this issue might be clarified by producing the individual isoforms recombinantly by site-directed mutagenesis and determining their activity; however, this would fall outside the scope of the present study.

Type I RIPs are mostly encoded by intron-less genes [39]. Therefore, genomic DNA was used as a PCR template. Even though the N- and C-terminal regions were not determined by PCR sequencing, it can be concluded from the PCR data that sapovaccarin-S1 and -S2 are also encoded by intron-less genes. The isoforms described here could be identified from intact protein mass, tryptic peptides as well as from the chromatogram of the PCR analysis. Given that the amino acid substitution at position 110 can be found in the genomic DNA, the isoforms could not have been created by alternative splicing, but had to have been encoded by two different genes.

RIPs from *Saponaria vaccaria* L. were first mentioned by Bolognesi et al. in 1995 [5]. The authors isolated one RIP by aqueous extraction and cation exchange chromatography. Its molecular mass of 28 kDa was determined by SDS-PAGE and gel filtration. In their study, they determined a pI of >9.5 and demonstrated its *N*-glycosidase activity. These results are essentially in agreement with ours. With the applied isolation and mass determination methods, Bolognesi et al. were not able to differentiate the two isoforms. Their assumption that the isolated protein fraction was composed of one protein, underlines the importance of accurate mass spectrometry methods for future research.

Thermostability of sapovaccarin-S1 and -S2 in comparison to gypsophilin-S and His-dianthin was studied by DSC. The Tm of sapovaccarin-S1 and -S2, gypsophilin-S and His-dianthin were determined as 69 °C, 65 °C and 66 °C, respectively. Thus, the isolated mixture of sapovaccarin-S1 and -S2 showed the highest thermal stability among proteins studied. Little data on the thermostability of RIPs has been published. Saporin-S6 and gelonin, a type I RIP from *Gelonium multiflorum*, were analyzed by infrared spectroscopy and two-dimensional correlation spectroscopy at neutral pH (50 mM sodium phosphate buffer, pH 7.4) in two different studies [40,41]. Tm of saporin-S6 was measured at 58 °C and gelonin's at 66 °C, indicating a moderately high thermostability [40,41]. These data are in accordance with the DSC data of our study. The moderately high thermostability as well as the high *N*-glycosylase activity demonstrated in this study show, that sapovaccarin-S1 and -S2 seem to exhibit potential for toxin moieties in immunotoxins.

4. Materials and Methods

4.1. Seed Material

The Canadian Carnation BioProducts Company, Saskatoon, S7H 3R2, Canada provided the seeds of *Saponaria vaccaria* L. (Caryophyllaceae). In addition to whole dried seeds, eight differently processed seed fractions were available: six sieve analysis fractions, an embryo-enriched fraction (EEF) and a perisperm-enriched fraction (PSF). The six sieve analysis fractions were obtained by collecting the particles which were stopped by the sieves (mesh size 40, 60, 80, 100 and 200) and the particles which have passed mesh size 200 (<200). The embryo-enriched fraction (EEF) and the perisperm-enriched fractions (PSF), which is the starch and hull fraction of the seed which remains after the embryo is removed, are achieved by separating the embryo from the rest of the seed using an impact and a roller mill and sieving.

4.2. Isolation of Sapovaccarin-S1 and -S2

The whole seeds, sieving fraction 40, EEF and PSF had to be ground using an electric mill (M20—Universalmühle, IKA, Staufen, Germany). The raw material of the remaining seed fractions was fine enough from the beginning. Each seed fraction was defatted twice with n-hexane (10 mL/g seed) at 4 °C for 30 min and dried at room temperature. The defatted seed fractions were extracted in PBS pH 7.4 (8 mL/g seed) by gentle stirring at 4 °C over 24 h and thereafter centrifuged at 8000× *g* for 20 min. The resulting supernatant is referred to as 'crude extract' and was purified by the following steps: Proteins were separated by fractionated ammonium sulfate precipitation at saturations of 30%, 60% and 90% ammonium sulfate. The precipitated proteins were resuspended in PBS and analyzed by SDS-PAGE (12.5%). The protein fractions were also tested for their *N*-glycosylase activity by measuring the released adenine by TLC densitometry at 260 nm [28]. The 90%-ammonium sulfate fractions showed the highest enzyme activity and were used for further purification by cation exchange chromatography using a prepacked SP Sepharose High Performance column (HiTrap SP XL 1 mL, GE Healthcare Europe, Freiburg, Germany) connected to an ÄKTA start protein purification system (GE Healthcare Europe, Freibug, Germany). Then, 7.0 mL of samples equilibrated with 50 mM HEPES, pH 7.0 were applied to the column and bound proteins were eluted from the column by 0.1 M, 0.2 M, 0.3 M, 0.4 M and 1 M NaCl (in 50 mM HEPES, pH 7.0) at a flow rate of 1 mL/min and detected at 280 nm.

4.3. Recombinant Expression of His-Dianthin and Isolation of Gypsophilin-S

N-terminally His-tagged dianthin 30 (His-dianthin) from the plant *Dianthus caryophyllus* L. was recombinantly expressed in E. coli NiCo21(DE3) (New England Biolabs, Ipswich, QLD, USA), purified by Ni-nitrilotriacetic acid affinity chromatography and analyzed by SDS-PAGE as described elsewhere [42]. Gypsophilin-S was isolated from the seeds of *Gypsophila elegans* M.Bieb. using ammonium sulfate precipitation and subsequent ion exchange chromatography. The isolation is described in detail elsewhere [29].

4.4. SDS-PAGE and Protein Quantification

12.5% SDS-polyacrylamide gels were used for SDS-PAGE using the Lämmli method. Protein bands were stained with Coomassie Brillant Blue G250 as described elsewhere [43]. Protein concentrations were determined by using a modified Bradford method (ROTI®Nanoquant, Carl Roth GmbH, Karlsruhe, Germany).

4.5. Protein Mass Spectrometry

Protein and peptide sequences were analyzed by matrix-assisted laser desorption time-of-flight mass spectrometry (MALDI-TOF-MS). All MALDI-TOF-MS measurements were performed with an Ultraflex-II TOF/TOF instrument (Bruker Daltonics, Bremen, Germany) equipped with a 200 Hz solid-state Smart beam^{TM} laser. Data were analyzed using the software FlexAnalysis provided with the instrument. Samples were applied by the dried-droplet technique. Peptides were generated by trypsin, LysC, chymotrypsin and AspN in-gel digestion following a protocol described elsewhere [44]. The mass fingerprints of the generated peptides were recorded in positive reflector mode (RP_PepMix) over an *m/z* range of 600–4000. α-cyano-4-hydroxycinnamic acid was used as matrix. Selected tryptic peptides got analyzed by tandem MS using the LIFT mode [45]. N-terminal c ions and C-terminal (z + 2) ions were generated from the intact and acetone precipitated protein using in-source decay (ISD). As a matrix, 1,5-diaminonaphthalene (1,5-DAN) was used. Spectra were recorded in positive reflector mode (RP_PepMix) over an m/z range of 600–6000.

For high resolution intact protein mass analysis by liquid chromatography-electrospray ionization mass spectrometry (LC-ESI-MS) the isolated protein sample was analyzed using the Ultimate 3000 liquid chromatography system connected to a Q Exactive HF mass spectrometer via the ion max source with HESI-II probe (Thermo Scientific, Waltham, MA, USA). The following MS source parameters were used: spray voltage 3.6 kV, capillary temperature 320 °C, sheath gas 10, auxiliary gas 4, S-lens RF level 60, intact protein mode on. For the analysis 7 µL of a 10 µM protein solution were desalted and concentrated by injection on a reversed-phase cartridge (MSPac DS-10, 2.1 mm × 10 mm, Thermo Scientific, Waltham, MA, USA) at 60 °C using buffer A (0.1% formic acid, 5% acetonitrile in water) at a constant flow rate of 22 µL/min for 3 min. This was followed by a short linear gradient of 5–95% buffer B (0.1% formic acid in 80% acetonitrile, 20% water) within 10 min followed by washing and re-equilibration. Full MS spectra were acquired using the following parameters: mass range m/z 600–2500, resolution 120,000, AGC target 3 × 106, µscans 5, maximum injection time 200 ms. MS raw data were analyzed using BioPharma Finder (version 3.2, Thermo Scientific, Waltham, MA, USA). First, an averaged spectrum over the chromatographic peak was generated followed by spectral deconvolution using a relative abundance threshold of 20% and the function 'consider overlaps' turned off.

4.6. DNA Extraction from the Seeds and Determination of the DNA Sequence by PCR

Peptide mass fingerprinting and subsequent MS/MS-analysis enabled the identification of first sections of the amino acid sequence of sapovaccarin-S1 and -S2. In order to complete the identified peptides to a full sequence, two PCRs were conducted. Primer pair A was designed based on the peptide mass fingerprint results and primer pair B was derived from the results of the first PCR round and the N-terminus determined by Bolognesi et al. (Table 2). The template DNA was extracted from 75 mg whole dried seeds which were frozen overnight, using the PureLink Plant Total DNA Purification kit (Life Technologies, Carlsbad, CA, USA). DNA concentrations were determined with the NanoDrop (Thermo Fisher Scientific, Waltham, MA, USA). The PCR was conducted using the Phusion High-Fidelity DNA Polymerase (New England Biolabs, Ipswich, QLD, USA). According to the instructions, 25 µL reactions using 64.8 ng of DNA template were prepared. The PCR followed a 3-step protocol with the following cycling steps: Initial denaturation 98 °C for 30 s, denaturation 98 °C for 30 s, annealing for 30 s (annealing temperature see Table 2), extension 72 °C for 22.8 s, final extension 72 °C for 10 min. Denaturation, annealing and extension steps were repeated for 35 cycles. PCR products were separated in a 1% agarose gel. The Monarch Genomic DNA purification Kit (New England Biolabs, Ipswich, QLD, USA) was used to extract the PCR products from the gel. For sequencing purposes extracted PCR product concentrations were determined. PCR products were prepared with corresponding primers and sent to LGC Genomics, Berlin, Germany for sequencing. The Expasy translation tool (https://web.expasy.org/translate/, accessed on 17 September 2021) was used to translate the DNA sequence to a protein sequence.

Table 2. Primer pairs used for sequencing of sapovaccarin-S1 and -S2.

Primer Pairs		Primer Sequence	Annealing Temperature (°C)
A	Forward	5′-AAT GCT AAG ATT ACA CAA GGG-3′	59
	Reverse	5′-GCC CAA ATA CAT AAG GAG TCC C-3′	
B	Forward	5′-CAT TAA ATC TCG CAA ATC C-3′	55
	Reverse	5′-GAC TCC ATC AAT TGA CGT TAC-3′	

4.7. Homology Modeling

Homology modeling was performed with MOE (version 2020.0901, Chemical Computing Group, Montreal, QC, Canada). The high resolution (1.4 Å) crystal structure of dianthin 30 with PDB ID 1RL0 served as a template). The target sequence and dianthin 30 exhibit 79% sequence identity and 95% sequence similarity. The target sequence and the sequence of dianthin 30 were aligned and checked for correct alignment.

4.8. Adenine-Releasing Assay

The *N*-glycosylase activity of different samples was determined by using the adenine-releasing assay as described elsewhere [28]. The assay is based on the cleavage of an adenine from an artificial substrate, a DNA oligonucleotide 5′-A30-3′ (A30). 10 µL of protein sample was mixed with 10 µg A30 (Metabion International AG, Planegg/Steinkirchen, Germany) and filled up to 50 µL with *N*-glycosylase buffer (50 mM sodium acetate, 100 mM KCl, pH 5). Deviating from the publication the mixtures were incubated at 37 °C over night. Samples (each 10 µL) were applied to a TLC 0.25 mm pre-coated silica gel 60 glass plate with fluorescent indicator UV254 (Macherey-Nagel, Düren, Germany) and developed by acetonitrile/water/ammonia (32%), (18:1.6:0.6). In addition, for quantification purposes different adenine amounts (0.125 µg, 0.25 µg, 0.5 µg and 1.0 µg) were applied on the plate. Released adenine was determined by TLC densitometry at 260 nm using the TLCs canner 4 (CAMAG, Berlin, Germany).

4.9. Cytotoxicity

To monitor the cytotoxicity of sapovaccarin-S1 and -S2 a label-free live-cell imaging system—the CytoSMART Omni system (CytoSMART Technologies B.V., Eindhoven, Netherlands) was used. The CytoSMART Omni system is an automated brightfield microscope, scanning the complete well surface that can be placed in the incubator. The cytotoxicity studies were performed with Huh-7 cells, a hepatocyte carcinoma cell line. The Huh-7 cell line was obtained from Dr. Mirko Pinotti, University of Ferrara. Then, 8000 cells/well were seeded in 96-well plates, each well containing 150 µL Minimum Essential Medium supplemented with 10% FBS, the plate was placed on the CytoSMART Omni device. Cells were cultured at 37 °C and 5% carbon dioxide. After 24 h sapovaccarin-S1 and -S2 were added (in 20 µL PBS each 3 wells) ranging from 0.1 to 1000 nM (final concentrations). Control cells were only treated with PBS. Image analysis and confluence calculation was performed using the CytoSMART image analysis software package.

4.10. Differential Scanning Calorimetry

The thermal stability of protein samples was investigated by differential scanning calorimetry (DSC). A NanoDSC (TA Instruments, New Castle, DE, USA) with capillary cells of 0.3 mL volume was used to carry out the calorimetric measurements. Successive heating and cooling buffer-buffer scans using PBS were repeated three times at a scanning rate of 1 °C/min and over a temperature range of 10–100 °C. During the measurements a total pressure of 3.0 atm was applied to the reference and the sample cell. Protein samples were prepared in PBS with a concentration of 0.4 mg/mL. Prior to measurements buffers and samples were degassed under vacuum for 15 min. A heating scan of each sample was recorded under the same conditions as the buffer-buffer scans. DSC data was analyzed by using the NanoAnalyze software (version 3.11.0, TA instruments New Castle, DE, USA). Buffer-buffer scans got subtracted from each sample scan.

4.11. Accession Numbers

The protein sequence data reported in this paper will appear in the UniProt Knowledgebase under the accession numbers Q7M1L6 and C0HM39.

Supplementary Materials: The following supporting information can be downloaded at: https://www.mdpi.com/article/10.3390/toxins14070449/s1. Figure S1: Cation exchange chromatogram of the isolation of sapovaccarin-S1 and -S2 from the 90% ammonium sulfate fraction of perisperm-enriched seed fraction (PSF). The Y-axis on the left side shows the absorbance at 280 nm in mAU. The Y-axis on the right side represents the composition of the elution buffer in %. 50 mM HEPES, pH 7.0 was used as starting buffer. The NaCl concentration in the elution buffer was gradually increased by adding 1 M NaCl in 50 mM HEPES, pH 7.0. The retention time in min is shown on the X-axis. Sapovaccarin-S1 and -S2 eluted at 0.3 M NaCl. Collected fractions are labeled with 1 to 5; Figure S2: SDS-PAGE (12.5%) of the cation exchange chromatography fractions of sapovaccarin-S1 and S2 using the PSF, Coomassie Brilliant Blue stain. I: Protein marker (in kDa); II: Fraction 1—flow through (6.3–17.5 min); III: Fraction 2 (32.0–35.0 min); IV: Fraction 3 (40.0–42.5 min); V: Fraction 4 (42.5–44.0 min); VI: Fraction 5—sapovaccarin-S1 and -S2 (44.0–46.9 min); VII: Sapovaccarin-S1 and -S2 fraction, concentrated 3 times; VIII: His-dianthin (0.66 µg); Figure S3: DNA sequence of sapovaccarin-S1 obtained by PCR analysis. Based on the tandem MS results, a pair of oligonucleotide primers was designed with the forward primer derived from the DNA sequence of gypsophilin-S (forward primer A) and the reverse primer derived from the C-terminal DNA region of saporin-S6 (reverse primer A). Using these primers and template DNA isolated directly from the seeds of *S. vaccaria* a PCR was performed that yielded a 351-bp PCR product which however did not cover the complete sequence. Therefore, a second primer pair was designed with the reverse primer based on the DNA of the first PCR round (reverse primer B) and the forward primer derived from the DNA of the N-terminus of dianthin 30 (forward primer B). The 449-bp PCR product overlapped in 72 bp with the first PCR product. Combining both sequences resulted in a 728-bp DNA sequence that was translated into a 242-amino acid sequence; Figure S4: In-source decay (ISD) analysis of sapovaccarin (type I RIP from *Saponaria vaccaria* L.). C-terminal ions (z + 2) according to the sequence given in the insert (z9 to z27) and additionally c22 to c28 according to the N-terminal sequence published by Bolognesi et al. are detected; Figure S5: Section of the MS1 spectrum of a tryptic digest of sapovaccarin highlighting the tryptic peptides pos. 102–126 of the isoforms S1 and S2; Figure S6: MS/MS spectra of the tryptic peptide pos. 102–126 of the sapovaccarin isoforms S1 (top) and S2 (bottom). Spectra were analyzed using the Mascot MS/MS search software and the corresponding peptide sequences and the matched b- and y-ions are indicated; Figure S7: Theoretical b- and y-ion series of the tryptic peptide pos. 102–126 of the sapovaccarin isoforms S1 (left) and S2 (right). B-ions differ by 30 Da starting from b9 and y-ions starting from y17; Figure S8: Sequence chromatogram of nucleotides encoding for amino acid positions 107 to 113. At nucleotide position 328 the guanine peak overlays an adenine peak. The codon change from GCG (sapovaccarin-S1) to ACG (sapovaccarin-S2) implicates an amino acid substitution at position 110 from alanine to threonine. Consistent with the results of the trypsin digest and the Orbitrap-based intact protein measurement, the guanine peak is more intense than the adenine peak.

Author Contributions: L.S. designed and performed experiments, analyzed data and wrote the main manuscript. C.W. and B.K. designed and performed mass spectrometry experiments, analyzed data, wrote the mass spectrometry parts of the manuscript, read and edited the manuscript; A.W. designed research, performed cytotoxicity experiments, analyzed data, read and edited the manuscript. All authors have read and agreed to the published version of the manuscript.

Funding: This research was funded by the Deutsche Forschungsgemeinschaft (DFG, German Research Foundation)—project number 422686308. We acknowledge support by the Open Access Publication Fund of the Freie Universität Berlin.

Institutional Review Board Statement: Not applicable.

Informed Consent Statement: Not applicable.

Data Availability Statement: Protein sequencing data have been deposited in UniProt Knowledgebase under the accession numbers Q7M1L6 and C0HM39.

Acknowledgments: For mass spectrometry (C.W. and B.K.), we would like to acknowledge the assistance of the Core Facility BioSupraMol supported by the Deutsche Forschungsgemeinschaft (DFG). For the provided seed material, we would like to acknowledge Canadian Carnation BioProducts Company.

Conflicts of Interest: The authors declare no conflict of interest.

References

1. Zhou, G.; Tang, L.; Wang, T.; Zhou, X.; Kou, Z.; Wang, Z.; Wu, J. Phytochemistry and pharmacological activities of *Vaccaria hispanica* (Miller) Rauschert: A review. *Phytochem. Rev.* **2016**, *15*, 813–827. [CrossRef]
2. Yun, Y.S.; Morita, H.; Takeya, K.; Itokawa, H. Cyclic peptides from higher plants. 34. Segetalins G and H, structures and estrogen-like activity of cyclic pentapeptides from Vaccaria segetalis. *J. Nat. Prod.* **1997**, *60*, 216–218. [CrossRef] [PubMed]
3. Yun, Y.S.; Shimizu, K.; Morita, H.; Takeya, K.; Itokawa, H.; Shirota, O. Triterpenoid saponin from Vaccaria segetalis. *Phytochemistry* **1998**, *47*, 143–144. [CrossRef]
4. Tian, M.; Huang, Y.; Wang, X.; Cao, M.; Zhao, Z.; Chen, T.; Yuan, C.; Wang, N.; Zhang, B.; Li, C.; et al. Vaccaria segetalis: A Review of Ethnomedicinal, Phytochemical, Pharmacological, and Toxicological Findings. *Front. Chem.* **2021**, *9*, 666280. [CrossRef] [PubMed]
5. Bolognesi, A.; Olivieri, F.; Battelli, M.G.; Barbieri, L.; Falasca, A.I.; Parente, A.; Del Vecchio Blanco, F.; Stirpe, F. Ribosome-inactivating proteins (RNA N-glycosidases) from the seeds of *Saponaria ocymoides* and *Vaccaria pyramidata*. *Eur. J. Biochem.* **1995**, *228*, 935–940. [CrossRef]
6. Schrot, J.; Weng, A.; Melzig, M.F. Ribosome-inactivating and related proteins. *Toxins* **2015**, *7*, 1556–1615. [CrossRef]
7. Endo, Y.; Tsurugi, K. Mechanism of action of ricin and related toxic lectins on eukaryotic ribosomes. *Nucleic Acids Symp. Ser.* **1986**, *17*, 187–190.
8. Brigotti, M.; Rambelli, F.; Zamboni, M.; Montanaro, L.; Sperti, S. Effect of alpha-sarcin and ribosome-inactivating proteins on the interaction of elongation factors with ribosomes. *Biochem. J.* **1989**, *257*, 723–727. [CrossRef]
9. Flexner, S. The Histological Changes Produced by Ricin and Abrin Intoxications. *J. Exp. Med.* **1897**, *2*, 197–216. [CrossRef]
10. Olsnes, S.; Pihl, A. Different biological properties of the two constituent peptide chains of ricin, a toxic protein inhibiting protein synthesis. *Biochemistry* **1973**, *12*, 3121–3126. [CrossRef]
11. Timar, J.; McIntosh, D.P.; Henry, R.; Cumber, A.J.; Parnell, G.D.; Davies, A.J. The effect of ricin B chain on the intracellular trafficking of an A chain immunotoxin. *Br. J. Cancer* **1991**, *64*, 655–662. [CrossRef] [PubMed]
12. Sandvig, K.; Grimmer, S.; Lauvrak, S.U.; Torgersen, M.L.; Skretting, G.; van Deurs, B.; Iversen, T.G. Pathways followed by ricin and Shiga toxin into cells. *Histochem. Cell Biol.* **2002**, *117*, 131–141. [CrossRef] [PubMed]
13. Barbieri, L.; Battelli, M.G.; Stirpe, F. Ribosome-inactivating proteins from plants. *Biochim. Biophys. Acta* **1993**, *1154*, 237–282. [CrossRef]
14. Bolognesi, A.; Polito, L.; Scicchitano, V.; Orrico, C.; Pasquinelli, G.; Musiani, S.; Santi, S.; Riccio, M.; Bortolotti, M.; Battelli, M.G. Endocytosis and intracellular localisation of type 1 ribosome-inactivating protein saporin-s6. *J. Biol. Regul. Homeost. Agents* **2012**, *26*, 97–109.
15. Stirpe, F.; Williams, D.G.; Onyon, L.J.; Legg, R.F.; Stevens, W.A. Dianthins, ribosome-damaging proteins with anti-viral properties from *Dianthus caryophyllus* L. (carnation). *Biochem. J.* **1981**, *195*, 399–405. [CrossRef]
16. Stirpe, F.; Gasperi-Campani, A.; Barbieri, L.; Falasca, A.; Abbondanza, A.; Stevens, W.A. Ribosome-inactivating proteins from the seeds of *Saponaria officinalis* L. (soapwort), of *Agrostemma githago* L. (corn cockle) and of Asparagus officinalis L. (asparagus), and from the latex of *Hura crepitans* L. (sandbox tree). *Biochem. J.* **1983**, *216*, 617–625. [CrossRef]
17. Ferreras, J.M.; Citores, L.; Iglesias, R.; Jiménez, P.; Girbés, T.; Lord, J.M.; Hartley, M.R. Sambucus ribosome-inactivating proteins and lectins. In *Toxic Plant Proteins*; Springer: Berline, Germany, 2010; Volume 18, pp. 107–131.
18. Girbes, T.; Ferreras, J.M.; Arias, F.J.; Stirpe, F. Description, distribution, activity and phylogenetic relationship of ribosome-inactivating proteins in plants, fungi and bacteria. *Mini Rev. Med. Chem.* **2004**, *4*, 461–476. [CrossRef]
19. De Zaeytijd, J.; Chen, P.; Scheys, F.; Subramanyam, K.; Dubiel, M.; De Schutter, K.; Smagghe, G.; Van Damme, E.J. Involvement of OsRIP1, a ribosome-inactivating protein from rice, in plant defense against Nilaparvata lugens. *Phytochemistry* **2020**, *170*, 112190. [CrossRef]
20. Stirpe, F.; Barbieri, L.; Gorini, P.; Valbonesi, P.; Bolognesi, A.; Polito, L. Activities associated with the presence of ribosome-inactivating proteins increase in senescent and stressed leaves. *FEBS Lett.* **1996**, *382*, 309–312. [CrossRef]
21. Gilabert-Oriol, R.; Weng, A.; Mallinckrodt, B.; Melzig, M.F.; Fuchs, H.; Thakur, M. Immunotoxins constructed with ribosome-inactivating proteins and their enhancers: A lethal cocktail with tumor specific efficacy. *Curr. Pharm. Des.* **2014**, *20*, 6584–6643. [CrossRef]

22. Citores, L.; Iglesias, R.; Ferreras, J.M. Antiviral Activity of Ribosome-Inactivating Proteins. *Toxins* **2021**, *13*, 80. [CrossRef] [PubMed]
23. Krishnan, R.; McDonald, K.A.; Dandekar, A.M.; Jackman, A.P.; Falk, B. Expression of recombinant trichosanthin, a ribosome-inactivating protein, in transgenic tobacco. *J. Biotechnol.* **2002**, *97*, 69–88. [CrossRef]
24. Iglesias, R.; Perez, Y.; de Torre, C.; Ferreras, J.M.; Antolin, P.; Jimenez, P.; Rojo, M.A.; Mendez, E.; Girbes, T. Molecular characterization and systemic induction of single-chain ribosome-inactivating proteins (RIPs) in sugar beet (Beta vulgaris) leaves. *J. Exp. Bot.* **2005**, *56*, 1675–1684. [CrossRef] [PubMed]
25. Bertholdo-Vargas, L.R.; Martins, J.N.; Bordin, D.; Salvador, M.; Schafer, A.E.; Barros, N.M.; Barbieri, L.; Stirpe, F.; Carlini, C.R. Type 1 ribosome-inactivating proteins-entomotoxic, oxidative and genotoxic action on *Anticarsia gemmatalis* (Hubner) and Spodoptera frugiperda (J.E. Smith) (Lepidoptera: Noctuidae). *J. Insect Physiol.* **2009**, *55*, 51–58. [CrossRef]
26. Di Massimo, A.M.; Di Loreto, M.; Pacilli, A.; Raucci, G.; D'Alatri, L.; Mele, A.; Bolognesi, A.; Polito, L.; Stirpe, F.; De Santis, R. Immunoconjugates made of an anti-EGF receptor monoclonal antibody and type 1 ribosome-inactivating proteins from *Saponaria ocymoides* or *Vaccaria pyramidata*. *Br. J. Cancer* **1997**, *75*, 822–828. [CrossRef] [PubMed]
27. Shi, W.W.; Mak, A.N.; Wong, K.B.; Shaw, P.C. Structures and Ribosomal Interaction of Ribosome-Inactivating Proteins. *Molecules* **2016**, *21*, 1588. [CrossRef]
28. Weng, A. A novel adenine-releasing assay for ribosome-inactivating proteins. *J. Chromatogr. B Anal. Technol. Biomed. Life Sci.* **2018**, *1072*, 300–304. [CrossRef]
29. Kokorin, A.; Weise, C.; Sama, S.; Weng, A. A new type 1 ribosome-inactivating protein from the seeds of Gypsophila elegans M.Bieb. *Phytochemistry* **2019**, *157*, 121–127. [CrossRef]
30. Legname, G.; Bellosta, P.; Gromo, G.; Modena, D.; Keen, J.N.; Roberts, L.M.; Lord, J.M. Nucleotide sequence of cDNA coding for dianthin 30, a ribosome inactivating protein from *Dianthus caryophyllus*. *Biochim. Biophys. Acta* **1991**, *1090*, 119–122. [CrossRef]
31. Maras, B.; Ippoliti, R.; De Luca, E.; Lendaro, E.; Bellelli, A.; Barra, D.; Bossa, F.; Brunori, M. The amino acid sequence of a ribosome-inactivating protein from Saponaria officinalis seeds. *Biochem. Int.* **1990**, *21*, 831–838.
32. Lee-Huang, S.; Kung, H.F.; Huang, P.L.; Huang, P.L.; Li, B.Q.; Huang, P.; Huang, H.I.; Chen, H.C. A new class of anti-HIV agents: GAP31, DAPs 30 and 32. *FEBS Lett.* **1991**, *291*, 139–144. [CrossRef]
33. Fermani, S.; Falini, G.; Ripamonti, A.; Polito, L.; Stirpe, F.; Bolognesi, A. The 1.4 anstroms structure of dianthin 30 indicates a role of surface potential at the active site of type 1 ribosome inactivating proteins. *J. Struct. Biol.* **2005**, *149*, 204–212. [CrossRef] [PubMed]
34. Savino, C.; Federici, L.; Ippoliti, R.; Lendaro, E.; Tsernoglou, D. The crystal structure of saporin SO6 from Saponaria officinalis and its interaction with the ribosome. *FEBS Lett.* **2000**, *470*, 239–243. [CrossRef]
35. Moolten, F.; Zajdel, S.; Cooperband, S. Immunotherapy of experimental animal tumors with antitumor antibodies conjugated to diphtheria toxin or ricin. *Ann. N. Y. Acad. Sci.* **1976**, *277*, 690–699. [CrossRef]
36. Barbieri, L.; Valbonesi, P.; Bonora, E.; Gorini, P.; Bolognesi, A.; Stirpe, F. Polynucleotide:adenosine glycosidase activity of ribosome-inactivating proteins: Effect on DNA, RNA and poly(A). *Nucleic Acids Res.* **1997**, *25*, 518–522. [CrossRef]
37. Barbieri, L.; Valbonesi, P.; Righi, F.; Zuccheri, G.; Monti, F.; Gorini, P.; Samori, B.; Stirpe, F. Polynucleotide:Adenosine glycosidase is the sole activity of ribosome-inactivating proteins on DNA. *J. Biochem.* **2000**, *128*, 883–889. [CrossRef]
38. Ferreras, J.M.; Barbieri, L.; Girbes, T.; Battelli, M.G.; Rojo, M.A.; Arias, F.J.; Rocher, M.A.; Soriano, F.; Mendez, E.; Stirpe, F. Distribution and properties of major ribosome-inactivating proteins (28 S rRNA N-glycosidases) of the plant *Saponaria officinalis* L. (Caryophyllaceae). *Biochim. Biophys. Acta* **1993**, *1216*, 31–42. [CrossRef]
39. Bolognesi, A.; Bortolotti, M.; Maiello, S.; Battelli, M.G.; Polito, L. Ribosome-Inactivating Proteins from Plants: A Historical Overview. *Molecules* **2016**, *21*, 1627. [CrossRef]
40. Sanchez, M.; Scire, A.; Tanfani, F.; Ausili, A. The thermal unfolding of the ribosome-inactivating protein saporin-S6 characterized by infrared spectroscopy. *Biochim. Biophys. Acta* **2015**, *1854*, 1357–1364. [CrossRef]
41. Scire, A.; Tanfani, F.; Ausili, A. A Spectroscopic Study on Secondary Structure and Thermal Unfolding of the Plant Toxin Gelonin Confirms Some Typical Structural Characteristics and Unravels the Sequence of Thermal Unfolding Events. *Toxins* **2019**, *11*, 483. [CrossRef]
42. Gilabert-Oriol, R.; Weng, A.; Trautner, A.; Weise, C.; Schmid, D.; Bhargava, C.; Niesler, N.; Wookey, P.J.; Fuchs, H.; Thakur, M. Combinatorial approach to increase efficacy of Cetuximab, Panitumumab and Trastuzumab by dianthin conjugation and co-application of SO1861. *Biochem. Pharm.* **2015**, *97*, 247–255. [CrossRef] [PubMed]
43. Neuhoff, V.A.; Arold, N.; Taube, D.; Erhardt, W. Improved staining of proteins in polyacrylamide gels including isoelectric focusing gels with clear background at nanogram sensitivity using Coomassie Brillant Blue G-250 and R-250. *Electrophoresis* **1988**, *9*, 255–262. [CrossRef] [PubMed]
44. Shevchenko, A.; Wilm, M.; Vorm, O.; Mann, M. Mass spectrometric sequencing of proteins silver-stained polyacrylamide gels. *Anal. Chem.* **1996**, *68*, 850–858. [CrossRef]
45. Suckau, D.; Resemann, A.; Schuerenberg, M.; Hufnagel, P.; Franzen, J.; Holle, A. A novel MALDI LIFT-TOF/TOF mass spectrometer for proteomics. *Anal. Bioanal. Chem.* **2003**, *376*, 952–965. [CrossRef] [PubMed]

toxins

MDPI

Article

Isolation, Characterization and Biological Action of Type-1 Ribosome-Inactivating Proteins from Tissues of *Salsola soda* L.

Nicola Landi [1,†], Sara Ragucci [1,†], Lucía Citores [2], Angela Clemente [1], Hafiza Z. F. Hussain [1], Rosario Iglesias [2], José M. Ferreras [2] and Antimo Di Maro [1,*]

[1] Department of Environmental, Biological and Pharmaceutical Sciences and Technologies (DiSTABiF), University of Campania Luigi Vanvitelli, Via Vivaldi 43, 81100 Caserta, Italy
[2] Department of Biochemistry and Molecular Biology and Physiology, Faculty of Sciences, University of Valladolid, E-47011 Valladolid, Spain
* Correspondence: antimo.dimaro@unicampania.it
† These authors contributed equally to this work.

Abstract: Ribosome-inactivating proteins (RIPs) are known as RNA N-glycosylases. They depurinate the major rRNA, damaging ribosomes and inhibiting protein synthesis. Here, new single-chain (type-1) RIPs named sodins were isolated from the seeds (five proteins), edible leaves (one protein) and roots (one protein) of *Salsola soda* L. Sodins are able to release Endo's fragment when incubated with rabbit and yeast ribosomes and inhibit protein synthesis in cell-free systems (IC_{50} = 4.83–79.31 pM). In addition, sodin 5, the major form isolated from seeds, as well as sodin eL and sodin R, isolated from edible leaves and roots, respectively, display polynucleotide:adenosine glycosylase activity and are cytotoxic towards the Hela and COLO 320 cell lines (IC_{50} = 0.41–1200 nM), inducing apoptosis. The further characterization of sodin 5 reveals that this enzyme shows a secondary structure similar to other type-1 RIPs and a higher melting temperature (Tm = 76.03 ± 0.30 °C) and is non-glycosylated, as other sodins are. Finally, we proved that sodin 5 possesses antifungal activity against *Penicillium digitatum*.

Keywords: antifungal activity; agretti; cytotoxicity; edible plants; protein purification; rRNA N-glycosylases

Key Contribution: Here, we reported the isolation and characterization of seven type-1 RIPs named sodins from *Salsola soda*, known for its edible leaves (agretti in Italian). Furthermore, we focused our attention on their biological and antifungal activities.

Citation: Landi, N.; Ragucci, S.; Citores, L.; Clemente, A.; Hussain, H.Z.F.; Iglesias, R.; Ferreras, J.M.; Di Maro, A. Isolation, Characterization and Biological Action of Type-1 Ribosome-Inactivating Proteins from Tissues of *Salsola soda* L.. *Toxins* **2022**, *14*, 566. https://doi.org/10.3390/toxins14080566

Received: 22 July 2022
Accepted: 17 August 2022
Published: 19 August 2022

1. Introduction

Ribosome-inactivating proteins (RIPs) are a group of ribotoxic enzymes which act on ribosomes in a highly specific and irreversible manner. They are N-glycosylases (EC 3.2.2.22) capable of hydrolyzing the N-glycosidic bond of a specific adenosine in the sarcin ricin loop (SRL) of major rRNA (A_{4324}, rat liver 28S rRNA numbering) [1]. The consequent formation of an apurinic site prevents the interaction of prokaryotic or eukaryotic elongation factors (EF-G or EF-2, respectively) with ribosomes, blocking mRNA-tRNA translocation and thus inhibiting protein synthesis and inducing the apoptotic pathway [2]. In addition, these enzymes possess polynucleotide:adenosine glycosylase (PNAG) activity on different polynucleotide substrates (e.g., viral RNA and herring sperm DNA) [3–5]. On the other hand, other enzymatic activities such as DNAse [6,7], RNAse [8], chitinase [9], phosphatase, lipase [10] and superoxide dismutase properties [11–13] have also been attributed to RIPs, although some authors ascribe these activities to possible contamination [4,14]. RIPs are mostly found in flowering plants [15,16], few are found in fungi [17] and bacteria [18] and one is found in alga [19].

Based on the presence or absence of a quaternary structure, there are two main groups of RIPs: single-chain proteins (type-1 RIPs), with a molecular weight of ~30 kDa and basic pI, and two-chain proteins (type-2 RIPs), with a molecular weight of ~60 kDa and neutral pI. The latter consist of an enzymatic active A-chain linked through a disulphide bridge to a lectinic B-chain, which allows for the entry into the cell. For this reason, type-2 RIPs are more toxic with respect to type-1 RIPs. In addition, a third group of non-canonical single-chain RIPs (type-3 RIPs) was found only in Poaceae, including JIP60 isolated from barley [20,21] and b-32 isolated from maize [22], made of a type-1-like N-terminal domain with N-glycosylase activity, covalently linked to a C-terminal domain with an unknown function [23].

Although their physiological role is still unknown, RIPs have a broad spectrum of activities, including antiviral, antibacterial and antifungal action, as well as anticancer properties [24–26]. Thus, the potential applications of RIPs span many fields, from agriculture (e.g., toxicity to pests and antifungal activity) [27] to biomedicine for the construction of antibody-RIPs conjugates (i.e., immunotoxins) against target cancer cells [28]. From the foregoing, it is clear that continuing the study of RIPs distribution in higher plants (including edible species) can contribute to expanding the availability on novel potential biotechnological and pharmacological tools.

Salsola soda L., commonly known as barilla plant or 'agretti' in Italy, is an annual, edible halophytic plant that is widespread in south Europe, mostly near the coast. In the past, the plant was used for the production of an impure sodium carbonate named 'barilla' from the sodium chloride in the soil (e.g., to make soap and glass) [29]. The plant tissues are rich in alkaloids, saponins, coumarins and sterols, as well as flavonoids with anti-inflammatory and antidiabetic potential [30]. According to the taxonomy, *S. soda* belongs to Caryophyllales [31], a plant order known as a source of RIPs [32]. The plant is an Amaranthacea with succulent leaves, small sessile hermaphrodite flowers and indehiscent fruits. It is known that the isoforms of these ribonucleolytic enzymes can be found in different tissues of the same plant [33–35] or in the same plant tissue [36]. However, little information about the distribution of RIPs in plant organs and tissues is available in the literature, especially with regard to edible plants, considering that they are often consumed raw [37].

In this framework, we report the purification and characterization of seven novel RIPs named sodins: five from the seeds, one from the roots and one from the edible leaves of *S. soda*. We describe the distribution of both N-glycosylase and PNAG activity in the different tissues of *S. soda*. Moreover, considering the high expression of sodins in the seeds, we further characterized the major form (i.e., sodin 5) by evaluating some structural features, the cytotoxic effect in cancer cell lines and the antifungal activity against *Penicillium digitatum*.

2. Results and Discussion

2.1. Purification of Type-1 RIPs from Salsola soda Seeds

The crude extract from *S. soda* seeds, obtained by homogenizing seeds (50 g) in 500 mL of phosphate-buffered saline, pH 7.2, possessed the ability to depurinate the hsDNA substrate (data not shown) [38]. Thus, to ascertain the presence of type-1 RIPs, the total crude extract was subjected to acid precipitation following a protocol for the extraction of basic proteins [34]. The supernatant was subjected to step-wise cation exchange chromatography, and eluted basic proteins were separated by gel-filtration. Pooled fractions with a molecular weight around 29 kDa were further subjected to cation exchange chromatography on the CM-Sepharose column, eluted with a linear NaCl gradient (0–0.17 M) in 5 mM Na-phosphate buffer. As shown in Figure 1A, five protein peaks with PNAG activity were detected, confirming the presence of various PNAG-enzymes (isoforms), which is a common feature in plant RIP expression [15].

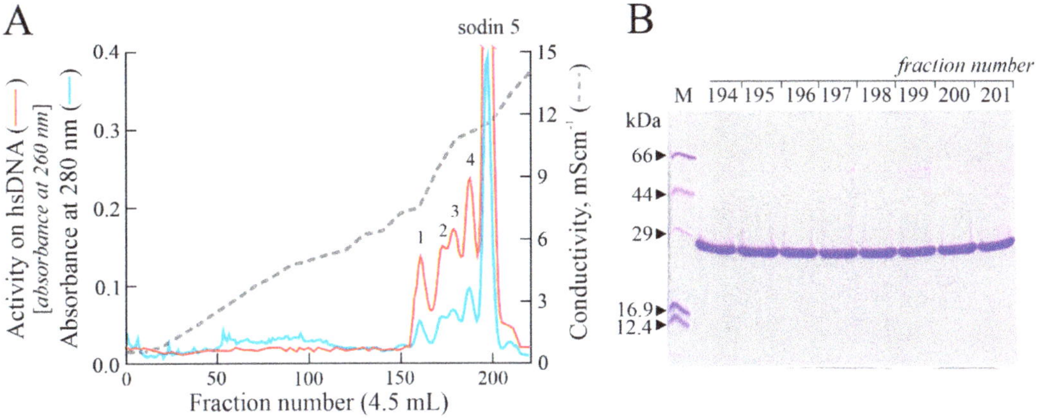

Figure 1. (**A**) Elution profile after cation exchange chromatography on the CM-Sepharose column, showing five peaks (peaks 1–4 and sodin 5) with PNAG activity (arbitrary units). (**B**) SDS-PAGE analysis of 194–201 fractions (5.0 µg) from sodin 5 obtained after cation exchange chromatography (**A**). M, molecular weight markers. SDS-PAGE in the presence of β-mercaptoethanol was carried out in 12% polyacrylamide separating gel and then stained with Coomassie brilliant blue.

In this framework, we decided to first characterize the main protein peak, eluted at a higher ionic strength (peak 5). In particular, the fractions from 194 to 201, corresponding to the main peak (hereafter, sodin 5), were evaluated by SDS-PAGE, showing a single protein band with an electrophoretic migration of ~29 kDa (Figure 1B). Thus, the fractions were pooled, dialyzed and used for further enzymatic and structural characterization.

The purification yield obtained by this procedure was of about 2.9 ± 0.15 mg/100 g of seeds for sodin 5.

2.2. Enzymatic and Structural Features of Sodin 5

In order to ascertain that the in vitro protein synthesis inhibition and the PNAG activity of sodin 5 were due to N-β-glycosylase action (characteristic enzymatic hallmark of RIPs from plants), we tested the β-fragment release by incubating the protein with rabbit ribosomes (Endo's assay) [32]. As shown in Figure 2A, sodin 5 is able to deadenylate rRNA from rabbit reticulocyte lysate, releasing the β-fragment after aniline treatment. Furthermore, as demonstrated by the PNAG activity assay, sodin 5 is ~1.8-fold less active than quinoin, a type-1 RIP isolated from *Chenopodium quinoa* seeds and used as a reference PNAG-enzyme (Figure 2B) [39].

In addition, sodin 5 purified from *S. soda* seeds inhibited protein synthesis in a rabbit reticulocyte lysate system, with an IC_{50} value of 4.83 pM (0.14 ng/mL). This value is similar to that of quinoin (IC_{50}= 5.08 pM; 0.15 ng/mL) and ~7.7-fold lower than that of saporin S6 (IC_{50} = 37 pM; 1.09 ng/mL) type-1 RIP isolated from *Saponaria officinalis* seeds [40]. On the other hand, the IC_{50} value of sodin 5 is of keen interest considering that type-1 RIPs have values of IC_{50} between 10 and 4000 pM [41].

The high amount of sodin 5 obtained from seeds of *S. soda* allowed us to perform a study on its secondary structure by Circular Dichroism (CD-) analysis. The far UV CD-spectrum of sodin 5 suggested that the periodic secondary structure of the protein is partially dominated by the α-helix, with a predicted percentage of ~30% (~25% β-strand) (Figure 3A). Therefore, these data show that sodin 5 has a secondary structure content similar to other RIPs, sharing a common 3D fold (RIP fold) consisting of a β-sheet N-terminal domain and an α-helix-rich C-terminal domain [42,43]. Subsequently, the thermal denaturation curve of sodin 5 was obtained using UV-spectroscopy by measuring the increment of absorbance at 278 nm, increasing the temperature. The melting temperature

(Tm) of sodin 5 was 76.03 ± 0.30 °C (Figure 3B). The thermal unfolding curve at pH 7.2 shows that this type-1 RIP is a highly stable protein. In particular, the Tm value of sodin 5 is higher than that of both quinoin (68.2 °C [39]) and saporin S6 (58.0 °C) [44]. Both quinoin and saporin S6 have been isolated from Caryophyllales, the same plant order of *S. soda*.

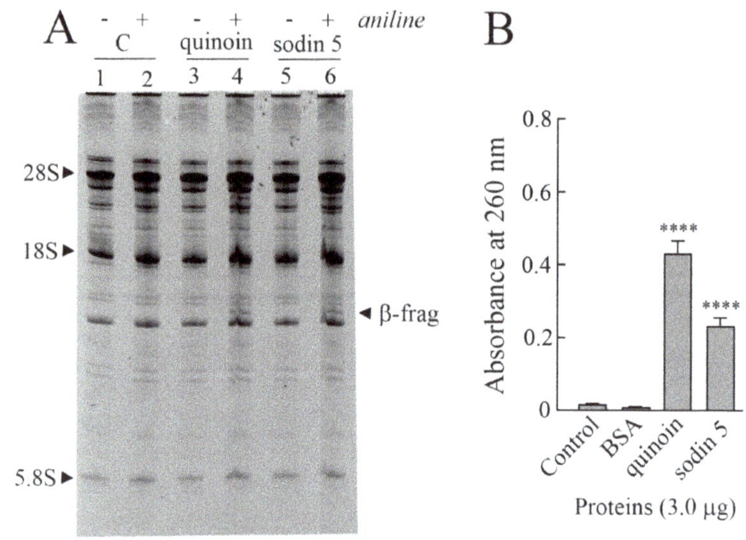

Figure 2. (**A**) rRNA N-glycosylase activity on rabbit ribosomes. Quinoin from *C. quinoa* seeds (3.0 μg; lanes 3 and 4) as a positive control and sodin 5 (3.0 μg; lanes 5 and 6) were incubated with ribosomes. Then, rRNA was extracted, treated with acid aniline and separated as described in the Materials and Methods section. (+) and (−) indicate with and without aniline treatment. 'β-frag' indicates the position of Endo's fragment released by the aniline treatment of rRNA from rabbit ribosomes. (**B**) Polynucleotide:adenosine glycosylase activity of BSA (negative control) or quinoin and sodin 5 type-1 RIPs. Proteins (3.0 μg) were assayed on salmon sperm DNA as described in the Materials and Methods section. The mean results ± SD of three experiments performed in triplicate are reported. The data were compared to the control and analyzed by one-way ANOVA with Dunnett's post hoc test (****, $p < 0.0001$).

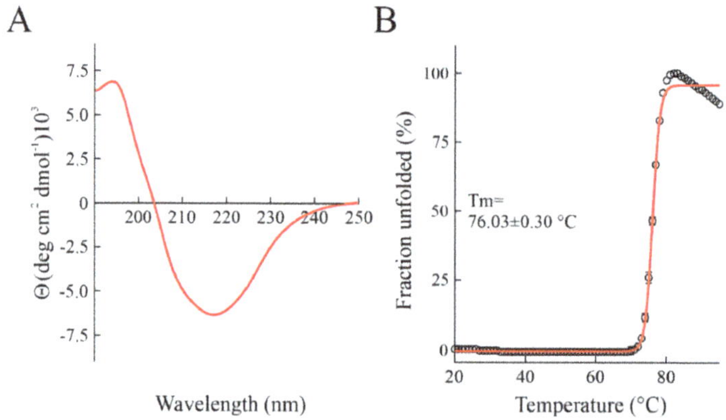

Figure 3. (**A**) Far-UV CD spectrum of sodin 5. (**B**) Thermal denaturation profile of sodin 5 (concentration: 0.15 mg mL^{-1}). The fraction unfolded at 278 nm is plotted as a function of temperature. The red line represents fit curve.

Finally, considering that several type-1 RIPs are N-glycosylated, a specific analysis for glycoproteins detection was carried out. When analyzed by SDS-PAGE and sugar staining, sodin 5 appeared to be non-glycosylated (Figure S1).

2.3. *Minor Forms of Type-1 RIPs from Salsola soda Seeds*

To obtain information on minor peaks 1–4 (Figure 1A), showing PNAG activity and eluted at a lower ionic strength with respect to sodin 5, the fractions of each protein peak were analytically re-chromatographed by FPLC on an AKTA Purifier System from cation exchange chromatography using a Source 15S PE 4.6/100 column (Figure S2A). Each eluted peak displayed a single band of ~29 kDa by SDS-PAGE analysis (Figure S1A).

The pooled fractions of peaks 1–4, (hereafter sodins 1–4) were able to release the β-fragment similarly to sodin 5 as a consequence of the RIPs action (Figure S3). Therefore, sodins 1–4, with a molecular weight of ~29 kDa and specific N-β-glycosylase activity, can be considered type-1 RIPs. Moreover, among type-1 RIPs from *S. soda* seeds, sodin 1 displayed the higher PNAG activity (~2.3-fold more active than sodin 5), while sodins 2, 3 and 4 are, respectively, ~1.3-, 2.0- and 1.8-fold less active than sodin 5 (Figure S2B). These data agreed with the documented different ability of type-1 RIPs to deadenylate nucleic acid substrates [3].

In addition, the specific analysis for glycoproteins detection after SDS-PAGE was carried out. Similar to sodin 5, the analysis shows that these enzymes are non-glycosylated (Figure S1B).

The purification yield obtained by this procedure was of about 0.37 ± 0.01, 0.45 ± 0.02, 0.47 ± 0.01 and 0.67 ± 0.02 mg/100 g of seeds for sodins 1–4, respectively.

2.4. *Type-1 RIPs from Edible Leaves and Roots of Salsola soda*

In order to evaluate the RIPs distribution in the tissues of *S. soda*, the same protocol used for the purification of sodins from *S. soda* seeds was also applied on the edible leaves and roots of this plant. The approach, coupled with the detection of enzymatic activity, allowed for the purification of two different type-1 RIPs from *S. soda* edible leaves and roots, respectively. However, considering the lower number of raw basic proteins, after cation step-wise chromatography and gel-filtration (see Section 4.2), the pools of basic proteins from the edible leaves and roots with a molecular weight of 29 kDa were separately subjected to analytical cation exchange chromatography using a Source 15S PE 4.6/100 column on AKTA Purifier System (Figure 4A).

The single protein peaks from edible leaves and roots were analyzed by SDS-PAGE to verify the purity and integrity. As reported in Figure 4B, both protein peaks showed the presence of a single band with an electrophoretic migration of ~29 kDa. These two-novel type-1 RIPs from edible leaves and roots were named sodin eL and sodin R, respectively.

The purification yield obtained by this procedure was of about 17.5 ± 0.61 μg/100 g of edible leaves and 27.8 ± 0.87 μg/100 g of roots for sodin eL and sodin R, respectively, confirming the low number of type-1 RIPs in edible leaves and roots compared to the quantity found in seeds.

In addition, the specific analysis for glycoproteins detection after SDS-PAGE shows that these enzymes are non-glycosylated, like both sodin 5 and sodins 1–4 from *S. soda* seeds (Figure S4).

Finally, the N-β-glycosylase action (characteristic enzymatic hallmark of RIPs from plants) of sodin eL and sodin R has been tested. As shown in Figure 5A, both proteins release the β-fragment by incubating the protein with rabbit ribosomes (Endo's assay) following aniline treatment, similarly to sodin 5 from *S. soda* seeds. In addition, type-1 RIPs from *S. soda* roots and edible leaves also displayed PNAG activity. In particular, as shown in Figure 5B, sodin eL and sodin R displayed a PNAG activity that was ~2.2- and 2.9-fold higher than that of sodin 5.

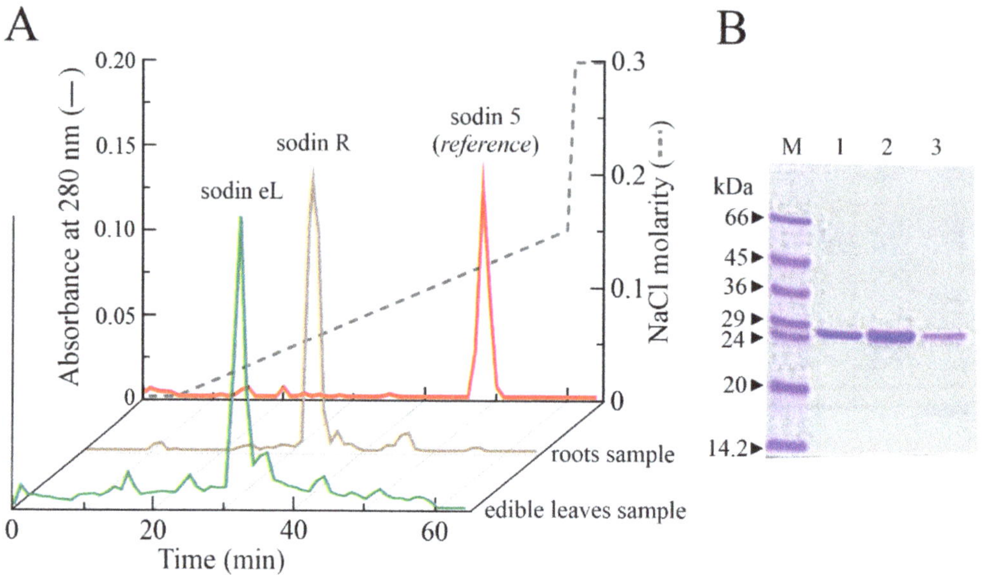

Figure 4. (**A**) Elution profile after FPLC on an AKTA Purifier System from cation exchange chromatography using a Source 15S PE 4.6/100 column, showing a single protein peak from *S. soda* roots and edible leaves, named sodin R and sodin eL, respectively. The elution profile of sodin 5 from *S. soda* seeds was reported as a reference chromatographic profile. (**B**) SDS-PAGE analysis of fractions (5.0 µg) from sodin 5, sodin R and sodin eL (lanes 1, 2 and 3, respectively), obtained after Source 15S chromatography. M, molecular weight markers. SDS-PAGE in the presence of β-mercaptoethanol was carried out in 12% polyacrylamide separating gel and then stained with Coomassie brilliant blue.

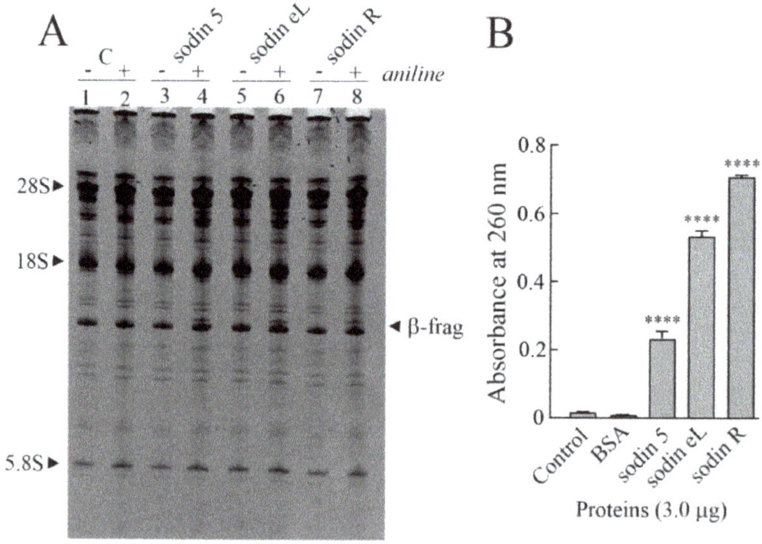

Figure 5. (**A**) rRNA N-glycosylase activity assayed on rabbit ribosomes. Sodin 5 (3.0 µg; lanes 3 and 4) as a positive control and sodin eL (3.0 µg; lanes 5 and 6) or sodin R (3.0 µg; lanes 7 and 8) were incubated with ribosomes. Then, rRNA was extracted, treated with acid aniline and separated as

described in the Materials and Methods section. (+) and (−) indicate with and without aniline treatment. 'β-frag' indicates the position of Endo's fragment released by the aniline treatment of rRNA from rabbit ribosomes. (**B**) Polynucleotide:adenosine glycosylase activity of BSA (negative control) or sodin 5, sodin eL and sodin R type-1 RIPs. Proteins (3.0 μg) were assayed on salmon sperm DNA as described in the Materials and Methods section. The mean results ± SD of three experiments performed in triplicate are reported. Data were compared to the control and analyzed by one-way ANOVA with Dunnett's post hoc test (****, $p < 0.0001$).

In addition, since the rRNA N-glycosylase activity might play a role in plant defense, we assayed the effect of sodin 5, sodin eL and sodin R, as well as quinoin, on ribosomes from yeasts (*Saccharomyces cerevisiae*) homologous to ribosomes from the putative fungal pathogens of plants. As shown in Figure 6, these RIPs displayed rRNA N-glycosylase activity on yeast ribosomes, as indicated by the release of a diagnostic β-fragment identical to that reported for BE27 from *Beta vulgaris* L. (sugar beet) and type-1 RIPs from *Phytolacca dioica* L. [45,46].

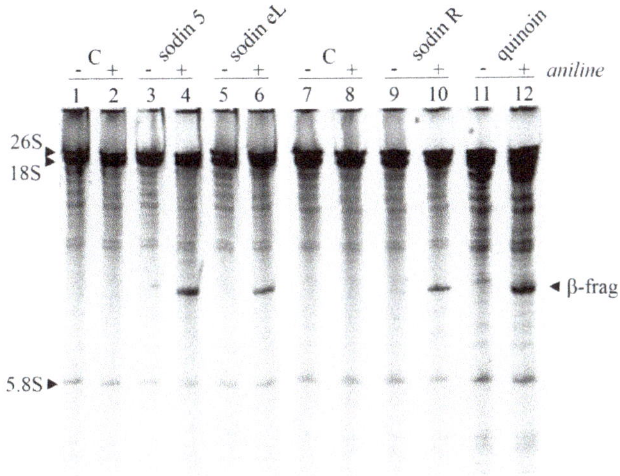

Figure 6. rRNA N-glycosylase activity assayed on yeast ribosomes. rRNA N-glycosylase activity was analyzed as reported in the Materials and Methods section. Each lane contained 5 μg of RNA isolated from either untreated (control) or RIP-treated ribosomes from yeast. (+) and (−) indicate with and without aniline treatment. 'β-frag' indicates the position of Endo's fragment released by the aniline treatment of rRNA from yeast ribosomes.

Finally, sodin eL and sodin R inhibited protein synthesis in a rabbit reticulocyte lysate system, with IC_{50} values of 79.31 pM (2.3 ng/mL) and 65.52 pM (1.9 ng/mL), respectively. These values are ~15-fold higher with respect to the IC_{50} of sodin 5 isolated from *S. soda* seeds.

2.5. Cytotoxic Effects of Sodins from S. soda Tissues in Cell Cultures

RIPs are cytotoxic toward several human cell lines (normal and malignant), although type-1 RIPs are usually less cytotoxic than type-2 RIPs due to the lack of a B-chain, which improves the entry of the A-chain in the cells. Indeed, typical IC_{50} values of toxic type-2 RIPs on cultured animal cells are in the range of 0.3–17,000 pM, while IC_{50} values of type-1 RIPs are in the range of 170–3300 nM [41]. Thus, we decided to verify the cytotoxic effects of sodins or quinoin on both HeLa and COLO 320 cell lines.

Table 1 lists the IC_{50} values (concentration of protein causing the death of 50% of cells) of sodin 5, sodin eL and sodin R from *S. soda* seeds, edible leaves and roots, respectively, compared with the IC_{50} of quinoin. Type-1 RIPs from both *S. soda* tissues and *C. quinoa* seeds were toxic to HeLa and COLO 320 cells, exhibiting IC_{50} values ranging from 0.41 to 1200 nM.

Table 1. Cytotoxicity of sodins and quinoin. HeLa or COLO 320 cells were grown in RPMI 1640 medium and incubated with different RIP concentrations for 48 or 72 h, and cell viability was evaluated by a colorimetric assay, as indicated in Section 4.6 of the Materials and Methods section. When reported, cells were pre-treated with Z-VAD for 3 h (see Materials and Methods) and then incubated with different RIP concentrations. Data represent the mean of three experiments performed in triplicate.

Type-1 RIP	HeLa 48 h		COLO 48 h	COLO 72 h
	–	+Z-VAD Pretreatment	–	
quinoin	1.9×10^{-9}	1.0×10^{-7}	1.0×10^{-6}	3.9×10^{-7}
sodin 5	2.0×10^{-9}	2.5×10^{-7}	1.2×10^{-6}	3.3×10^{-7}
sodin eL	1.3×10^{-9}	–	–	$>1.2 \times 10^{-7}$
sodin R	4.1×10^{-10}	–	–	1.6×10^{-7}

The most sensitive were HeLa cells, with IC_{50} values from 0.41 to 2.0 nM after 48 h of treatment, while COLO 320 cells have values between 160 - >120 nM after 72 h of treatment (Figure 7A). These data agree with those previously reported for other type-1 RIPs, such as type-1 RIPs isolated from *P. dioica*, which exhibit IC_{50} values ranging from 1.0 to 1000 nM against the same cell lines [45]. On the other hand, the cytotoxicity of sodin 5 and quinoin is similar, with IC_{50} values 10^3-fold higher for COLO 320 cells with respect to Hela cells.

There are important differences in toxicity among type-1 RIPs based on their capability to reach the ribosomes of target cells. Based on the above studies, sodins and quinoin display a considerable toxicity against HeLa cells for type-1 RIPs. To see if sodin 5 and sodin R can reach and inactivate the ribosomes after being endocytosed, we analyzed the rRNA from HeLa cells after 48 h of RIP treatment. Figure 7B displayed that the ribosomes were depurinated, as proved by the detection of a diagnostic β-fragment following RNA treatment with acid aniline. Thus, both sodin 5 and sodin R can reach the ribosomes of target cells, inhibiting protein synthesis.

Several studies highlight that RIP cytotoxicity in the cells is associated with their ability to induce apoptosis [12]. Apoptosis might be a consequence of the ribotoxic stress induced by the RIP after entry into the cytosol, or both processes could run in parallel. Apoptosis is characterized by cell shrinkage, nuclear condensation, changes in the cell membrane and mitochondria, DNA fragmentation into 200 base oligomers and protein degradation by caspases. In this framework, in order to ascertain if the observed cytotoxic effects of both sodin 5 and quinoin were mediated via apoptosis, we evaluated the sensitivity to the pan-caspase inhibitor Z-VAD or the cleavage of chromosomal DNA into oligonucleosomal fragments, considered a late-stage apoptosis hallmark. In particular, HeLa cells were pre-treated and maintained in 100 μM Z-VAD for 48 h, and the cell viability was determined for different sodin 5 and quinoin concentrations. As shown in Figure 7A, the presence of Z-VAD improved cell survival. In particular, in the presence of Z-VAD, viability increased from 14 to 49% in 4.3×10^{-7} M sodin 5-treated cells and from 14 to 46% in 5.7×10^{-7} M quinoin-treated cells. On the other hand, when COLO 320 cells were treated for 72 h with RIP concentrations close to their IC_{50}, the breakdown of the nuclear DNA into oligonucleosomal fragments was clearly observed (Figure 7C).

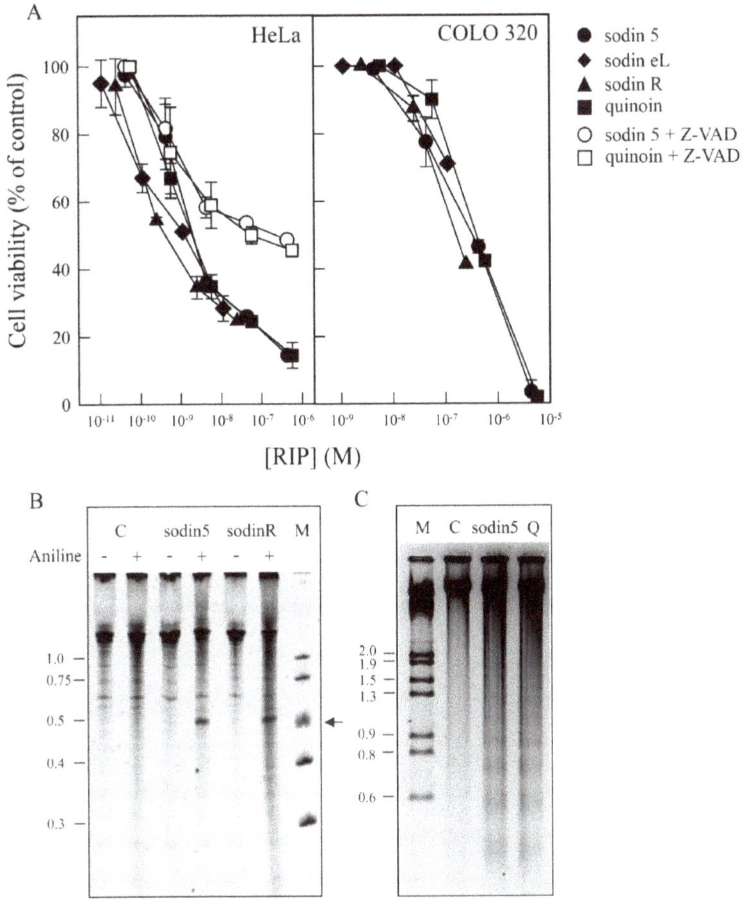

Figure 7. Induction of cytotoxicity and apoptosis on HeLa and COLO 320 cells by sodins and quinoin (**A**). Effect of sodins or quinoin on the viability of HeLa (left panel) and COLO 320 (right panel) cells. Cells were grown in RPMI 1640 medium and incubated with different type-1 RIP concentrations for 48 h (HeLa) and 72 h (COLO 320), and cell viability was evaluated by a colorimetric assay, as indicated in Section 4.6 of the Materials and Methods section. To investigate the effect of Z-VAD on the viability of HeLa cells, the cells were preincubated for 3 h with Z-VAD and then incubated with different concentrations of sodin 5 or quinoin for 48 h, and cell viability was evaluated. Data represent the mean ± SD of two experiments performed in duplicate. (**B**) rRNA N-glycosylase activity of sodin 5 and sodin R on RNA from HeLa cells. rRNA N-glycosylase activity was evaluated as reported in the Materials and Methods section. Each lane contained 2.0 μg of RNA isolated from either untreated cells (C, control) or cells incubated with 8 nM of sodin 5 or 5 nM of sodin R for 48 h. The arrow indicates the RNA fragment released as a result of RIP action upon the acid aniline treatment. Numbers indicate the size of the standards (M) in nucleotides. (**C**) Effect of sodin 5 and quinoin on internucleosomal DNA fragmentation. COLO 320 cells were incubated in the absence (C, control) or presence of 0.4 μM of sodin 5 or 0.6 μM of quinoin (Q) for 72 h. The DNA was isolated, and 4.0 μg was electrophoresed, as indicated in Section 4.7. The numbers indicate the corresponding size of the standards (M) (λDNA HindIII/EcoRI) in Kb. (+) and (−) indicate with and without aniline treatment.

Overall, our data suggested that the apoptotic pathway was implicated in the cell death mediated by sodin 5 and quinoin, as already proved for other type-1 RIPs [41].

2.6. Effect of Sodin 5 and Quinoin on the Growth of P. digitatum

A potential role for RIPs as plant defense proteins has been proposed based on their enzymatic activity, which can act by either inactivating pathogen ribosomes or their own ribosomes, causing cell death [27]. Antifungal activity has been attributed to several RIPs. In particular, a strong antifungal activity against *P. digitatum* has been described for the apoplastic type-1 RIP beetin 27 (BE27) from sugar beet and for the type-1 RIPs PD-S2 and dioicin 2 from *P. dioica* [45,47]. *P. digitatum* is a necrotrophic fungus responsible for the postharvest decay of citrus, an economically important crop worldwide. Therefore, we carried out experiments to evaluate the effects of sodin 5 from *S. soda* seeds and quinoin from *C. quinoa* seeds on the growth of *P. digitatum*. Thus, conidia of *P. digitatum* were grown in PDB medium for 24 h before exposure to different RIP concentrations, continuing the treatment for a further 46 h. As shown in Figure 8, sodin 5 and quinoin reduced the fungal growth in a concentration-dependent manner. Both RIPs induced a strong decrease in the growth at 40 µg/mL. Thus, 40, 10 and 4.0 µg/mL of sodin 5 resulted in 70%, 34% and 6% growth inhibition, respectively, after 70 h of growth. Similar results were obtained for quinoin, with 61%, 27% and 10% growth inhibition at the same concentrations. Sodin 5 and quinoin added from the beginning to the conidia as starting material inhibited fungal growth to the same extent as RIPs added at 24 h (once conidial germination occurred; data not shown), suggesting that sodin 5 and quinoin affect mycelial growth rather than conidial germination. As shown in Figures 2 and 6, sodin 5 and quinoin exhibited rRNA N-glycosylase activities against mammalian and fungal ribosomes, respectively. Thus, the antifungal activities of sodin 5 and quinoin against *P. digitatum* could be mediated by the inhibition of protein synthesis together with the ability to cross the membrane and enter into the fungal cells, as has been postulated for other type-1 RIPs [45,47].

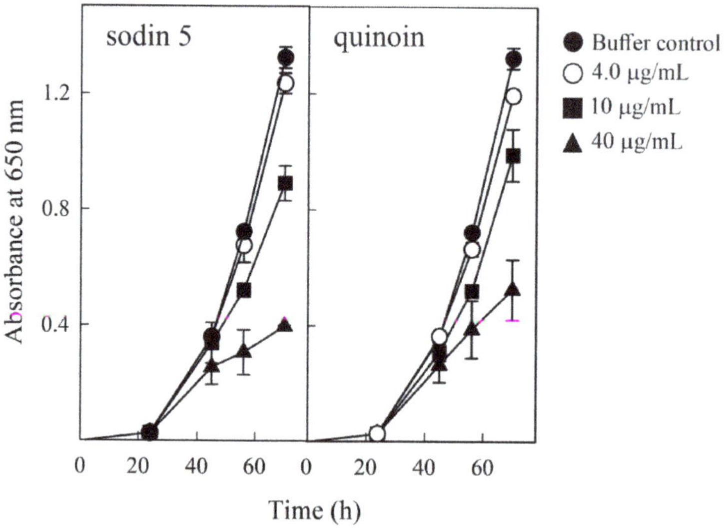

Figure 8. Antifungal activity of sodin 5 (left panel) and quinoin (right panel) against *Penicillium digitatum*, measured in a microtiter plate bioassay. Conidia of *P. digitatum* were grown in Potato Dextrose Broth (PDB) for 24 h before exposure to different RIP concentrations. Fungal growth was followed for 70 h and measured as an increase in absorbance at 650 nm. The curves represent the buffer control or different amounts (µg/mL) of both toxins. The mean results ± SE of two experiments performed in triplicate are reported.

3. Conclusions

In conclusion, we have isolated seven type-1 RIPs from the different tissues of *S. soda* ('agretti' in Italian): five type-1 RIPs from seeds (sodins), one from edible leaves (sodin eL) and one from roots (sodin R). All these enzymes are able to release the β-fragment following incubation with rabbit or yeast ribosome and exhibit PNAG activity.

Sodin 5, the major form expressed in seeds (2.9 ± 0.15 mg/100 g of seeds), with respect to other type-1 RIPs from *S. soda* tissues, exhibits an α+β structure typical of type-1 RIPs with a high melting temperature (Tm = 76.03 ± 0.30 °C) and is non-glycosylated, as the other six sodins. Furthermore, sodin 5, sodin eL and sodin R show cytotoxic effects towards the HeLa and COLO 320 cell lines, inducing apoptosis. In addition, since fungi are among the most important plant pathogens, we tested the antifungal properties of both sodin 5 and quinoin against *P. digitatum*, finding that both RIPs possess concentration-dependent antifungal activity.

Overall, this research aims to revisit RIPs in edible plants in light of their possible use as antiviral, antifungal and antipathogenic tools in agri-food, overcoming the preconception about transgenic plants, as these enzymes are physiologically present in edible plants.

4. Materials and Methods

4.1. Materials

The chemicals for chromatography were previously reported [39,48,49]. Single-stranded salmon sperm DNA was obtained from Sigma-Aldrich (St. Louis, MO, USA). Quinoin from the seeds of *C. quinoa* and PD-L4 from the leaves of *P. dioica* were isolated as previously reported [34,39]. The nuclease-treated rabbit reticulocyte lysate system was purchased from Promega (Madison, WI, USA).

The medium and the other chemicals were from Sigma Chemical Co. (St. Louis, MO, USA). The RPMI 1640 medium, fetal bovine serum (FBS), penicillin, streptomycin and trypsin were purchased from GIBCO BRL (Barcelona, Spain). The Z-VAD-fmk (pan-caspase inhibitor carbobenzoxy-valyl-alanyl-aspartyl-[O-methyl]-fluoromethylketone) named Z-VAD was purchased from R&D Systems (Abingdon, UK).

Buffer A: 5 mM Na-phosphate, pH 7.2, containing 0.14 M NaCl; buffer B: 10 mM Na-acetate, pH 4.0; and buffer C: 5 mM Na-phosphate, pH 7.2.

4.2. Purification of Type-1 RIPs from Seeds, Roots and Edible Leaves of S. soda

Type-1 RIPs from *S. soda* were purified with the same protocol used for quinoin, type-1 RIP from *C. quinoa* seeds, as reported by Landi et al., 2021 [39]. Briefly, the crude extract in buffer A was first subjected to acid precipitation at pH 4.0 and cation step-wise chromatography using a SP-Streamline resin [column L × I.D. 20 cm × 30 mm, flow rate 3.0 mL/min; Cytiva, Buccinasco (MI) Italy]. Subsequently, the basic proteins, eluted with 1.0 M NaCl in buffer C, were gel-filtrated [HiLoad® 26/60 Superdex® column L × I.D. 60 cm × 26 mm, flow rate 2.5 mL/min (range 100–10 kDa); Cytiva] to separate the proteins by molecular weight, and then, basic proteins with a molecular weight of about 29 kDa were subjected to cation exchange chromatography on CM-Sepharose fast flow (Cytiva; column L × I.D. 25 cm × 16 mm) equilibrated in buffer C and eluted with a NaCl gradient up to 0.17 M (buffer C, 500 mL, buffer C containing 0.17 M NaCl, 500 mL; total volume 1 L using a peristaltic pump).

However, when the number of basic proteins after gel-filtration was lower (less than 200 μg), CM-Sepharose chromatography was replaced by FPLC on an AKTA Purifier System (Amersham Pharmacia; Milan, Italy) using a Source 15S PE 4.6/100 column, equilibrated in buffer C and eluted by a linear gradient from 0 to 50% of buffer C containing NaCl 0.3 M over 60 min (flow rate 1.0 mL/min). The same chromatographic step (Source 15S column) was carried out for minor forms of type-1 RIPs from the seeds of *S. soda* after cation exchange chromatography on the CM-Sepharose column.

4.3. Enzymatic Assays

4.3.1. rRNA N-Glycosylase Activity of RIPs on Rabbit Ribosomes

The rRNA N-glycosylase assay was conducted as previously described [45]. Rabbit reticulocytes lysate (40 μL) was incubated with RIP (3.0 μg) at 37 °C for 1 h. After treatment, the RNA was extracted by phenolization, treated with 1 M aniline acetate (pH 4.5) and precipitated with cold ethanol. Purified RNA was analyzed by polyacrylamide gel in denaturing conditions [7 M urea/5% acrylamide (*w*/*v*)] and stained with ethidium bromide.

4.3.2. rRNA N-Glycosylase Activity of RIPs on Yeast Ribosomes

The preparation of the 30,000 g (S30) supernatants from yeast was performed as described elsewhere [46]. The rRNA N-glycosylase activity was assayed in 50 μL samples of S30 supernatant from yeast, which was incubated with 5.0 μg of sodin 5, 0.7 μg of sodin eL, 1.5 μg of sodin R or 5.0 μg of quinoin for 1 h at 30 °C. After treatment, the RNA was extracted with phenol and treated with aniline for 10 min at 23 °C. The RNA samples were separated on a polyacrylamide gel in denaturing conditions [7 M urea/5% acrylamide (*w*/*v*)] and stained with Gel Red nucleic acid staining [50].

4.3.3. Polynucleotide: Adenosine Glycosylase Activity on Salmon Sperm DNA

The adenine release was measured as previously reported [45], incubating salmon sperm DNA (10 μg) with RIPs (3.0 μg) in 300 μL 50 mM magnesium acetate (pH 4.0) containing 100 mM KCl, at 30 °C for 1 h. After incubation, the DNA was precipitated with cold ethanol and centrifuged. Adenine release was determined spectrophotometrically, reading the supernatant at 260 nm. On the other hand, to evaluate arbitrary units of PNAG activity on single fractions from *S. soda* seeds after CM-Sepharose chromatography, an equal volume was tested.

4.3.4. Cell-Free Protein Synthesis Inhibition

The effect of RIPs on protein synthesis was determined through a coupled transcription-translation in vitro assay using a rabbit reticulocytes lysate system, as described elsewhere [51]. Samples of RIPs were diluted and added to the reaction mixture as previously described [51]. Three experiments were conducted in duplicate, and IC_{50} (concentration that inhibits 50% protein synthesis) values were calculated by linear regression.

4.4. Analytical Procedures

The proteins' homogeneity was evaluated by SDS-PAGE with a Mini-Protean II (Bio-Rad; Milan, Italy) using a 6% stacking and 12% separating polyacrylamide gel under reducing conditions; a precision plus protein kit (Bio-Rad) was used as the reference proteins. The protein concentration was determined by a Pierce BCA Protein Assay Kit (Life Technologies Italia Fil., Monza, Italy). The glycosylation analysis was performed in gel after SDS-PAGE by using the Pro-Q™ Emerald 300 Glycoprot Probes Kombo (Life Technologies Italia). Glycosylated proteins were visualized by a ChemiDocTM XRS system.

4.5. Circular Dichroism and Thermal Stability Determination

The far-UV CD spectrum of sodin 5 was determined at 25 °C on a Jasco J-815 dichrograph [Jasco Europe, Cremella (LC) Italy]. A protein concentration of 0.15 mg/mL (5.15 μM) in 10 mM Na-phosphate, pH 7.2 (path-length quartz cuvette of 0.1 cm), was used for the far-UV spectrum measurements. DichroWeb (online analysis for protein Circular Dichroism spectra; http://dichroweb.cryst.bbk.ac.uk/html/home.shtml (accessed on 8 June 2022); [52]) was used to estimate the percentages of secondary structural elements.

Protein (~0.15 mg/mL) in 10 mM sodium phosphate, pH 7.2, was subjected to heat-induced denaturation, as previously reported [39].

4.6. Cell Viability Assays

The COLO 320 (human colon adenocarcinoma) and HeLa cell lines used in this study were obtained from the European Collection of Cell Cultures (ECACC). The cells were grown in RPMI 1640 medium (GIBCO BRL, Barcelona, Spain) supplemented with 10% fetal bovine serum (FBS), 100 U mL^{-1} penicillin and 0.1 mg mL^{-1} streptomycin under 5% CO_2 at 37 °C. Cell viability was determined as previously reported [45]. The concentration of RIPs causing a 50% reduction in viability (IC_{50}) was calculated by linear regression analysis. Sodin 5 and quinoin toxicity was also evaluated using HeLa cells pre-treated with 100 μM of the pan-caspase inhibitor Z-VAD. The reagent was added to cells 3 h before RIP administration, and the cell viability was determined for different RIP concentrations.

4.7. DNA Fragmentation Analysis

COLO 320 cells (1 × 10^6/plate) were incubated for 72 h in the presence of RIP (~0.5 μM). After treatment, cells were harvested by centrifugation (1000× *g* for 5 min). The pellets were lysed in 50 mM Tris Cl, pH 8.0, containing 10 mM EDTA and 0.5% SDS, and the DNA was isolated following the manufacturer's instructions [Genomic Prep Cells and Tissue DNA Isolation Kit (GE Healthcare, Madrid, Spain)]. DNA electrophoresis was carried out as previously reported [45].

4.8. Antifungal Activity Measurements

The growth inhibition assays of sodin 5 from *S. soda* seeds and quinoin from *C. quinoa* against *P. digitatum* were performed in 96-well microtiter plates. The conidia of *P. digitatum* (100 spores/well), obtained as indicated [47], were incubated at 26 °C in 150 μL PDB medium for 24 h to allow for conidia germination. The incubation was continued in the presence or in the absence of different RIP concentrations for a further 46 h. Fungal growth was followed for 70 h and measured as an increase in absorbance at 650 nm. Fungal growth was monitored spectrophotometrically using a microtiter plate reader (ELISA reader Multiskan) after 0, 24, 45, 56 and 70 h of incubation. The absorbance of cultures without cells was subtracted as the background.

Supplementary Materials: The following supporting information can be downloaded at: https://www.mdpi.com/article/10.3390/toxins14080566/s1, Figure S1: SDS-PAGE analysis and in-gel staining for sugars of sodins isolated from *S. soda* tissues; Figure S2: Protein purification and polynucleotide:adenosine glycosylase activity of sodins 1–4; Figure S3: rRNA N-glycosylase activity of sodins from *S. soda* seeds assayed on rabbit ribosomes; Figure S4: staining for sugars of sodins isolated from *S. soda* tissues after SDS-PAGE.

Author Contributions: N.L., S.R., A.C. and H.Z.F.H. conducted the protein purification, characterization and enzymatic assays; L.C., R.I. and J.M.F. evaluated IC_{50} and performed cytotoxic activities; A.D.M. was responsible for the conceptualization, data analysis, writing and funding acquisition; S.R. and A.D.M. were responsible for the review and editing. All authors have read and agreed to the published version of the manuscript.

Funding: This work was supported by the University of Campania 'Luigi Vanvitelli', by the project 'Sviluppo di Nutraceutici da Fonti Naturali—BIONUTRA', PON Ricerca e Innovazione 2014–2020 of the Campania region (code ARS01_01166) and the Grant VA033G19 (Consejería de Educación, Junta de Castilla y León) to the GIR ProtIBio.

Institutional Review Board Statement: Not applicable.

Informed Consent Statement: Not applicable.

Data Availability Statement: The data presented in this study are available in this article.

Acknowledgments: The abnegation of all authors has made this study possible without dedicated funds considering the chronic difficulties afflicting the Italian research. The Authors are grateful for IT support to 'Maurizio Muselli', DiSTABiF.

Conflicts of Interest: The authors have no conflict of interest to declare.

Abbreviations

CD, circular dichroism; IC_{50}, concentration that inhibits 50% of protein synthesis or reduces 50% of cell viability; PDB, potato dextrose broth; PNAG, polynucleotide:adenosine glycosylase; RIPs, ribosome-inactivating proteins; SRL, sarcin ricin loop; Tm, melting temperature.

References

1. Endo, Y.; Huber, P.W.; Wool, I.G. The ribonuclease activity of the cytotoxin alpha-sarcin. The characteristics of the enzymatic activity of alpha-sarcin with ribosomes and ribonucleic acids as substrates. *J. Biol. Chem.* **1983**, *258*, 2662–2667. [CrossRef]
2. Shi, X.; Khade, P.K.; Sanbonmatsu, K.Y.; Joseph, S. Functional role of the sarcin-ricin loop of the 23S rRNA in the elongation cycle of protein synthesis. *J. Mol. Biol.* **2012**, *419*, 125–138. [CrossRef]
3. Barbieri, L.; Valbonesi, P.; Bonora, E.; Gorini, P.; Bolognesi, A.; Stirpe, F. Polynucleotide:adenosine glycosidase activity of ribosome-inactivating proteins: Effect on DNA, RNA and poly(A). *Nucleic Acids Res.* **1997**, *25*, 518–522. [CrossRef] [PubMed]
4. Barbieri, L.; Valbonesi, P.; Righi, F.; Zuccheri, G.; Monti, F.; Gorini, P.; Samorí, B.; Stirpe, F. Polynucleotide:Adenosine glycosidase is the sole activity of ribosome-inactivating proteins on DNA. *J. Biochem.* **2000**, *128*, 883–889. [CrossRef] [PubMed]
5. Barbieri, L.; Gorini, P.; Valbonesi, P.; Castiglioni, P.; Stirpe, F. Unexpected activity of saporins. *Nature* **1994**, *372*, 624. [CrossRef]
6. Ruggiero, A.; Chambery, A.; Di Maro, A.; Mastroianni, A.; Parente, A.; Berisio, R. Crystallization and preliminary X-ray diffraction analysis of PD-L1, a highly glycosylated ribosome inactivating protein with DNase activity. *Protein Pept. Lett.* **2007**, *14*, 407–409. [CrossRef]
7. Aceto, S.; Di Maro, A.; Conforto, B.; Siniscalco, G.G.; Parente, A.; Delli Bovi, P.; Gaudio, L. Nicking activity on pBR322 DNA of ribosome inactivating proteins from *Phytolacca dioica* L. leaves. *Biol. Chem.* **2005**, *386*, 307–317. [CrossRef]
8. Mock, J.W.; Ng, T.B.; Wong, R.N.; Yao, Q.Z.; Yeung, H.W.; Fong, W.P. Demonstration of ribonuclease activity in the plant ribosome-inactivating proteins alpha- and beta-momorcharins. *Life Sci.* **1996**, *59*, 1853–1859. [CrossRef]
9. Shih, N.R.; McDonald, K.A.; Jackman, A.P.; Girbés, T.; Iglesias, R. Bifunctional plant defence enzymes with chitinase and ribosome inactivating activities from *Trichosanthes kirilowii* cell cultures. *Plant Sci.* **1997**, *130*, 145–150. [CrossRef]
10. Lombard, S.; Helmy, M.E.; Piéroni, G. Lipolytic activity of ricin from Ricinus sanguineus and *Ricinus communis* on neutral lipids. *Biochem. J.* **2001**, *358*, 773–781. [CrossRef]
11. Li, X.-D.; Chen, W.-F.; Liu, W.-Y.; Wang, G.-H. Large-scale preparation of two new ribosome-inactivating proteins–cinnamomin and camphorin from the seeds of *Cinnamomum camphora. Protein Expr. Purif.* **1997**, *10*, 27–31. [CrossRef] [PubMed]
12. Stirpe, F. Ribosome-inactivating proteins. *Toxicon* **2004**, *44*, 371–383. [CrossRef] [PubMed]
13. Peumans, W.J.; Hao, Q.; Van Damme, E.J. Ribosome-inactivating proteins from plants: More than RNA N-glycosidases? *FASEB J.* **2001**, *15*, 1493–1506. [CrossRef] [PubMed]
14. Day, P.J.; Lord, J.M.; Roberts, L.M. The deoxyribonuclease activity attributed to ribosome-inactivating proteins is due to contamination. *Eur. J. Biochem.* **1998**, *258*, 540–545. [CrossRef] [PubMed]
15. Bolognesi, A.; Bortolotti, M.; Maiello, S.; Battelli, M.G.; Polito, L. Ribosome-Inactivating Proteins from Plants: A Historical Overview. *Molecules* **2016**, *21*, 1627. [CrossRef] [PubMed]
16. Di Maro, A.; Citores, L.; Russo, R.; Iglesias, R.; Ferreras, J.M. Sequence comparison and phylogenetic analysis by the Maximum Likelihood method of ribosome-inactivating proteins from angiosperms. *Plant Mol. Biol* **2014**, *85*, 575–588. [CrossRef]
17. Landi, N.; Hussain, H.Z.F.; Pedone, P.V.; Ragucci, S.; Di Maro, A. Ribotoxic Proteins, Known as Inhibitors of Protein Synthesis, from Mushrooms and Other Fungi According to Endo's Fragment Detection. *Toxins* **2022**, *14*, 403. [CrossRef]
18. O'Loughlin, E.V.; Robins-Browne, R.M. Effect of Shiga toxin and Shiga-like toxins on eukaryotic cells. *Microbes Infect.* **2001**, *3*, 493–507. [CrossRef]
19. Liu, R.S.; Yang, J.H.; Liu, W.Y. Isolation and enzymatic characterization of lamjapin, the first ribosome-inactivating protein from cryptogamic algal plant (*Laminaria japonica* A). *Eur. J. Biochem.* **2002**, *269*, 4746–4752. [CrossRef]
20. Becker, W.; Apel, K. Isolation and characterization of a cDNA clone encoding a novel jasmonate-induced protein of barley (*Hordeum vulgare* L.). *Plant Mol. Biol.* **1992**, *19*, 1065–1067. [CrossRef]
21. Chaudhry, B.; Müller-Uri, F.; Cameron-Mills, V.; Gough, S.; Simpson, D.; Skriver, K.; Mundy, J. The barley 60 kDa jasmonate-induced protein (JIP60) is a novel ribosome-inactivating protein. *Plant J.* **1994**, *6*, 815–824. [CrossRef] [PubMed]
22. Walsh, T.A.; Morgan, A.E.; Hey, T.D. Characterization and molecular cloning of a proenzyme form of a ribosome-inactivating protein from maize. Novel mechanism of proenzyme activation by proteolytic removal of a 2.8-kilodalton internal peptide segment. *J. Biol. Chem.* **1991**, *266*, 23422–23427. [CrossRef]
23. Lapadula, W.J.; Sánchez Puerta, M.V.; Juri Ayub, M. Revising the taxonomic distribution, origin and evolution of ribosome inactivating protein genes. *PLoS ONE* **2013**, *8*, e72825. [CrossRef] [PubMed]
24. Akkouh, O.; Ng, T.B.; Cheung, R.C.; Wong, J.H.; Pan, W.; Ng, C.C.; Sha, O.; Shaw, P.C.; Chan, W.Y. Biological activities of ribosome-inactivating proteins and their possible applications as antimicrobial, anticancer, and anti-pest agents and in neuroscience research. *Appl. Microbiol. Biotechnol.* **2015**, *99*, 9847–9863. [CrossRef]
25. Rotondo, R.; Ragucci, S.; Castaldo, S.; Oliva, M.A.; Landi, N.; Pedone, P.V.; Arcella, A.; Di Maro, A. Cytotoxicity Effect of Quinoin, Type 1 Ribosome-Inactivating Protein from Quinoa Seeds, on Glioblastoma Cells. *Toxins* **2021**, *13*, 684. [CrossRef]
26. Citores, L.; Iglesias, R.; Ferreras, J.M. Antiviral Activity of Ribosome-Inactivating Proteins. *Toxins* **2021**, *13*, 80. [CrossRef]

27. Zhu, F.; Zhou, Y.-K.; Ji, Z.-L.; Chen, X.-R. The Plant Ribosome-Inactivating Proteins Play Important Roles in Defense against Pathogens and Insect Pest Attacks. *Front. Plant Sci.* **2018**, *9*, 146. [CrossRef]
28. Pizzo, E.; Di Maro, A. A new age for biomedical applications of Ribosome Inactivating Proteins (RIPs): From bioconjugate to nanoconstructs. *J. Biomed. Sci.* **2016**, *23*, 54. [CrossRef] [PubMed]
29. Hammer, K.; Pignone, D.; Cifarelli, S.; Perrino, P. Barilla (*Salsola soda*, Chenopodiaceae). *Econ. Bot.* **1990**, *44*, 410–412. [CrossRef]
30. Iannuzzi, A.M.; Moschini, R.; De Leo, M.; Pineschi, C.; Balestri, F.; Cappiello, M.; Braca, A.; Del-Corso, A. Chemical profile and nutraceutical features of *Salsola soda* (agretti): Anti-inflammatory and antidiabetic potential of its flavonoids. *Food Biosci.* **2020**, *37*, 100713. [CrossRef]
31. Murshid, S.S.A.; Atoum, D.; Abou-Hussein, D.R.; Abdallah, H.M.; Hareeri, R.H.; Almukadi, H.; Edrada-Ebel, R. Genus Salsola: Chemistry, Biological Activities and Future Prospective-A Review. *Plants* **2022**, *11*, 714. [CrossRef] [PubMed]
32. Ragucci, S.; Bulgari, D.; Landi, N.; Russo, R.; Clemente, A.; Valletta, M.; Chambery, A.; Gobbi, E.; Faoro, F.; Di Maro, A. The Structural Characterization and Antipathogenic Activities of Quinoin, a Type 1 Ribosome-Inactivating Protein from Quinoa Seeds. *Int. J. Mol. Sci.* **2021**, *22*, 8964. [CrossRef] [PubMed]
33. Chow, T.P.; Feldman, R.A.; Lovett, M.; Piatak, M. Isolation and DNA sequence of a gene encoding alpha-trichosanthin, a type I ribosome-inactivating protein. *J. Biol. Chem.* **1990**, *265*, 8670–8674. [CrossRef]
34. Di Maro, A.; Valbonesi, P.; Bolognesi, A.; Stirpe, F.; De Luca, P.; Siniscalco Gigliano, G.; Gaudio, L.; Delli Bovi, P.; Ferranti, P.; Malorni, A.; et al. Isolation and characterization of four type-1 ribosome-inactivating proteins, with polynucleotide:adenosine glycosidase activity, from leaves of *Phytolacca dioica* L. *Planta* **1999**, *208*, 125–131. [CrossRef] [PubMed]
35. Bolognesi, A.; Polito, L.; Lubelli, C.; Barbieri, L.; Parente, A.; Stirpe, F. Ribosome-inactivating and adenine polynucleotide glycosylase activities in *Mirabilis jalapa* L. tissues. *J. Biol. Chem.* **2002**, *277*, 13709–13716. [CrossRef]
36. Parente, A.; De Luca, P.; Bolognesi, A.; Barbieri, L.; Battelli, M.G.; Abbondanza, A.; Sande, M.J.; Gigliano, G.S.; Tazzari, P.L.; Stirpe, F. Purification and partial characterization of single-chain ribosome-inactivating proteins from the seeds of *Phytolacca dioica* L. *Biochim. Biophys. Acta-Gene Struct. Expr.* **1993**, *1216*, 43–49. [CrossRef]
37. Barbieri, L.; Polito, L.; Bolognesi, A.; Ciani, M.; Pelosi, E.; Farini, V.; Jha, A.K.; Sharma, N.; Vivanco, J.M.; Chambery, A.; et al. Ribosome-inactivating proteins in edible plants and purification and characterization of a new ribosome-inactivating protein from *Cucurbita moschata*. *Biochim. Biophys. Acta-Gen. Subj.* **2006**, *1760*, 783–792. [CrossRef]
38. Di Maro, A.; Chambery, A.; Daniele, A.; Casoria, P.; Parente, A. Isolation and characterization of heterotepalins, type 1 ribosome-inactivating proteins from *Phytolacca heterotepala* leaves. *Phytochemistry* **2007**, *68*, 767–776. [CrossRef]
39. Landi, N.; Ruocco, M.R.; Ragucci, S.; Aliotta, F.; Nasso, R.; Pedone, P.V.; Di Maro, A. Quinoa as source of type 1 ribosome inactivating proteins: A novel knowledge for a revision of its consumption. *Food Chem.* **2021**, *342*, 128337. [CrossRef]
40. Stirpe, F.; Gasperi-Campani, A.; Barbieri, L.; Falasca, A.; Abbondanza, A.; Stevens, W.A. Ribosome-inactivating proteins from the seeds of *Saponaria officinalis* L. (soapwort), of *Agrostemma githago* L. (corn cockle) and of *Asparagus officinalis* L. (asparagus), and from the latex of *Hura crepitans* L. (sandbox tree). *Biochem. J.* **1983**, *216*, 617–625. [CrossRef]
41. Stirpe, F.; Gilabert-Oriol, R. Ribosome-Inactivating Proteins: An Overview. In *Plant Toxins*; Carlini, C.R., Ligabue-Braun, R., Gopalakrishnakone, P., Eds.; Springer: Dordrecht, The Netherlands, 2017; pp. 153–182. [CrossRef]
42. Monzingo, A.F.; Collins, E.J.; Ernst, S.R.; Irvin, J.D.; Robertus, J.D. The 2.5 A structure of pokeweed antiviral protein. *J. Mol. Biol.* **1993**, *233*, 705–715. [CrossRef] [PubMed]
43. Savino, C.; Federici, L.; Ippoliti, R.; Lendaro, E.; Tsernoglou, D. The crystal structure of saporin SO6 from *Saponaria officinalis* and its interaction with the ribosome. *FEBS Lett.* **2000**, *470*, 239–243. [CrossRef]
44. Sánchez, M.; Scirè, A.; Tanfani, F.; Ausili, A. The thermal unfolding of the ribosome-inactivating protein saporin-S6 characterized by infrared spectroscopy. *Biochim. Biophys. Acta-Proteins Proteom.* **2015**, *1854*, 1357–1364. [CrossRef] [PubMed]
45. Iglesias, R.; Citores, L.; Ragucci, S.; Russo, R.; Di Maro, A.; Ferreras, J.M. Biological and antipathogenic activities of ribosome-inactivating proteins from *Phytolacca dioica* L. *Biochim. Biophys. Acta-Gen. Subj.* **2016**, *1860*, 1256–1264. [CrossRef]
46. Iglesias, R.; Citores, L.; Di Maro, A.; Ferreras, J.M. Biological activities of the antiviral protein BE27 from sugar beet (*Beta vulgaris* L.). *Planta* **2015**, *241*, 421–433. [CrossRef]
47. Citores, L.; Iglesias, R.; Gay, C.; Ferreras, J.M. Antifungal activity of the ribosome-inactivating protein BE27 from sugar beet (*Beta vulgaris* L.) against the green mould Penicillium digitatum. *Mol. Plant Pathol.* **2016**, *17*, 261–271. [CrossRef]
48. Landi, N.; Pacifico, S.; Ragucci, S.; Iglesias, R.; Piccolella, S.; Amici, A.; Di Giuseppe, A.M.A.; Di Maro, A. Purification, characterization and cytotoxicity assessment of Ageritin: The first ribotoxin from the basidiomycete mushroom *Agrocybe aegerita*. *Biochim. Biophys. Acta-Gen. Subj.* **2017**, *1861*, 1113–1121. [CrossRef]
49. Di Maro, A.; Terracciano, I.; Sticco, L.; Fiandra, L.; Ruocco, M.; Corrado, G.; Parente, A.; Rao, R. Purification and characterization of a viral chitinase active against plant pathogens and herbivores from transgenic tobacco. *J. Biotechnol.* **2010**, *147*, 1–6. [CrossRef]
50. Iglesias, R.; Citores, L.; Ferreras, J.M. Ribosomal RNA N-glycosylase Activity Assay of Ribosome-inactivating Proteins. *Bio-Protocol* **2017**, *7*, e2180. [CrossRef]

51. Ferreras, J.M.; Citores, L.; Iglesias, R.; Jiménez, P.; Souza, A.M.; Gayoso, M.J.; Girbés, T. Occurrence and new procedure of preparation of nigrin, an antiribosomal lectin present in elderberry bark. *Food Res. Int.* **2011**, *44*, 2798–2805. [CrossRef]
52. Miles, A.J.; Ramalli, S.G.; Wallace, B.A. DichroWeb, a website for calculating protein secondary structure from circular dichroism spectroscopic data. *Protein Sci.* **2022**, *31*, 37–46. [CrossRef] [PubMed]

toxins

Article

Structure and Biological Properties of Ribosome-Inactivating Proteins and Lectins from Elder (*Sambucus nigra* L.) Leaves

Rosario Iglesias [1,†], Rosita Russo [2,†], Nicola Landi [2], Mariangela Valletta [2], Angela Chambery [2], Antimo Di Maro [2], Andrea Bolognesi [3], José M. Ferreras [1,*] and Lucía Citores [1,*]

1 Department of Biochemistry and Molecular Biology and Physiology, Faculty of Sciences, University of Valladolid, E-47011 Valladolid, Spain
2 Department of Environmental, Biological and Pharmaceutical Sciences and Technologies (DiSTABiF), University of Campania 'Luigi Vanvitelli', Via Vivaldi 43, 81100 Caserta, Italy
3 Department of Experimental, Diagnostic and Specialty Medicine-DIMES, Alma Mater Studiorum-University of Bologna, Via S. Giacomo 14, 40126 Bologna, Italy
* Correspondence: josemiguel.ferreras@uva.es (J.M.F.); lucia.citores@uva.es (L.C.)
† These authors contributed equally to this work.

Abstract: Ribosome-inactivating proteins (RIPs) are a group of proteins with rRNA N-glycosylase activity that catalyze the removal of a specific adenine located in the sarcin–ricin loop of the large ribosomal RNA, which leads to the irreversible inhibition of protein synthesis and, consequently, cell death. The case of elderberry (*Sambucus nigra* L.) is unique, since more than 20 RIPs and related lectins have been isolated and characterized from the flowers, seeds, fruits, and bark of this plant. However, these kinds of proteins have never been isolated from elderberry leaves. In this work, we have purified RIPs and lectins from the leaves of this shrub, studying their main physicochemical characteristics, sequences, and biological properties. In elderberry leaves, we found one type 2 RIP and two related lectins that are specific for galactose, four type 2 RIPs that fail to agglutinate erythrocytes, and one type 1 RIP. Several of these proteins are homologous to others found elsewhere in the plant. The diversity of RIPs and lectins in the different elderberry tissues, and the different biological activities of these proteins, which have a high degree of homology with each other, constitute an excellent source of proteins that are of great interest in diagnostics, experimental therapy, and agriculture.

Keywords: anticancer agents; galactose; lectin; nanoLC–tandem mass spectrometry (nLC-MS/MS); protein synthesis (inhibition); ribosome-inactivating protein (RIP); ricin; sugar binding

Key Contribution: This work contributes to expanding our knowledge of the family of RIPs and RIP-related lectins produced by *Sambucus nigra*, with eight new proteins found in leaves: one type 2 RIP and two related lectins that are specific for galactose, four type 2 RIPs with deficient sugar binding domains and one type 1 RIP. This knowledge is important for the potential medical and biotechnological use of these proteins.

Citation: Iglesias, R.; Russo, R.; Landi, N.; Valletta, M.; Chambery, A.; Di Maro, A.; Bolognesi, A.; Ferreras, J.M.; Citores, L. Structure and Biological Properties of Ribosome-Inactivating Proteins and Lectins from Elder (*Sambucus nigra* L.) Leaves. *Toxins* **2022**, *14*, 611. https://doi.org/10.3390/toxins14090611

Received: 29 July 2022
Accepted: 29 August 2022
Published: 1 September 2022

1. Introduction

Ribosome-inactivating proteins (RIPs) are a group of proteins with rRNA N-glycosylase activity (EC 3.2.2.22) that catalyze the elimination of a specific adenine located in the sarcin–ricin loop (SRL) present in the large rRNA of eukaryotes and prokaryotes [1,2]. The elimination of this adenine (A4324 in rat ribosomes, or the equivalent in other organisms) inactivates ribosomes, which leads to the irreversible inhibition of protein synthesis and, therefore, cell death [1,2]. RIPs have been classified according to their structure as type 1 RIPs, consisting of a polypeptide chain with N-glycosylase activity, and type 2 RIPs, formed by two polypeptide chains, an A (active) chain with enzymatic activity, and a B (binding) chain with lectin activity that can bind to receptors on the surface of cells, facilitating the entry of RIP [2]. Some type 2 RIPs, such as ricin, are extremely toxic, while others have low toxicity; this is because the binding of the B chain to oligosaccharides present on the surface of cells is less effective, and because once internalized, the RIP follows an intracellular pathway different from ricin [1,3,4]. The toxicity of type 1 RIPs is lower, as they lack the lectin part and are, therefore, unable to bind to cells as type 2 RIPs do. Although the structure, activity, and mode of action of RIPs are known, their biological function is unclear. It has been proposed that these proteins could play an important role in the defense of plants against viruses, fungi, and insects [5,6].

Because of their diverse activities, RIPs, alone or as part of a conjugate, are good candidates for developing selective antiviral and anticancer agents [1,6]. Conjugates consist of a component directed against the target, such as an antibody, lectin, or growth factor, attached to a toxic component. RIPs have been used as the toxic component in several conjugates that have been tested in experimental therapies against various malignancies [3,7]. In agriculture, RIPs have been shown to increase resistance against viruses, fungi, and insects in transgenic plants [5,6].

RIPs are present in many angiosperm plants, both monocotyledonous and dicotyledonous, although in some plant families it is more common to find RIPs than in others; therefore, there are families such as Poaceae, Euphorbiaceae, Cucurbitaceae, Caryophyllaceae, Amaranthaceae, and Phytolaccaceae where several species with RIPs have been found, and other families where they have never been found [2]. Some species contain a wide variety of diverse RIPs, such as rice [8], or species of the genus *Phytolacca* [9,10].

The case of the genus *Sambucus* is unique, since more than 40 RIPs and lectins have been isolated and characterized to varying degrees from species belonging to this taxon [3,11]. Type 2 RIPs isolated from *Sambucus* have the peculiarity that, although they are enzymatically more active than ricin, they lack the high toxicity of ricin to cells and animals [11]. The presence in the same species of type 2 RIPs (heterodimeric and tetrameric), lectins (monomeric and homodimeric) structurally related to the above, together with type 1 RIPs, make the genus *Sambucus* an ideal model with which to study these proteins. Although these kinds of proteins can be found in other species, most of them have been obtained from elderberry (*Sambucus nigra* L.). Type 1 RIPs, type 2 RIPs (heterodimeric and tetrameric), and B-chain-related lectins have been obtained from the bark, seeds, flowers, and fruits, but the proteins from the leaves of this species have never been isolated. In this work we have isolated and characterized the most abundant RIPs and lectins in elderberry leaves by investigating their main physicochemical and structural properties, including amino acid sequence. We further studied their most important biological properties, and through in silico experiments, we explored potential mechanisms of sugar binding to these proteins.

2. Results

2.1. Isolation of RIPs and Lectins from Elderberry Leaves

Species of the genus *Sambucus* are one of the best sources for the isolation of RIPs and related lectins. These proteins have been found in *Sambucus ebulus* L. (dwarf elder), *S. nigra* L. (European elder), *S. sieboldiana* (Miq.) Blume ex Schwer. (Japanese elder), and *S. racemosa* L. (red elder) [3,11]. The most frequently used species, and the one with the greatest variety of these proteins, is *S. nigra*, in which RIPs and lectins have been isolated and characterized from bark, fruits, seeds, flowers, and pollen. Although RIPs from leaves of this species have not yet been isolated, in data banks the sequences of three proteins obtained from *S. nigra* leaves cDNA can be found: a type 2 RIP with an amino acid sequence similar to that of nigrin b from bark (named nigrin l), a monomeric lectin with a sequence similar to that of SNAIV from fruits (named SNAlm), and a homodimeric lectin with a sequence resembling that of SELld from the leaves of *S. ebulus* (named SNAld). Therefore, we aimed to isolate and characterize the RIPs and lectins from elderberry leaves. For this purpose, we optimized a standard RIP purification procedure [12] previously used to isolate RIPs and lectins from elderberry bark [13] and dwarf elder leaves [14]. A schematic overview of the procedures used to purify the RIPs and lectins from *S. nigra* leaves is shown in Figure 1.

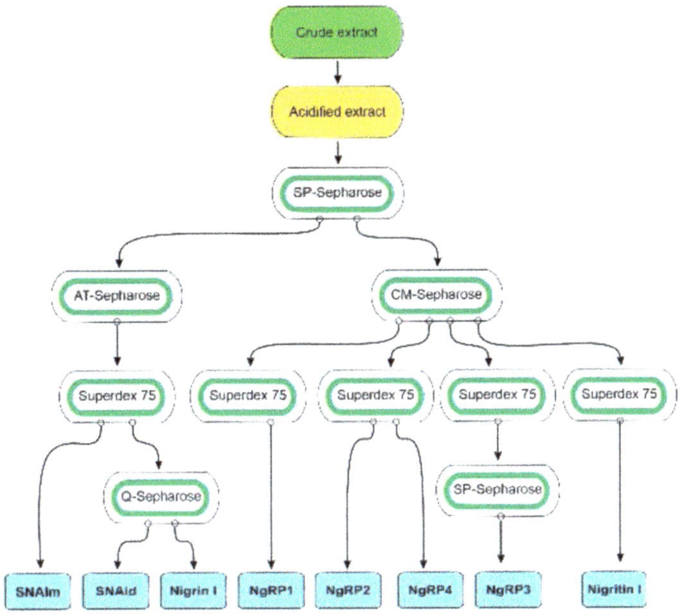

Figure 1. Schematic overview of the procedures used to purify the RIPs and lectins from *S. nigra* leaves. All steps of purification are detailed in the Materials and Methods.

A great difficulty in carrying out this purpose is the extraordinary variability that these proteins present in the leaves. For this reason, two purifications were derived from the leaves of *S. nigra*. Two crude extracts were prepared from 700 and 540 g of leaves. Then, an acidified crude extract was obtained and subjected to ion exchange SP-Sepharose chromatography. After washing the column with sodium acetate (pH 4.5), the bound protein was eluted first with 5 mM sodium phosphate (pH 6.66), and then with sodium chloride. The fraction eluted with sodium phosphate displayed both protein synthesis inhibitory activity and erythrocyte agglutination ability, while the fraction eluted with sodium chloride inhibited protein synthesis but failed to agglutinate erythrocytes (data not shown). The yield of proteins was different in the two preparations: in the first, a higher amount of protein was obtained by eluting with sodium phosphate (1.4-fold higher); in the second, a better yield was obtained by eluting with sodium chloride (threefold higher). Therefore, we used the first preparation for purifying proteins eluted with sodium phosphate and the second preparation for purifying proteins eluted with NaCl.

The fraction eluted with sodium phosphate from the SP-Sepharose column was further purified by affinity chromatography using an acid-treated (AT)-Sepharose column. After washing with buffer, the D-galactose-binding proteins were eluted from the column with 0.2 M lactose, concentrated, and subjected to chromatography using a Superdex 75 HiLoad column (Figure 2a). The first peak contained nigrin l and SNAld, and the second contained SNAlm. The fractions of the first peak were dialyzed and subjected to Q-Sepharose chromatography. Proteins were eluted from the column with an NaCl gradient yielding two peaks (Figure 2b). The fractions of the first peak contained nigrin l and those of the second peak contained SNAld. The estimated yields of this preparation were 2, 12, and 1.4 mg per 100 g of leaves for nigrin l, SNAlm, and SNAld, respectively.

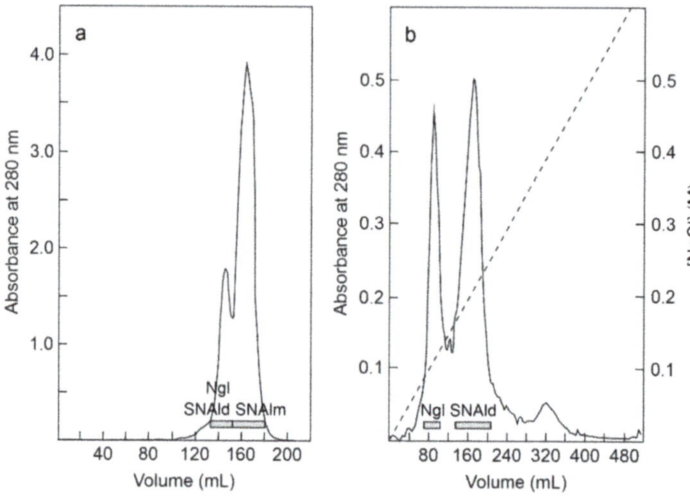

Figure 2. Purification of nigrin l, SNAlm, and SNAld: (**a**) The fraction eluted with lactose from the AT-Sepharose column was concentrated and chromatographed using Superdex 75 HiLoad. The fractions of the first peak contained nigrin l (Ngl) and SNAld, and those of the second peak contained SNAlm (horizontal bars); (**b**) the fractions of the first peak shown on panel (**a**) were dialyzed and subjected to chromatography using Q-Sepharose that was eluted with an NaCl gradient (dashed line), as indicated in the Materials and Methods. The fractions of the first peak contained nigrin l (Ngl) and those of the second peak contained SNAld (horizontal bars).

The fraction eluted with sodium chloride from the SP-Sepharose column was dialyzed and subjected to cation exchange chromatography using CM-Sepharose with a linear gradient of NaCl. As shown in Figure 3a, CM-Sepharose chromatography resolved several protein peaks. The fractions of the peaks were analyzed by electrophoresis in the presence and absence of 2-mercaptoethanol, and their effect on protein synthesis was tested. Further purification of the peaks yielded five proteins that strongly inhibited protein synthesis. Four of these new proteins corresponded to type 2 RIPs, and were named nigrin-Related Proteins 1–4 (nigrin-RPs 1–4). A new type 1 RIP was also found, and we named it nigritin l. For the purification of these new RIPs, the fractions derived via CM-Sepharose chromatography were collected as shown in Figure 3a, and subjected to subsequent chromatographies separately. Thus, nigrin-RP1 was purified by chromatography using Superdex 75 (Figure 3b). The protein obtained contained traces of nigrin l that were eliminated by chromatography using AT-Sepharose, as indicated in the Materials and Methods section. Nigrin-RP2 and nigrin-RP4 were also purified by chromatography using Superdex 75 (Figure 3c) and the trace contaminants were removed by re-chromatographing the proteins in the same column. To purify nigritin l, the fractions indicated in Figure 3a were subjected to chromatography using Superdex 75 (Figure 3d) and the contaminant traces were removed by re-chromatographing the protein on the same column. Finally, nigrin-RP3 was purified by chromatography using Superdex 75 (Figure 3e) followed by chromatography using SP-Sepharose with an NaCl gradient (Figure 3f). The estimated yields of the current preparation were 2.2, 2.6, 0.29, 0.03 and 1.5 mg per 100 g of leaves for nigrin-RPs 1–4 and nigritin l, respectively.

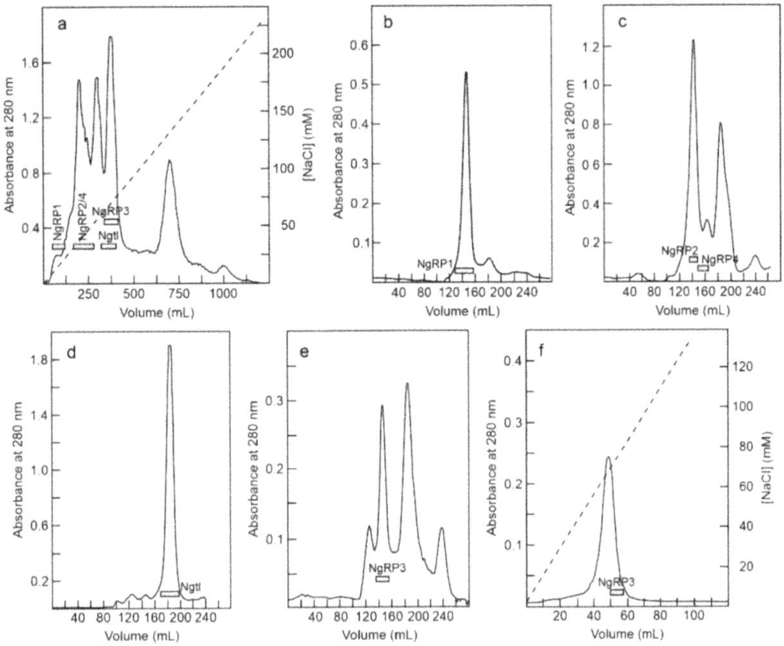

Figure 3. Purification of nigrin-Related Proteins 1–4, and nigritin l: (**a**) The proteins eluted with NaCl in SP-Sepharose were dialyzed and subjected to cation exchange chromatography using CM-Sepharose, as described in the Materials and Methods. The protein was eluted with a linear gradient of NaCl (dashed line). The fractions indicated by horizontal bars were separately pooled, concentrated, and subjected to molecular exclusion chromatography using Superdex 75 HiLoad; (**b**) Purification of nigrin-RP1. The fractions marked with NgRP1 in panel **a** were concentrated and subjected to chromatography using Superdex 75 HiLoad. The fractions indicated with the horizontal bar (NgRP1) were pooled, and the traces of nigrin l were eliminated by chromatography using AT-Sepharose, as indicated in the Materials and Methods; (**c**) Purification of nigrin-RP2 and nigrin-RP4. The fractions indicated with NgRP2/4 in panel (**a**) were concentrated and subjected to chromatography using Superdex 75 HiLoad. The fractions indicated with the horizontal bars (NgRP2 and NgRP4) were pooled and the contaminant traces were removed by re-chromatographing the proteins on the same column; (**d**) Purification of nigritin l. The fractions marked with Ngtl on panel (**a**) were concentrated and subjected to chromatography using Superdex 75 HiLoad. The fractions marked with the horizontal bar (Ngtl) were pooled and the contaminant traces were removed by re-chromatographing the proteins in the same column; (**e**,**f**) Purification of nigrin-RP3. The purification of nigrin-RP3 was conducted using the first preparation, with fractions equivalent to those indicated with NgRP3 in panel (**a**), which were concentrated and subjected to chromatography using Superdex 75 HiLoad (**e**). The fractions indicated with NgRP3 in panel **e** were dialyzed and subjected to chromatography using SP-Sepharose (**f**), which was eluted with an NaCl gradient (dashed line), as described in the Materials and Methods.

2.2. Characterization of RIPs and Lectins from Elderberry Leaves

Purified proteins from elderberry leaves were analyzed by electrophoresis on polyacrylamide gels in the presence of SDS (SDS-PAGE), and in the absence or presence of 2-mercaptoethanol. As shown in Figure 4, in the absence of a reductant, all of the proteins exhibited molecular weights between 50 and 65 kDa except for the lectin, SNAlm, and the type 1 RIP, nigritin l, which exhibited molecular weights of about 32.4 and 25.5 kDa, respectively. In the presence of 2-mercaptoethanol, all of the proteins produced bands with molecular weights between 25 and 35 kDa. Therefore, they are all dimeric proteins except

SNAlm and nigritin l, which are monomeric proteins. However, we found that reduction with 2-mercaptoethanol also induced changes in the apparent molecular weight of nigritin l, reducing the mobility of the protein to a molecular weight of about 27.5 kDa, suggesting the presence of intrachain disulfide bonds.

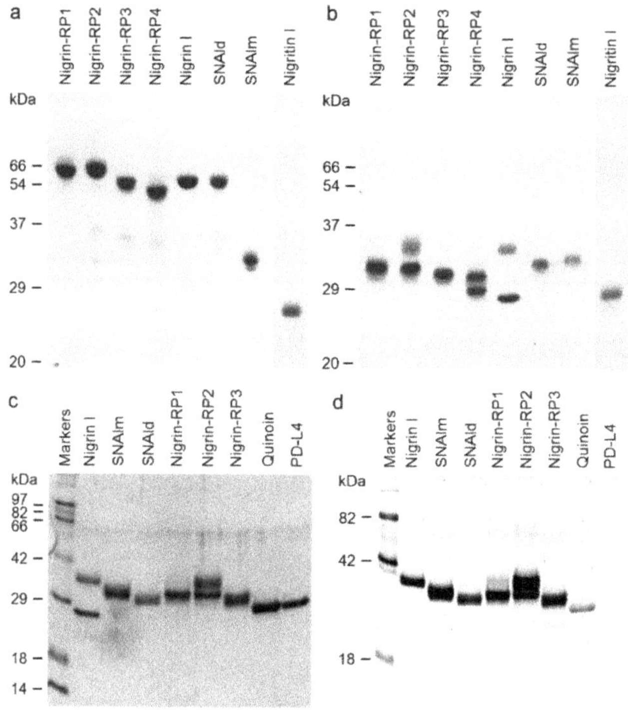

Figure 4. Analysis of purified proteins from elderberry leaves by electrophoresis on polyacrylamide gels: SDS-PAGE of the isolated proteins without (**a**) or with (**b**) 2-mercaptoethanol was carried out on 12% polyacrylamide separating gel and then stained with Coomassie brilliant blue. Samples of 5 µg of each protein were loaded on the gel except for nigritin l, for which 3 µg were loaded. The numbers indicate the corresponding size of the standards in kDa; (**c**,**d**) Sugar staining of RIPs and lectins from *S. nigra* leaves after SDS-PAGE. Each lane contained 5 µg of protein. SDS-PAGE was carried out on a 12% polyacrylamide separating gel in the presence of 2-mercaptoethanol, followed by Coomassie blue staining (**c**), or in-gel glycan detection (**d**) using the Pro-Q Emerald 300 glycoprotein staining kit. Stained glycoproteins were visualized by UV transillumination. Markers: CandyCane™ glycoproteins (**d**) and Coomassie blue (**c**) molecular weight standards.

The type 2 RIPs, nigrin l, and nigrin-RPs 1–4 are heterodimeric proteins consisting of a catalytic chain (A chain) and a lectin chain (B chain), both linked through a disulfide bond. In the presence of 2-mercaptoethanol, the apparent molecular weight values obtained for these proteins were 27.3 and 33.7 kDa for the two chains of nigrin l, 30.7 kDa and 33.7 kDa for the two chains of nigrin-RP2, and 27.9 kDa and 29.8 kDa for the two chains of nigrin-RP4. On the other hand, nigrin-RP1 is composed of two subunits of 30.6 kDa and nigrin-RP3 of two subunits of 30.2 kDa. The homodimeric lectin SNAld with an apparent molecular weight of 61.8 kDa contained only a homogeneous protein band of 30.9 kDa in the presence of 2-mercaptoethanol.

Many RIPs and lectins obtained from different species of *Sambucus* are glycoproteins; thus, we studied whether the purified proteins were glycosylated. Figure 4 compares proteins stained with Coomassie blue (Figure 4c) and those detected with a glycoprotein staining kit (Figure 4d). Quinoin (a glycosylated type 1 RIP) and PDL4 (a non-glycosylated type 1 RIP) are shown as controls. It can be observed that all the RIPs and lectins tested are strongly glycosylated, and, in those in which the A chain can be distinguished from the B chain, the latter is the most glycosylated. Thus, while the nigrin l B chain stained for carbohydrate, nigrin l A chain did not show any staining. On the contrary, nigrin-RP2 contained sugar chains on both subunits.

RIPs are potent inhibitors of protein synthesis in eukaryotes, as they are enzymes capable of inactivating ribosomes catalytically. Therefore, in mammalian cell-free systems, they usually show values of IC_{50} (concentration that inhibits 50% protein synthesis) in the ng/mL range. Accordingly, we tested all the proteins isolated from elderberry leaves in a coupled transcription–translation in vitro assay using a rabbit reticulocyte lysate system, finding the following values of IC_{50}: 0.36, 0.75, 3.0, 0.35, 0.3 and 6.5 ng/mL for nigrin l, nigrin-RPs 1–4, and nigritin l, respectively (Figure S1). It is worth mentioning that, although all are good inhibitors of protein synthesis, there are great differences among the RIPs displaying IC_{50} values that differ up to 20 times. As expected, the lectins SNAlm and SNAld that lack a catalytic chain did not inhibit protein synthesis up to the maximum tested concentration of 1 µg/mL.

Type 2 RIPs consist of two chains, one being the enzymatic chain and the other being a lectin able to recognize sugars, mostly galactose residues. Due to the lectin activity, type 2 RIPs promote human erythrocyte agglutination. Among proteins isolated from *S. nigra* leaves, only nigrin l and the lectins SNAlm and SNAld can be purified by affinity chromatography using AT-Sepharose 6 B, which exposed galactose residues (Figure 1). The type 2 RIPs, nigrin-RPs 1–4, were not retained on AT-Sepharose (data not shown). We therefore studied the agglutination capacity of nigrin-RPs 1–4 and found that even a concentration of 200 µg/mL did not have any effect on human erythrocytes. Under the same conditions, nigrin l, SNALm, and SNALld agglutinated human erythrocytes at concentrations as low as 12.5, 40, and 6.2 µg /mL, respectively. Therefore, nigrin RPs 1–4, unlike nigrin l, may lack functional sugar binding domains. Similar type 2 RIPs have been previously described in *S. nigra* (SNLRPs 1 and 2) and *S. ebulus* (ebulin-RP) [14,15].

2.3. rRNA N-Glycosylase, Adenine Polynucleotide Glycosylase, and DNA Nicking Activities

RIPs are enzymes that irreversibly inactivate ribosomes because of their N-glycosylase activity (EC 3.2.2.22). The enzyme catalyzes the hydrolysis of the N-glycosidic bond between adenine number 4324 and its ribose in rat ribosomes (or equivalent adenine in sensitive ribosomes of other organisms) [1,2]. This activity can be evidenced by detecting, by means of a polyacrylamide gel electrophoresis, the RNA fragment released (Endo's fragment or diagnostic fragment) when the apurinic RNA is incubated in the presence of acid aniline [16]. As shown in Figure 5a, the type 2 RIPs, nigrin l and nigrin-RPs 1–4, and the type 1 RIP, nigritin l, cause, after treatment with acid aniline, the release of the diagnostic fragment of 460 nucleotides from rabbit reticulocyte ribosomes. As shown in Figure 5b, the type 2 RIPs can also depurinate yeast ribosomes, releasing a fragment of 368 nucleotides in the presence of acid aniline. Prokaryotic ribosomes are not sensitive to most RIPs. This is the case, for example, for ricin, volkensin [17], and other type 2 RIPs obtained from different species of the genus *Sambucus* [11]. However, they are sensitive to some type 1 RIPs, such as those obtained from *Pytolacca dioica* or *Beta vulgaris* [18,19]. As shown in Figure 5c, BE27 (obtained from the leaves of *B. vulgaris*) releases the diagnostic fragment from ribosomes of *Micrococcus lysodeikticus*, while the type 1 RIP, nigritin l, and the type 2 RIPs (Figure 5c and data not shown, respectively) from elderberry leaves do not, indicating that prokaryote ribosomes are not sensitive to these proteins.

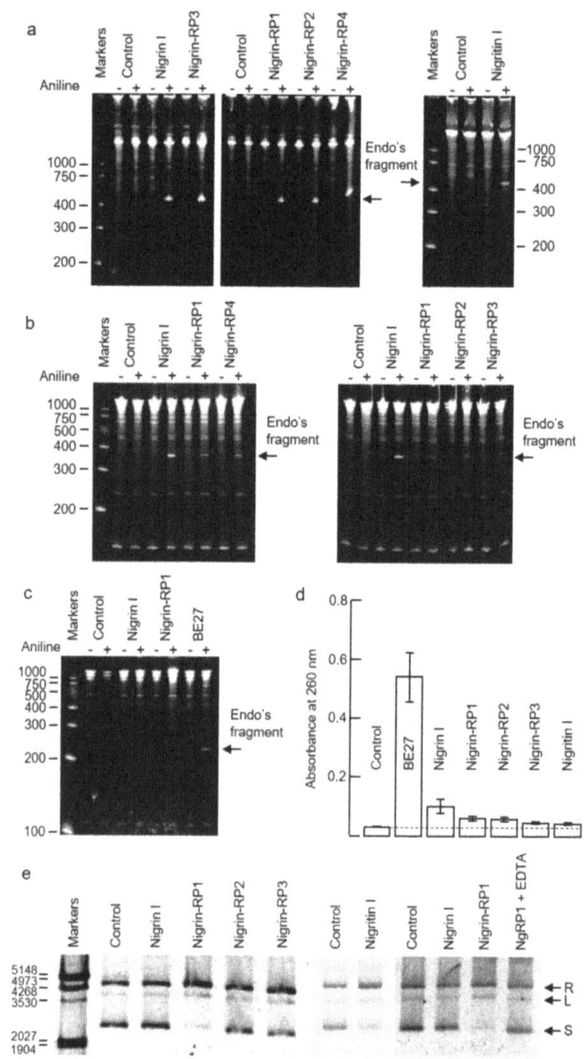

Figure 5. Enzymatic activities of RIPs from *S. nigra* leaves: (**a**–**c**) rRNA N-glycosylase activity in animal, yeast, and bacterial ribosomes. The rRNA-glycosylase activity was tested as indicated in the Materials and Methods. Each lane contained 5 µg of RNA isolated from either untreated (control) or RIP-treated ribosomes from rabbit reticulocyte lysate (**a**), the yeast *Saccharomyces cerevisiae* (**b**), and 1 µg of RNA isolated from the bacterium *Micrococcus lysodeikticus* (**c**). The arrows indicate the RNA fragment (Endo's fragment) released as a result of the action of RIP after treatment with acid aniline (+). The numbers indicate the size of the markers in nucleotides; (**d**) Adenine polynucleotide glycosylase activity (APG) on DNA from salmon sperm. APG activity of 5 µg of RIP was assayed on salmon sperm DNA as described in the Materials and Methods, and the absorbance of the released adenine was measured at 260 nm. Data represent the mean of two duplicate experiments ± SE; (**e**) Nicking activity on pCR2.1 DNA. Samples comprising 200 ng/10 µL of plasmid DNA were incubated with 5 µg of RIP. Nigrin-RP1 (NgRP1) was also incubated in the presence of EDTA. R, L, and S indicate relaxed, linear, and supercoiled forms of pCR2.1, respectively. The numbers indicate the size of the markers in bp.

Some RIPs are also capable of removing more than one adenine from rRNA [20], and many of them can depurinate not only rRNA, but also other polynucleotide substrates, such as DNA, poly(A), mRNA, tRNA, and viral RNA. Therefore, the names adenine polynucleotide glycosylase (APG) and polynucleotide–adenosine glycosylase (PNAG) have been proposed for RIPs [21]. RIPs display different APG activities on DNA and RNA, all of which are capable of depurinating DNA from herring and salmon sperm [18,21]. However, this APG activity on DNA varies markedly between different RIPs. Generally, type 2 RIPs, such as ricin and kirkiin, have low activities [22], while some type 1 RIPs, such as those obtained from *Pytolacca dioica* or *Beta vulgaris*, possess very high activities [18,19]. Figure 5d shows the APG activities of RIPs from elderberry leaves on DNA compared with BE27 activity, measured as the absorbance at 260 nm produced by adenines released from salmon sperm DNA. As shown, RIPs from elderberry leaves have low activities, similar to ricin, stenodactylin, and kirkiin, compared to BE27 activity. Of these, the most active is nigrin l, which is twice as active as the others. However, none of the elderberry RIPs showed activity on tobacco mosaic virus RNA (data not shown).

Some RIPs exhibit topoisomerase (nicking) activity on plasmid DNA, transforming supercoiled DNA into the relaxed form; we tested whether elderberry RIPs possessed this activity, and found that only nigrin-RP1 and the type 1 RIP, nigritin l, were able to convert the supercoiled PCR 2.1 DNA forms into the relaxed forms (Figure 5e). Such activity was dependent of Mg^{2+} ions because it was inhibited by EDTA (Figure 5e).

Therefore, elderberry leaves contain mainly a type 1 RIP (nigritin l), which displayed rRNA N-glycosylase activity, a type 2 RIP with N-glycosylase activity and lectin activity (nigrin l), four type 2 RIPs with N-glycosylase activity that fail to agglutinate erythrocytes (nigrin-RPs 1–4), and a monomeric (SNAlm) and a dimeric lectin (SNAld).

2.4. Peptide Mapping of RIPs from S. nigra Leaves by High-Resolution MS/MS

The proteins isolated from *S. nigra* leaves, nigrin l, nigrin-RPs 1–3, SNALm, and SNAld were characterized via a peptide mapping approach based on high-resolution nanoLC–tandem mass spectrometry (Figure 6). Nigrin-RP4 was not studied due to the low levels obtained via the purification process. Similarly, nigritin l was not further investigated due to the lack of sequence data available for database searches. Preliminary optimization of sample preparation steps for MS analyses was performed to define the conditions for the reduction, alkylation, and tryptic hydrolysis of *S. nigra* proteins. An extensive step of protein denaturation and disulfide bridge reduction performed with 20 mM DTT at 95 °C was needed to obtain a significant yield of tryptic peptides suitable for MS analysis. Then, free cysteinyl residue alkylation with iodoacetamide (IAA) and enzymatic proteolysis with trypsin were both performed before nanoLC–ESI–MS/MS analyses on a Q Exactive Orbitrap mass spectrometer. A data-dependent acquisition mode was used, during which higher-energy collisional dissociation (HCD) MS/MS spectra were obtained for the five most intense mass peaks in each scan, allowing for accurate amino acid sequencing of tryptic peptides. By this approach, for the monomeric lectin, 18 peptide spectral matches (PSMs) were mapped on the SNAlm sequence (AC: AAN86132). Amino acid sequences of peptides obtained by high-resolution tandem mass spectrometry are reported in Table S1. Representative MS/MS spectra of peptides are reported in Figure S2. Regarding the dimeric lectin, nine PSMs were mapped on the SNAld sequence (AC: AAN86131, Table S2 and Figure S3).

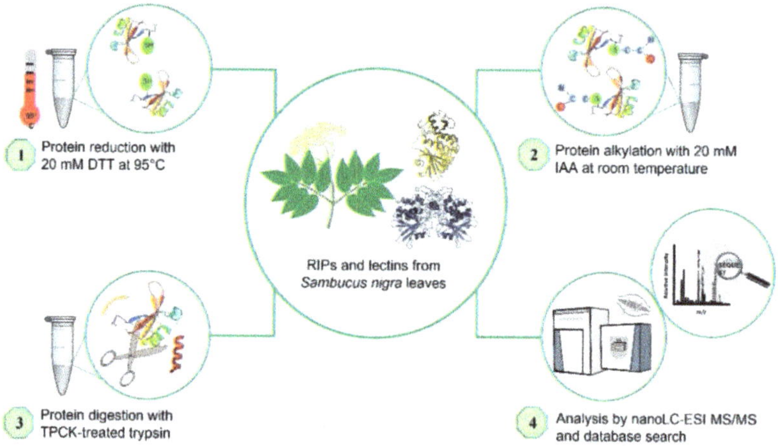

Figure 6. Schematic overview of the experimental workflow used for the peptide mapping of proteins from *Sambucus nigra* leaves by high-resolution nanoLC–MS/MS.

A high number of MS/MS spectra (53 PSMs) of the type 2 RIP, nigrin l, were mapped on the A chain and B chain of nigrin l (AC: AAN86130, Table S3 and Figure S4).

For both nigrin-RP 1 and 2, sequenced peptides were mapped on SNLRP2 A and B chains (AC: AAC49672, Tables S4 and S5) (23 and 24 PSMs, respectively). On the contrary, nigrin-RP3 was identified as SNLRP1 (AC: AAC49673, Table S6) (27 PSMs). Representative MS/MS spectra of peptides from nigrin-RPs 1, 2, and 3 are reported in Figures S5–S7.

Peptide mapping of proteins purified from *S. nigra* leaves by high-resolution MS/MS identifies the proteins obtained by affinity chromatography using AT-Sepharose with three sequences obtained from *S. nigra* leaves cDNA: the type 2 RIP, nigrin l, and the lectins SNAlm and SNAld (accession numbers AAN86130, AAN86132, and AAN86131, respectively). On the other hand, nigrin-RP3 would be a type 2 RIP homologous to SNLRP1 from elderberry bark (accession number AAC49673), and nigrin-RPs 1 and 2 would be homologous to the type 2 RIP, SNLRP2, also found in the bark of elderberry (accession number AAC49672). Alternatively, nigrin-RPs 1 and 2 could be the same protein with different degrees of glycosylation (see Figure 4d). Evidence against this hypothesis is the fact that these two proteins exhibit different protein synthesis inhibitory activities (IC_{50} fourfold higher for nigrin-RP2) and different behaviors against the supercoiled plasmid (see Figure 5e). Figure 7 graphically presents these data by comparing the sequences obtained by mass spectrometry with the sequences found in the data banks. It is noteworthy that neither SNLRP1 nor SNLRP2 agglutinate erythrocytes [15].

2.5. *Carbohydrate Binding Properties of Nigrin l, SNAlm, and SNAld*

Nigrin l, SNAlm, and SNAld showed hemagglutination activities in human erythrocytes (Section 2.2). To elucidate the sugar binding specificities of these proteins, hemagglutination inhibition with various monosaccharides and disaccharides was carried out (Table 1). The results show that the agglutination produced by the three proteins was inhibited by D-galactose and lactose (β-D-galactopyranosyl-(1→4)-D-glucose). In none of the three proteins was an affinity for D-glucose, D-fructose, D-mannose, or L-fucose observed at the maximum sugar concentration tested. The protein showing the highest affinity for galactose was SNAlm, whereas SNAld showed the lowest affinity. In all cases, lactose inhibited agglutination at a concentration four times lower than galactose.

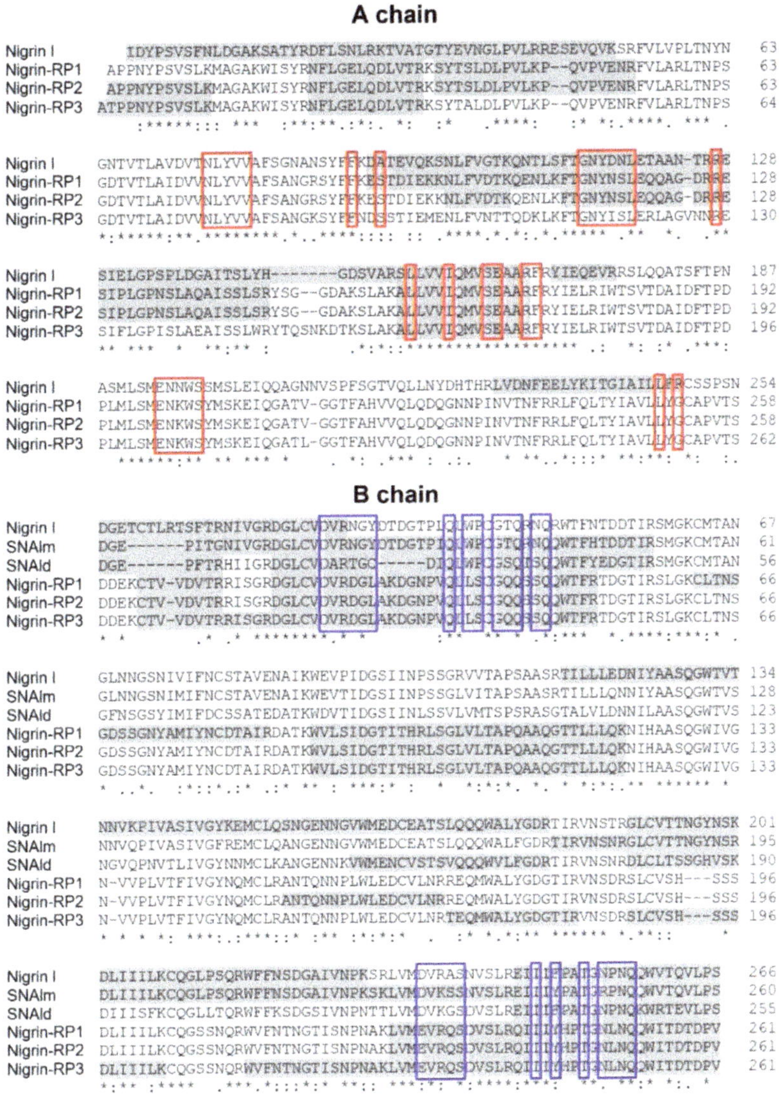

Figure 7. Comparison of the sequences of RIPs and lectins from leaves of *S. nigra* obtained by peptide mapping via high-resolution nanoLC–MS/MS mass spectrometry with sequences deposited in the data banks. The gray shaded sequences were obtained by peptide mapping via high-resolution MS/MS mass spectrometry and match sequences obtained from the data banks with the access numbers AAN86130 (nigrin l), AAN86132 (SNAlm), AAN86131 (SNAld), AAC49672 (nigrin-RP1 and 2), and AAC49673 (nigrin-RP3). The red boxes indicate the amino acids that possibly form the catalytic pocket, and the blue boxes indicate the amino acids that possibly form the 1α and 2γ sugar binding sites. Identical residues (*), conserved substitutions (:), and semiconserved substitutions (.) are reported.

Table 1. Inhibition of the hemagglutination activity of nigrin l, SNAlm, and SNAld by sugars.

	Minimum Concentration Inhibiting Hemagglutination (mM)		
Carbohydrates [1]	**Nigrin l**	**SNAlm**	**SNAld**
D-galactose	50	6.25	100
Lactose	12.5	1.56	25

[1] No inhibition of hemagglutination at the maximum sugar concentration tested (200 mM) was observed with the following sugars: D-glucose, D-fructose, D-mannose, and L-fucose.

2.6. Structural Analysis of Nigrin l, SNAlm, and SNAld

Given the availability of the complete amino acid sequence, it was possible to predict the three-dimensional structures of nigrin l, SNAlm, and SNAld with a computational model using the potentials of deep learning and neural networks [23]. The best three-dimensional models obtained for nigrin l, SNAlm, and SNAld are shown in Figure 8, and all showed local Distance Difference Test (lDTT) values that were much higher than 90%, which makes them suitable for characterizing the binding sites [23]. As described for ricin and other type 2 RIPs [24,25], the A chain of nigrin l consists of three folding domains. The first domain includes the N-terminal, and is composed of six antiparallel β-sheets and two α-helices in the order aAbcdeBf. The second domain consists of five α-helices (helices from C to G). The last domain consists of two α-helices and two antiparallel β-sheets in a α-helix–β-fork–α-helix motif (HghI). Similar to other type 2 RIPs, the B chain is made up of two homologous globular lectin domains arising from gene duplication, which are made up exclusively of β-sheets. Each domain consists of four homologous subdomains (1λ, 1α, 1β, and 1γ for lectin 1; 2λ, 2α, 2β, and 2γ for lectin 2). The subdomains 1λ and 2λ are responsible for the linking to the A chain and the interconnection between the two domains of the B chain, respectively. The subdomains 1α, 1β, and 1γ are arranged in a trefoil structure. This arrangement is also present in lectin 2 with the subdomains 2α, 2β, and 2γ. The 1α and 2γ subdomains contain the two D-galactose binding sites. SNAlm and SNAld have structures similar to nigrin l, but both lack the A chain. In addition, SNAld has an additional cysteine (Cys 23) that allows it to form a dimer with another identical chain.

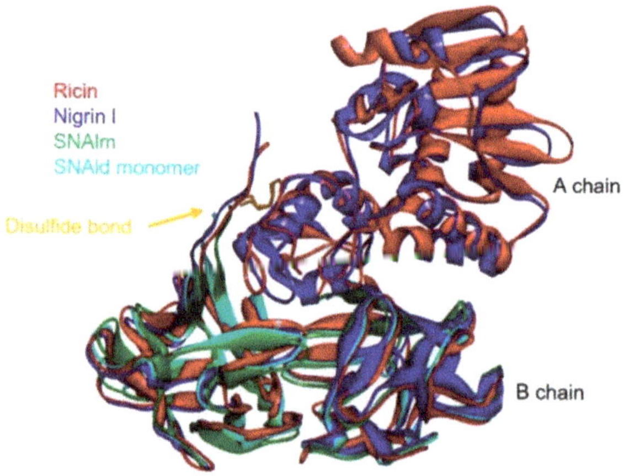

Figure 8. Comparison of the structures of ricin, nigrin l, SNAlm, and the monomer of SNAld. The three-dimensional structural modeling was carried out using AlphaFold2 software, and the figure was generated using Discovery Studio 2021. The arrows indicate the position of the disulfide bond linking A and B chains.

2.7. Molecular Docking

The availability of 3D models of the proteins that allow studies at the molecular level encouraged us to study how D-galactose binds to the 1α and 2γ sites of nigrin l, SNAlm, and SNAld. For this purpose, we carried out a molecular docking study using Autodock 4.2, and compared the results with those already published for ricin [26].

As shown in Figure 9, the sequences of the binding sites for sugars of nigrin l, SNAlm, and SNAld are similar to those of ricin. Seven of the fourteen amino acids in the binding pocket of the 1α site are identical, as are seven out of twelve amino acids at the 2γ site. This is consistent with the fact that all these proteins are specific for D-galactose and lactose. However, there are differences with the sugar binding sites of ricin that may influence the affinity of these sites for sugars. The 1α site is identical in nigrin l and SNAlm. In addition, these sites are relatively similar to that of ricin. In all three proteins, the binding of β-D-galactopyranose is the result of the C–H–π interaction between the aromatic rings of tryptophan (W37, W39, and W33 in ricin, nigrin l, and SNAlm, respectively) and the apolar face of the pyranosic ring of galactose. The C–H groups of carbons 3, 4, 5, and 6 are oriented towards the aromatic rings of tryptophan, allowing the π interaction of the electron cloud with the aliphatic protons of sugar that carry a positive partial charge. The polar face of galactose forms hydrogen bonds with five amino acids located on the other side of the binding pocket, four of which (aspartic, arginine, glutamine, and asparagine) are identical in all three proteins. The 1α site of the SNAld is very different. This is likely influenced by the presence within this site of the cysteine (C23) that forms the disulfide bond with the other subunit. Nevertheless, both galactose and lactose can bind at the 1α site of SNAld, although galactose adopts a different arrangement that can affect the affinity for this sugar. In this case, carbons 1, 5, and 6 are oriented towards the aromatic rings of tryptophan, and only coincide, on the other side, the amino acids aspartic and arginine.

At the 2γ site, the amino acids that provide aromatic rings are tyrosine (in ricin and SNAlm) and phenylalanine (in nigrin l and SNAld). However, these residues seem to not affect the orientation of the pyranosic ring at this site, as it is virtually identical. The C–H groups of carbons 4 and 6 are oriented towards the aromatic ring, allowing the interaction with the apolar face of D-galactose, while the polar face forms hydrogen bonds with various amino acids, of which at least four are identical in the three proteins. In ricin, the orientation of galactose is slightly different since the C–H groups of carbons 3, 4, and 5 are oriented towards the aromatic ring, while on the other side, only two amino acids (aspartic and asparagine) coincide.

Similar to D-galactose, lactose can also bind to the 1α and 2γ sites of nigrin l, SNAlm, and SNAld (Figures S8 and S9). However, in all cases, lactose inhibited the agglutination of erythrocytes at a concentration four times lower than D-galactose (Table 1). This is not in accordance with the estimated free energy of binding data provided by Autodock 4.2 (Table S7). One possible explanation is that lactose contains D-galactose in the β-pyranosic form, allowing effective binding to aromatic amino acids at sugar binding sites.

It is worth mentioning that peptide mapping identifies the sequences of the binding sites of nigrin-RPs 1–3 (Figure 7), which match with the sequences of SNLRP1 and SNLRP2 [15]. These are proteins from the bark of *S. nigra* that, similar to nigrin-RPs, do not agglutinate erythrocytes. This was initially attributed to the fact that the 1α and 2γ sites were inactive because amino acid substitutions at these sites prevented carbohydrate binding [15]; however, SNLRP has subsequently been reported to interact with N-acetylglucosamine oligomers, as well as with many glycan structures containing N-acetylglucosamine [27].

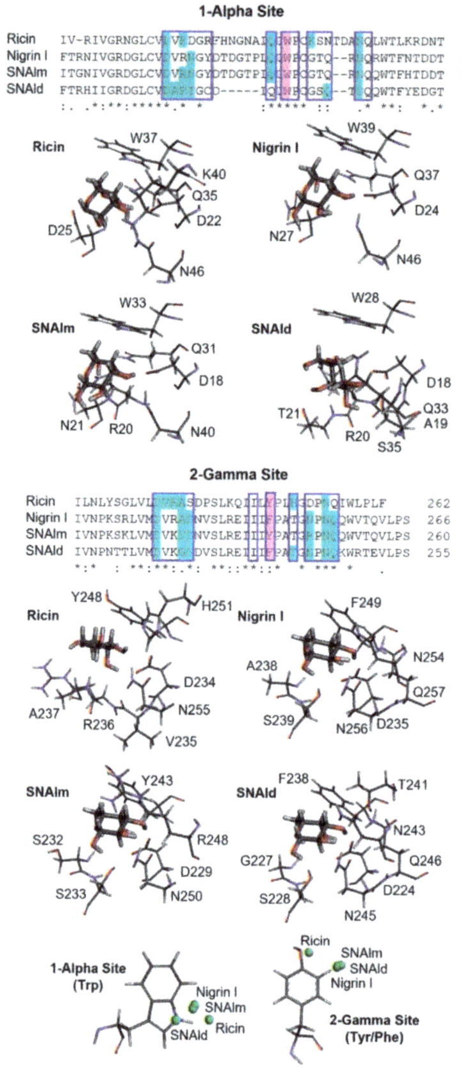

Figure 9. Three-dimensional models of the galactose binding sites of ricin, nigrin l, SNAlm, and SNAld. The alignment of the sequences of subdomains 1α and 2γ of ricin, nigrin l, SNAlm, and SNAld are represented. The binding pockets of sites 1α and 2γ are indicated by blue boxes. Amino acids involved in galactose binding by C–H–π interactions or hydrogen bonds are colored purple and cyan, respectively. Identical residues (*), conserved substitutions (:), and semiconserved substitutions (.) are reported. The galactose binding sites of ricin (PDB 2AAI), nigrin l, SNAlm, and SNAld complexed with β-D-galactopyranose (thick sticks) are represented. The amino acids that bind the galactose molecule, either by C–H–π interactions or both conventional and nonconventional hydrogen bonds, are represented by thin sticks. At the bottom, the positions of the centroids of the pyranosic ring of D-galactose with respect to the aromatic rings of sites 1α and 2γ of ricin, nigrin l, SNAlm, and SNAld are represented by green balls.

2.8. Cytotoxic Effect of RIPs from Elderberry Leaves in Cell Cultures

Type 1 RIPs consisting of a single enzymatic (A) chain usually display lower toxicity than type 2 RIPs consisting of a binding (B) chain with lectin activity linked to the enzymatic A chain. The carbohydrate binding domains of the B chain recognize glycosylated receptors on the cell surface, facilitating the entry of the A chain into the cell. Ricin is a well-known example of a highly toxic type 2 RIP. However, all type 2 RIPs found in the genus *Sambucus* are considered nontoxic type 2 RIPs since, despite the high enzymatic activity on ribosomes comparable to that of ricin, they show much lower toxicity to cells and animals [3,4,11]. Figure 10a shows the toxicities of RIPs from *S. nigra* leaves towards HeLa and COLO 320 cells. The RIPs from *S. nigra* leaves were toxic to HeLa and COLO 320 cells, exhibiting IC_{50} (concentration of protein causing the death of 50% cells) values ranging from 19 to >1460 nM. In all cases, the cytotoxicity of these RIPs was much less than that exerted by ricin, which affects viability, with IC_{50} values several orders of magnitude lower (0.14–0.6 pM) [14]. The most sensitive were HeLa cells, showing IC_{50} values ranging from 19 to 580 nM, while COLO 320 cells exhibited values between 19 and >1460 nM after 48 h of treatment. When comparing the type 2 RIPs nigrin l, a galactose-binding protein, and nigrin-RPs 1–3, nigrin l was the most active toxin, with IC_{50} values of 19 nM for COLO 320 and HeLa cells. Nigrin-RPs 1–3 displayed very low toxicity, especially on COLO 320 cells, comparable to that of the type 1 RIP, nigritin l. The IC_{50} values obtained from treated HeLa cells were 130, 340, 580, and 280 nM for nigrin-RPs 1–3 and nigritin l, respectively. The lack of sugar binding activity of nigrin-RPs could play a role in their low cytotoxicity. HeLa cells treated with nigrin l, nigrin-RPs, and nigritin l exhibited morphological features characteristic of apoptosis, such as cell rounding and blebbing (data not shown). Several studies reported that the cytotoxicity of RIPs is associated, in addition to rRNA damage, with their ability to induce apoptosis [28]. To investigate the capability of nigrin l to reach the cytosol and inactivate the ribosomes after being endocytosed, we analyzed the ribosomal RNA from HeLa cells treated with the RIP for 48 h. Figure 10b shows that the ribosomes were depurinated, releasing the diagnostic fragment after treatment of the RNA with acid aniline, indicating that nigrin l was able to reach the ribosomes to inhibit protein synthesis. Apoptosis might be a consequence of the ribotoxic stress induced by the RIP after entry into the cytosol, or both processes could run in parallel. To determine whether the observed cytotoxic effects of nigrin l were also mediated via apoptosis in COLO 320 cells, we evaluated the breakdown of the nuclear DNA into oligonucleosomal fragments. As shown in Figure 10c, when COLO cells were treated with 40 nM nigrin l for 72 h, the cleavage of the chromosomal DNA was clearly observed. Thus, our data suggest that the apoptotic pathway was involved in the cell death mediated by RIPs from *S. nigra* leaves.

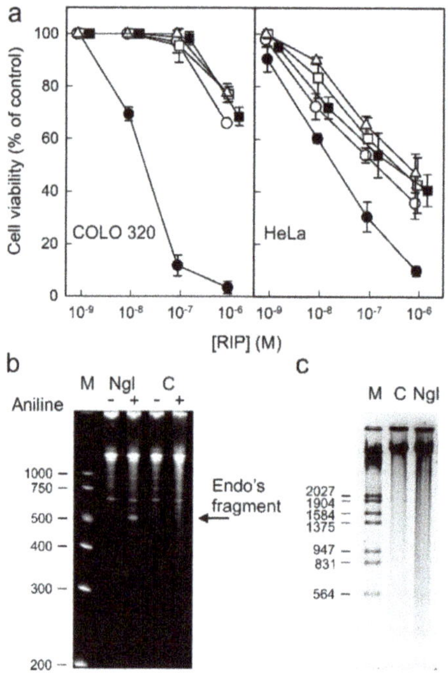

Figure 10. Cytotoxicity of nigrin l, nigrin-RPs 1–3, and nigritin l on HeLa and COLO 320 cells: (**a**) Effect of nigrin l (•), nigrin-RP1 (○), nigrin-RP2 (□), nigrin-RP3 (△), and nigritin l (■) on viability of COLO 320 (left panel) and HeLa (right panel) cells. Cells were incubated with different concentrations of RIPs for 48 h, and cell viability was evaluated by a colorimetric assay, as indicated in the Materials and Methods. Data represent the mean ± SD of two experiments performed in duplicate; (**b**) N-glycosylase activity of nigrin l on rRNA from HeLa cells. rRNA N-glycosylase activity was assayed as indicated in the Materials and Methods. Each lane contained 2 µg of RNA isolated from either untreated cells (C, control) or cells incubated with 40 nM of nigrin l for 48 h. The arrow indicates the RNA fragment released as a result of RIP action upon acid aniline treatment. Numbers indicate the size of the standards (M) in nucleotides; (**c**) Effect of nigrin l on internucleosomal DNA fragmentation. COLO 320 cells were incubated in the absence (C, control) or presence of 40 nM of nigrin l for 72 h. The DNA was isolated and 4.0 µg was electrophoresed, as indicated in Section 5.3.15. The numbers indicate the corresponding size of the standards (M) (λDNA HindIII/EcoRI) in pb.

3. Discussion

Species of the genus *Sambucus* are one of the best sources of ribosome-inactivating proteins. From different tissues, type 1 RIPs, type 2 RIPs (heterodimeric and tetrameric), and lectins (monomeric and dimeric) related to the B chain of type 2 RIPs have been obtained [3,11]. Type 2 RIPs isolated from *Sambucus* are peculiar in that they lack the toxicity of other type 2 RIPs, such as ricin [29], abrin [30], stenodactylin [28], and kirkiin [22]. For example, nigrin b is 50,000 times less toxic than ricin to HeLa cells, and 1,500 times less toxic to mice [31]. Therefore, they have been used as the enzymatic component of immunotoxins and other conjugates directed against tumor cells [3,11]. Due to their specificity for the α2,6-linked sialic acid, SNAI from the bark of *S. nigra* and SSA from the bark of *S. sieboldiana* are used in highly diverse biomedical applications, such as diagnosis by ELISA [32,33], histochemistry [34,35], confocal fluorescence microscopy [36], microarrays [37], and new therapeutic strategies [38]. Elderberry RIPs have also been used in agriculture to obtain transgenic plants resistant to viruses [39,40] and insects [41,42].

Most of these proteins have been obtained from *S. nigra*. Thus, from the bark of this species, a galactose-specific type 2 RIP (nigrin b or SNAV) [43,44]; a sialic-acid-specific type 2 RIP (SNAI') [45]; three type 2 RIPs that do not agglutinate erythrocytes, and that could be specific for N-acetylglucosamine (SNLRP1, SNLRP2, and basic nigrin b) [15,27,46]; a sialic-acid-specific tetrameric type 2 RIP (SNAI) [47]; and a galactose-specific monomeric lectin (SNAII) [48] have been purified. From the fruits two type 1 RIPs (nigritins f1 and f2) [49], a galactose-specific type 2 RIP (nigrin f) [50], a sialic-acid-specific tetrameric type 2 RIP (SNAIf) [51], and a galactose-specific monomeric lectin (SNAIV or SNAIVf) have been obtained [52]. From the seeds, nigrin s (type 2 RIP) [53] and SNAIII (monomeric lectin) [54], both specific for galactose, have been purified. Finally, the presence of SNAflu-I, a tetrameric type 2 RIP reported as being specific for galactose, has been described in the flowers of *S. nigra* [55,56]. Some of them can be considered isoforms; for example, SNAI from the bark and SNAIf from fruits have amino acid identities of 95%, and can therefore be considered as tissue-specific isoenzymes.

However, even though elderberry has been the subject of intense research for the search and study of RIPs and lectins since 1984 [57], leaf proteins have never been purified, despite crude leaf extracts exhibiting very potent activities in both protein synthesis inhibition in rabbit reticulocyte lysate (IC_{50} = 50 ng/mL) and hemagglutination (minimum concentration agglutinating erythrocytes = 1 mg/mL) (data not shown). One of the reasons for this is the difficulty of isolating a considerable number of proteins with similar characteristics, and which also present great developmental variations in their expression in this tissue (data not shown). In spite of this, we proposed to investigate the presence of RIPs and lectins in the leaves in order to isolate and characterize them to enhance our knowledge of these proteins.

We found three proteins that are specific for galactose: nigrin l (a heterodimeric type 2 RIP with an A chain of 27.3 kDa and a B chain of 33.7 kDa), SNAlm (a 32.4 kDa monomeric lectin), and SNAld (a homodimeric lectin with two identical 30.9 kDa subunits) (Figure 4a,b). These proteins could be identified by peptide mapping with three sequences obtained from *S. nigra* leaves cDNA. The sequenced peptides matched with the sequences with access numbers AAN86130 (nigrin l), AAN86132 (SNAlm), and AAN86131 (SNAld), with coverages of 59, 53, and 38%, respectively (Figure 7). Nigrin l can be considered an isoenzyme of nigrin b from the bark since they have an amino acid identity of 98.4%. Likewise, SNAlm can be considered an isoform of SNAIV (or SNAIVf) from fruits, with which it shares 90.2% of the amino acids. However, no dimeric lectins have been found in other elderberry tissues; thus, the most related protein is SELld, found in the leaves of *S. ebulus* [58], with which it presents a homology of 89.9%. Therefore, these dimeric lectins appear to be unique to the leaves. It is also noteworthy that in leaves, no tetrameric type 2 RIPs corresponding to SNAI from the bark, SNAIf from the fruits, or SNAflu-I from flowers were found, tissues in which this type of structure is among the predominant [13,51,55–57].

We also found four heterodimeric type 2 RIPs that fail to agglutinate erythrocytes (nigrin-RPs 1, 2, 3, and 4). Nigrin-RPs 1 and 3 are heterodimers whose A and B subunits have a similar molecular weight of about 30 kDa, whereas nigrin-RPs 2 and 4 are heterodimers whose A and B subunits have molecular weights of 30.7 and 33.7 kDa for nigrin-RP2 and 27.9 and 29.8 kDa for nigrin-RP4 (Figure 4a,b). All these proteins are strongly glycosylated, a characteristic they share with most RIPs and lectins from the genus *Sambucus* [14,46,49,50,59,60]. Peptide mapping identified nigrin-RP3 as an isoform of SNLRP1 from the bark, and nigrins-RPs 1 and 2 as SNLRP2 isoforms (Figure 7). Unfortunately, not enough nigrin-RP4 was obtained to perform peptide mapping. The data indicate that nigrin-RPs 1 and 2 could be isoforms with different amino acid sequences or proteins with the same sequence, but different states of glycosylation. In favor of the latter hypothesis is the fact that no differences in the amino acid sequences were found in peptide mapping (Figure 7). Moreover, the B chain of nigrin-RP2 is strongly glycosylated (Figure 4d), suggesting the occurrence of a different glycosylation pattern for the B chain of nigrin-RP2 compared to that of nigrin-RP1 (Figure 4d). This could also explain the different behaviors of the two proteins when interacting with the CM-Sepharose chromatography column (Figure 3a) and in SDS-PAGE (Figure 4). However, the evidence that the two proteins display different enzymatic activities, with nigrin-RP1 being more active, both as an inhibitor of protein synthesis and in nicking activity, leads us to hypothesize that they are probably proteins with different amino acid sequences. Finally, we also found a type 1 RIP, nigritin l, which we have not been able to map. However, based on its behavior in chromatography columns, electrophoresis, and its enzymatic activities (for example, its nicking activity), it could very likely correspond to an isoform of nigritin f1, a protein isolated from fruits of *S. nigra* [49].

One of the most important features of RIPs to consider is their enzymatic properties, both for their possible biological role and for their biotechnological applications. RIPs are inhibitors of protein synthesis using rRNA N-glycosylase activity, which catalyzes the elimination of a specific adenine located in the sarcin–ricin loop (SRL) that is present in the large rRNA of eukaryotes and prokaryotes [1,2]. All the RIPs from elderberry leaves were shown as strong inhibitors of protein synthesis in rabbit reticulocyte lysate, being the most potent nigrin l, nigrin-RP3 and nigrin-RP4 which presented an IC_{50} 20-fold lower than nigritin l. Since the sequence of the active site is very similar in all proteins (Figure 7), the difference could be attributed to small differences in the active site, but mainly to differences in the amino acid sequence of the A chain that binds to ribosomal proteins [61,62]. All RIPs in this study showed rRNA N-glycosylase activity, not only in rabbit reticulocyte ribosomes, but also in yeast ribosomes (Figure 5a,b). However, unlike BE27, none showed activity against bacterial ribosomes (Figure 5c). This is often considered an advantage for biotechnological applications because it facilitates their cloning and expression in bacteria.

Although RIPs are classified as rRNA N-glycosylases, one very important enzymatic activity of these proteins is the depurination of nucleic acids (adenine polynucleotide glycosylase or APG activity) [21]. Different RIPs show different APG activities on DNA and RNA; however, all of them can depurinate the DNA of herring and/or salmon sperm, and some of them can also depurinate various types of RNA [18,19,21]. Although this activity has only been demonstrated in vitro, it could be important for apoptotic activity against animal cells or antiviral activity [19]. All RIPs tested showed APG activity on salmon sperm DNA (Figure 5d), similar to that of ricin and kirkiin [22], although much lower than that of BE27 [19], since type 1 RIPs usually have an APG activity on DNA 10-fold higher than type 2 RIPs [21]. Of all the RIPs in elderberry leaves, the one that displayed the highest activity was nigrin l, which showed a twofold higher activity with respect to the other proteins.

Some RIPs show endonuclease (nicking) activity on the DNA of supercoiled plasmids that produces relaxed and sometimes linear plasmids. This ability may be necessary for these proteins to perform different biological functions, including resistance to pathogenic microorganisms or viruses [18,19]. Only nigrin-RP1 and nigritin l promoted the conversion of supercoiled DNA into relaxed forms. This activity was dependent on magnesium ions, as it was inhibited by EDTA, according to what has been reported for other RIPs [18,19]. Regarding this activity, nigrin-RP1 and nigrin-RP2 showed very different behaviors, so the differences between both proteins must be more than just a difference in the glycosylation pattern. Many RIPs do not present this activity, which is related to a different configuration in the structure near the active site that allows the accommodation of the supercoiled DNA of the plasmids [19,63].

Nigrin l, SNAlm, and SNAld were revealed to be lectins with affinities for galactose (Table 1), the affinity of SNAlm for both galactose and lactose being significantly higher. This contrasts with some elderberry RIPs that have affinities for sialic acid, and appear not to be present in elderberry leaves. Due to the arrangement of its hydroxyl groups, β-D-galactopyranose has two faces, a polar face and an apolar face. In several RIPs such as ricin, abrin, stenodactylin, and kirkiin, the binding of β-D-galactopyranose to the 1α and 2γ sites is the result of the C–H–π interaction between the aromatic ring(s) of an amino acid (tryptophan, phenylalanine, tyrosine, or histidine) and the apolar face of the pyranosic ring of galactose, and hydrogen bonds between the hydroxyl groups of the polar face and amino acids located on the other side of the pocket of binding [24,25].

The recent emergence of programs that predict the structure of proteins at the atomic level [23,64] has allowed us to conduct in silico experiments to predict how galactose and lactose bind to these proteins. A surprising finding is that the 1α site of the SNAld is fully functional, although the subdomain in which it is located is also used for the linking of the two monomers that form the dimeric protein (Figure 9). In SNAld, the 1α site has undergone numerous changes, including the deletion of five amino acids, which allows it to keep away the 1α site of the other subunit without affecting the galactose binding capacity (Figure S10). The way galactose binds to the 1α site of nigrin l and SNAlm is practically identical, and very similar to that of ricin, although some of the amino acids with which it forms hydrogen bonds change, which could affect the affinity for sugars (Figure 9). However, the orientation of galactose at the 1α site of SNAld is different, caused by what we have discussed above, which does not seem to affect the affinity for monosaccharide with respect to the other two proteins. In the case of the 2γ site, the way galactose binds to the three elderberry proteins is practically identical, and somewhat different from that of ricin (Figure 9). It has been suggested that the difference between toxic RIPs, such as ricin, and nontoxic RIPs such as ebulin l or nigrin b, lies in the 2γ site, which binds galactose in a different way, resulting in a decreased affinity for membrane glycoproteins containing galactose and, as a consequence of the deficient binding, an intracellular pathway different from that of ricin [3,11]. The most notable change at this site is the substitution of tyrosine for phenylalanine, which seemed to suggest that this change alone could modify the way galactose binds. In the case of ebulin l, this prevents the binding of lactose, which explains the fact that this protein has less affinity for the α-lactose-agarose gel [65]. Nigrin b, which has a 98.4% homology with nigrin l, also has less affinity for the AT-Sepharose matrix [13]; however, we have not observed that the binding of galactose to the 2γ site of nigrin l is affected by glucose as part of lactose (Figure S9). Nor does the change from tyrosine to phenylalanine seem to be decisive because SNAlm contains tyrosine and does not exhibit a different mode of galactose binding from that of nigrin l. However, this different affinity could be explained by the different way galactose binds to this site with respect to ricin (Figure 9).

Peptide mapping identified the sequence of the binding sites of nigrin-RPs l, 2, and 3 with the sequences of SNLRP1 and SNLRP2, type 2 RIPs that fail to agglutinate erythrocytes and do not recognize galactose [15]. In these proteins, the most prominent change occurs at the 1α site, which lacks an aromatic amino acid. The 2γ site contains tyrosine; however, several changes occur in the amino acids that form hydrogen bonds with galactose, which could affect the monosaccharide binding to this site. On the other hand, ebulin-RP is a heterodimeric type 2 RIP present in *S. ebulus* leaves that displays rRNA N-glycosylase activity but lacks functional sugar binding domains [14]. Moreover, the B chain of ebulin-RP shares a strong homology (83.14%) with the B chains of SNLRPs, and thus with those of nigrin-RPs. It has been proposed, based on molecular docking, that the loss of lectin activity of ebulin-RP may be due to the presence of inactive 1α and 2γ sites, and this could also be the case for nigrin-RPs from leaves.

We have performed a study of the cytotoxic activity of *S. nigra* leaf RIPs, including the type 2 RIPs (nigrin l and nigrin-RPs 1–3) and the type 1 nigritin l, towards COLO 320 and HeLa cells. First, our data confirmed that the elderberry type 2 RIPs are much less toxic than ricin, which affects the viability of these cells, with IC_{50} five–six orders of magnitude lower [14]. The reason for the different toxicities among type 2 RIPs is not clear. Type 2 RIPs from *Sambucus* leaves, inactivate ribosomes in vitro with higher efficiency than ricin, therefore, the different toxicities could be better attributed to differences between the B chains, which are responsible for the interaction with cellular membranes, than to the enzymatic A chains. In this line, it has been shown that the type 2 RIPs from *Sambucus*, ebulin l and nigrin b, bind to cells to a lesser extent than ricin [4,31]. Furthermore, these RIPs have lower affinities for galactose than ricin [4,13,65]. The differential affinity of

RIPs from *Sambucus* for galactosides on cell surfaces may determine its intracellular fate and possibly its cytotoxicity. Thus, it has been reported that, unlike ricin, type 2 RIPs from *Sambucus* follow a weakly productive Golgi-independent pathway to the cytosol to intoxicate the cells [4,14,31]. An important observation when comparing the toxicity of nigrin l, nigrin-RPs, and nigritin l towards COLO 320 and HeLa cells is that nigrin l was the most active toxin. Nigrin l is a galactose binding lectin, whereas nigrin-RPs 1–3 failed to bind to the affinity matrix of AT-Sepharose and to agglutinate erythrocytes. Therefore, the low cytotoxicity of nigrin-RPs compared to nigrin l could be related to deficient sugar binding domains, which is the major difference with nigrin l. According to this, the toxicity of nigrin-RPs is comparable to that of the type 1 RIP nigritin l that lacks a B chain. Several studies underline that the cytotoxicity of RIPs is associated with their ability to induce apoptosis [28]. We also found that RIPs from *S. nigra* leaves induced apoptosis. HeLa cells treated with the RIPs showed apoptotic morphological features and, in COLO 320 cells, nigrin l treatment led to oligonucleosomal DNA fragmentation. One of the unresolved questions is whether rRNA damage leading to the inhibition of protein synthesis is solely responsible for this RIP-induced apoptosis. On the one hand, the ribosome may not be the only substrate of RIP action and, on the other hand, in the case of type 2 RIPs, apoptosis might also be induced by lectin binding to specific glycosylated proteins on the cell membrane, leading to activation of cell death factor receptors.

4. Conclusions

This work contributes to expanding our knowledge of the family of RIPs and RIP-related lectins produced by *S. nigra*, with eight new proteins found in the leaves. Our results demonstrate the presence in this tissue of a type 2 RIP and two related lectins specific for galactose, four type 2 RIPs with deficient sugar binding domains, and one type 1 RIP.

Although initially only SNAI has been used, with the discovery and study of other proteins of the genus *Sambucus*, its use has been extended to other proteins, such as SSA, ebulin l, nigrin b, and SNAI'. The study of RIPs and related lectins of elderberry leaves completes the knowledge of this type of proteins in this species, and opens new perspectives, not only in the study of the biological functions attributed to them, but also in their use in biomedicine and agriculture.

New techniques such as protein structure prediction based on deep learning and neural networks, and peptide mapping based on high-resolution nanoLC–tandem mass spectrometry, can be very useful tools to further advance our knowledge of these types of proteins.

5. Materials and Methods

5.1. Materials

The sources of the chemicals were described previously [16]. Leaves from elder were harvested at Cobos de Cerrato (Palencia, Spain) in early summer. CM-Sepharose FF, Q-Sepharose FF, CM-Sepharose FF, Sepharose 6 B, and Superdex−75 HiLoad 26/60 columns were purchased from GE Healthcare (Barcelona, Spain). The acid-treated Sepharose (AT-Sepharose) was prepared as described in [66], treating the Sepharose 6 B with 0.1 N HCl at 50 °C for 3 h. The gel was then washed with water (Elix 5, Millipore) until a neutral pH was obtained, and stored in water at 4 °C until it was used. Tosyl phenylalanyl chloromethyl ketone (TPCK)-treated trypsin was purchased from Merk Life Science S.r.l. (Milan, Italy). Acetonitrile (CH3CN), formic acid (FA), and water (LC–MS grade) were from Fisher Scientific Italia (Milan, Italy). Century™-Plus RNA Markers were purchased from Fisher Scientific (Madrid, Spain). Rabbit reticulocyte lysate system (nuclease-treated) was purchased from Promega Biotech Iberica S.L. (Alcobendas, Madrid, Spain).

5.2. Cell Lines and Culture

COLO 320 (human colon carcinoma) and HeLa (human cervix epithelioid carcinoma) cells were obtained from the European Culture Collection (ECACC) and grown in RPMI

1640 medium (GIBCO BRL, Barcelona, Spain) supplemented with 10% fetal bovine serum (FBS), 100 U/mL penicillin, and 0.1 mg/mL streptomycin, under 5% CO_2 at 37 °C.

5.3. Methods

5.3.1. Preparation of Crude and Acidified Extracts and Chromatography Using SP-Sepharose

Two extracts were prepared from 700 and 540 g of *S. nigra* leaves. The leaves were crushed with dry ice in a crusher (Sammic Cutter K−52) and extracted with eight volumes of PBS (5 mM sodium phosphate, pH 7.5, 0.14 M NaCl) overnight at 4 °C. The crude extract was clarified by filtering it through a nylon mesh and then centrifuging it for 30 min at 9000 rpm in a JA−10 rotor (12,900× *g*) at 2 °C. The crude extract was acidified by adding glacial acetic acid to a pH of 4.05, and clarified again by filtering it through a nylon mesh and centrifuging it under the same conditions. The acidified crude extract was loaded onto a SP-Sepharose FF column (25 × 5 cm = 491 mL) equilibrated with 10 mM sodium acetate (pH 4.5). The column was washed at a flow rate of 7 mL/min with the same buffer until the absorbance at 280 nm of eluent dropped to almost zero. Proteins were first eluted with 5 mM sodium phosphate (pH 6.66) and then with the same buffer containing 1 M NaCl. Protein elution was controlled by measuring the absorbance at 280 nm, and the fractions eluted with sodium phosphate and NaCl were collected separately.

5.3.2. Purification of Nigrin l, SNAlm, and SNAld

The protein eluted with sodium phosphate from the first preparation (1116 mg) was supplemented with 0.28 M NaCl and subjected to chromatography using AT-Sepharose (25 × 5 cm = 137 mL) equilibrated with 5 mM sodium phosphate (pH 7.5) containing 0.28 M NaCl. The column was kept at 0 °C and three identical chromatographies were performed, each loading 372 mg of protein. The column was washed at a flow rate of 4.5 mL/min with sodium phosphate 5 mM (pH 7.5) containing 0.28 M NaCl until the absorbance at 280 nm dropped to almost zero, and was eluted with the same buffer containing 0.2 M lactose. The eluate from the three chromatographies (218 mg of protein) was combined and concentrated up to 17 mL using an Amicon YM10 membrane. Next, three aliquots were prepared, each of which was subjected to molecular exclusion chromatography using a Superdex 75 HiLoad 26/60 (60 × 2.6 cm = 319 mL) column equilibrated with PBS at a flow rate of 2 mL/min. Fractions of 5 mL were collected, and their composition was determined by SDS-PAGE in the presence and absence of 2-mercaptoethanol. Fractions containing SNAlm were collected, dialyzed against water, frozen, and freeze dried; 84 mg of lyophilized protein was obtained. Fractions containing nigrin l and SNAld were collected, concentrated, and re-chromatographed on the same column to remove traces of SNAlm, dialyzed. Subsequently, these fractions, in 5 mM sodium phosphate (pH 7.5), were subjected to anion exchange chromatography using Q-Sepharose FF (5 × 1.6 cm = 10 mL) equilibrated with 5 mM sodium phosphate (pH 7.5) at a flow rate of 4 mL/min. After loading the sample and washing with sodium phosphate buffer, the protein was eluted with a linear gradient of 600 mL of NaCl from 0 to 0.6 M. Fractions of 8 mL were collected, and the fractions containing nigrin l were pooled together, dialyzed against water, frozen, and freeze dried. The SNAld was further purified by repeating the same chromatography procedure; 14 mg of nigrin l and 10 mg of SNAld were obtained.

5.3.3. Purification of Nigrin-RPs 1, 2, 3, and 4, and Nigritin l

The protein eluted with NaCl from the second preparation (1820 mg) was dialyzed and subjected to cation exchange chromatography using CM-Sepharose FF (7 × 2.6 cm = 37 mL) equilibrated with 5 mM sodium phosphate (pH 6.66) at a flow rate of 7 mL/min. After loading the sample and washing with sodium phosphate, the protein was eluted with a linear gradient of 1596 mL of NaCl from 0 to 0.3 M. Fractions of 10.5 mL were collected, which were tested via protein synthesis, and analyzed by SDS-PAGE. Fractions containing nigrin-RPs 1–4 and nigritin l were placed together, concentrated using an Amicon YM10 membrane, and separately subjected to molecular exclusion chromatographies through a Superdex 75 HiLoad 26/60 column equilibrated with PBS at a flow rate of 2 mL/min. Fractions of 5 mL were collected, and their composition was determined by SDS-PAGE in the presence and absence of 2-mercaptoethanol. Nigrin-RP1 was subjected to chromatography using AT-Sepharose (5 × 4 cm = 20 mL) to remove traces of nigrin l; the fraction not retained by the AT-Sepharose was dialyzed against water and freeze dried, obtaining 12 mg of lyophilized protein. Both the fractions containing nigrin-RP2 and those containing nigrin-RP4 were re-chromatographed using a Superdex 75 HiLoad column to eliminate traces contaminating the other protein, obtaining 14 and 0.15 mg of nigrin-RP2 and nigrin-RP4, respectively. The purification of nigrin-RP3 was conducted using the first preparation. After molecular exclusion chromatography, the protein was dialyzed and purified to homogeneity by chromatography using SP-Sepharose FF (4.5 × 1 cm = 3.5 mL) equilibrated with 5 mM sodium phosphate (pH 7.5) at a flow rate of 1 mL/min. After loading the sample and washing with sodium phosphate, the protein was eluted with a linear gradient of 140 mL of NaCl from 0 to 0.2 M. Fractions of 2 mL were collected and the fractions containing nigrin-RP3 were placed together, dialyzed against water, frozen, and freeze dried, obtaining 2 mg of lyophilized protein. Nigritin l was re-chromatographed using a Superdex 75 HiLoad column to remove traces of contaminants, dialyzed, and freeze dried, obtaining 8 mg of lyophilized protein.

5.3.4. Analytical Procedures

Protein concentrations were determined using the spectrophotometric method of Kalb and Bernlohr [67]. Analyses of proteins by SDS-PAGE were carried out as described elsewhere [68] using 12% acrylamide gels and the Hoefer™ MiniVE system (Thermo Fisher Scientific-ES, Madrid, Spain). The glycosylation analysis was performed in gel after SDS-PAGE with a Mini-Protean II system (Bio-Rad; Milan, Italy) using the Pro-Q emerald 300 Glycoprot Probes Kombo (Life Technologies, Monza, Italy). Glycosylated proteins were visualized using the ChemiDoc™ XRS system.

5.3.5. Assays of Cell-Free Protein Synthesis

The effect of RIPs on protein synthesis was determined through a coupled transcription–translation in vitro assay using a rabbit reticulocyte lysate system [13]. The reaction mixture contained 0.6 µL of rabbit reticulocyte lysate and 5.8 µL of a mixture of the following components: 4.6 U ribonuclease inhibitor, 2.3 U T7 RNA polymerase, 0.2 µg luciferase T7 plasmid, rNTPs (0.4 mM each), amino acids (2 µM each), 10 mM Tris–HCl (pH 7.8), 0.2 mM spermidine, 28 mM KCl, 1 mM $MgCl_2$, and nuclease-free water. The mixtures were incubated at 30 °C for 10 min and placed on ice. Then, 1.6 µL of either water or different protein concentrations were added and the sample mixture was incubated at 30 °C for 40 min. Subsequently, 25 µL water was added and mixed with 28 µL of Luciferase Assay Reagent (Promega, Alcobendas, Madrid, Spain) at room temperature. Luminescence was determined with a Junior LB 9509 luminometer (Berthold Technologies GmbH & Co. KG, Bad Wildbad, Germany). Three experiments were conducted in duplicate, and IC_{50} (concentration that inhibits 50% protein synthesis) values were calculated by linear regression.

5.3.6. rRNA N-Glycosylase Assays on Rabbit Reticulocyte, Yeast, Bacterium Lysates, and HeLa Cells

rRNA N-glycosylase assays were conducted as described elsewhere [16]. Rabbit reticulocyte lysate (40 µL) was incubated with 5 µg of RIP at 30 °C for 1 h. N-glycosylase activity on *Saccharomyces cerevisiae* ribosomes was assayed in 50 µL samples of S−30 lysate from yeast in 10 mM Tris–HCl buffer (pH 7.6) containing 10 mM KCl, 10 mM magnesium acetate, and 6 mM 2-mercaptoethanol, which was incubated with 5 µg of RIPs at 30 °C for 1 h. N-glycosylase activity on *Micrococcus lysodeikticus* ribosomes was assayed using 100 µL of bacterial lysate samples in 20 mM Tris–HCl buffer (pH 7.8), which were incubated with 5 µg of RIP at 30 °C for 1 h. After treatment, the RNA was extracted by phenolization, treated with 1 M aniline acetate (pH 4.5), and precipitated with ethanol. HeLa cells (1 × 10^6/plate) were incubated for 48 h in the presence of 40 nM of nigrin l. After treatment, cells were harvested by centrifugation at 1000× *g* for 5 min. The pellets were lysed, and the RNA was isolated following the instruction of the RNeasy Mini Kit (Qiagen GmbH, Hilden, Germany). RNA was treated with 1 M aniline acetate (pH 4.5) for 10 min at 0 °C and precipitated with ethanol. The RNAs were subjected to electrophoresis at 15 mA for 2 h (rabbit and HeLa cells) or 1 h 30 min (yeast and bacterium) in a 7 M urea/5% (*w*/*v*) polyacrylamide gel and stained with GelRed nucleic acid stain (Biotium Inc., Hayward, CA, USA) [16].

5.3.7. Adenine Polynucleotide Glycosylase Activity on Salmon Sperm DNA and Tobacco Mosaic Virus (TMV) RNA

The adenine release was measured according to the method reported elsewhere [69] with a few modifications. First, 10 µg of salmon sperm DNA was incubated with 5 µg of RIP in 300 µL of a reaction mixture containing 100 mM KCl and 50 mM magnesium acetate (pH 4), at 30 °C for 2 h. After incubation, the DNA was precipitated with ethanol at −80 °C for 3 h and centrifugated at 13,000 rpm for 15 min. Adenine released from RIP-treated DNA was determined in the supernatants spectrophotometrically at 260 nm. Analysis of the adenine polynucleotide glycosylase activity on tobacco mosaic virus (TMV) RNA was carried out as described elsewhere [18].

5.3.8. DNA Cleavage Experiments

Nicking activity experiments were performed as previously reported [18]. Each reaction contained 5 µg of RIP and 200 ng of pCR2.1 DNA in a final volume of 10 µL, comprised of 10 mM Tris–HCl, 5 mM $MgCl_2$, 50 mM NaCl, and 50 mM KCl, pH 7.8. Samples were incubated for 2 h at 30 °C, run on agarose gel (0.8%) in TAE buffer (0.04 M Tris, 0.04 M acetate, 1 mM EDTA, pH 8.0) and visualized by GelRed nucleic acid staining (Biotium Inc., Hayward, CA, USA).

5.3.9. Tryptic Digestion and Sample Preparation for MS/MS Analyses

Aliquots of protein samples (50 µg) were reduced with 20 mM dithiothreitol (DTT, 5 min at 95 °C) and alkylated with 20 mM iodoacetamide (IAA, 30 min, in the dark, at room temperature). Enzymatic hydrolyses were performed on reduced and alkylated samples by adding TPCK-treated trypsin with an enzyme/substrate (E/S) ratio of 1:200 (*w*/*w*) for 3 h, 1:100 for 16 h, and 1:50 for 4 h at 37 °C.

5.3.10. High-Resolution NanoLC–Tandem Mass Spectrometry

Mass spectrometry analyses on tryptic samples (500 fmol) were performed on a Q Exactive Orbitrap mass spectrometer equipped with an EASY-Spray nano-electrospray ion source (Thermo Fisher Scientific, Bremen, Germany) and coupled with a Thermo Scientific Dionex UltiMate 3000 RSLCnano system (Thermo Fisher Scientific). Solvent composition was 0.1% formic acid in water (solvent A) and 0.1% formic acid in acetonitrile (solvent B). Peptides were loaded on a trapping PepMap™100 µCartridge Column C18 (300 µm × 0.5 cm, 5 µm, 100 angstroms) and desalted with solvent A for 3 min at a flow rate of 10 µL/min. After trapping, eluted peptides were separated on an EASY-Spray analytical column (50 cm × 75 µm ID PepMap RSLC C18, 3 µm, 100 angstroms) and heated to 35 °C at a flow rate of 300 nL/min using the following gradient: 4% B for 3 min, from 4% to 55% B in 60 min, from 55% to 70% B in 10 min, and from 70% to 95% B in 2 min. Eluting peptides were analyzed on the Q-Exactive mass spectrometer operating in positive polarity mode with capillary temperature of 280 °C and a potential of 1.9 kV applied to the capillary probe. Full MS survey scan resolution was set to 70,000 with an automatic gain control (AGC) target value of 3 × 10^6 for a scan range of 375–1500 *m*/*z* and maximum ion injection time (IT) of 100 ms. The mass (*m*/*z*) 445.12003 was used as lock mass. A data-dependent top 5 method was operated, during which higher-energy collisional dissociation (HCD) spectra were obtained at 17,500 MS2 resolution with AGC target of 1×10^5 for a scan range of 200–2000 m/z, maximum IT of 55 ms, 2 m/z isolation width, and normalized collisional energy (NCE) of 27. Precursor ions targeted for HCD were dynamically excluded for 15 s. Full scans and Orbitrap MS/MS scans were acquired in profile mode, whereas ion trap mass spectra were acquired in centroid mode. Charge state recognition was enabled by excluding unassigned charge states.

5.3.11. Data Processing

The acquired raw files were analyzed with Proteome Discoverer 2.4 software (Thermo Fisher Scientific, Rockford, IL, USA) using the SEQUEST HT search engine. The HCD MS/MS spectra were searched against the whole UniProt_SwissProt KB database and against a homemade database including *S. nigra* RIP sequences deposited into the NCBI_GeneBank_NIH assuming trypsin (Full) as digestion enzyme with two allowed numbers of missed cleavage sites. The mass tolerances were set to 10 ppm and 0.02 Da for precursor and fragment ions, respectively. Oxidation of methionine (+15.995 Da) was set as dynamic modification, and carbamidomethylation of cysteine (+57.021 Da) as static modification. False discovery rates (FDRs) for peptide spectral matches (PSMs) were calculated and filtered using the Target Decoy PSM Validator Node in Proteome Discoverer. The Target Decoy PSM Validator Node specifies the PSM confidences based on dynamic score-based thresholds. It calculates the node-dependent score thresholds needed to determine the FDRs, which are provided as input parameters of the node. The Target Decoy PSM Validator was run with the following settings: maximum Delta Cn of 0.05, a strict target FDR of 0.01, a relaxed target FDR of 0.05, and validation based on q-value. The Protein FDR Validator Node in Proteome Discoverer was used to classify protein identifications based on q-value. Proteins with a q-value of <0.01 were classified as high-confidence identifications and proteins with a q-value of 0.01–0.05 were classified as medium-confidence identifications. Only proteins identified with medium or high confidence were retained, resulting in an overall FDR of 5%. All MS/MS spectra were manually validated to assign the best PSM to peptide sequences. When multiple PSM were mapped on the same peptide sequence, those with the highest accuracy were selected.

5.3.12. Hemagglutination Activity and Carbohydrate Binding Properties

Hemagglutination activity (HA) was assayed as described elsewhere [25]. The HA was determined in microtiter plates. Each well contained 50 µL of serial dilutions of the proteins and 50 µL of 1% erythrocyte suspension, and the plates were incubated for 1 h at room temperature. The minimum concentration of protein causing complete agglutination was visually evaluated. For hemagglutination inhibition assay, six sugars (D-glucose, D-galactose, D-fructose, D-mannose, L-fucose, and lactose) were tested for their ability to inhibit the HA of the proteins. Each well contained 50 µL of carbohydrates serially diluted, and 25 µL of the proteins at a concentration one titer higher than the HA dose. An equal volume of 2% erythrocyte suspension (25 µL) was added to each well and incubated for 1 h at room temperature. The minimum concentration of the tested sugars that completely inhibited HA activity was determined.

5.3.13. Protein Structure Studies and Graphical Representation

The structure of ricin (accession number 2 AAI) is available in the Protein Data Bank (https://www.rcsb.org/ (accessed on 26 April 2022). The three-dimensional structural modeling of nigrin l, SNAlm, and SNAld was carried out with the AlphaFold2 software following the instructions of the website https://colab.research.google.com/github/sokrypton/ColabFold/blob/main/AlphaFold2.ipynb#scrollTo=G4yBrceuFbf3 (accessed on 26 May 2022) [23]. The study representations and graphs of protein structures were constructed with the help of the Discovery Studio Visualizer suite (v21.1.0) (https://www.3dsbiovia.com/ (accessed on 26 April 2022). The SNAld dimer model was carried out on the SymmDock server (http://bioinfo3d.cs.tau.ac.il/SymmDock/ (accessed on 19 June 2022)) [70].

5.3.14. Molecular Docking

The structures of β-D-galactose (PubChem CID 439353) and β-lactose (PubChem CID 6134) are available in the PubChem database (https://pubchem.ncbi.nlm.nih.gov/ (accessed on 10 June 2022)) [71]. Docking was carried out using Autodock 4.2 (http://autodock.scripps.edu/ (accessed on 15 October 2021)) [72], as described elsewhere [25]. The docking of D-galactose was performed on a grid of 120 × 120 × 120 points, with the addition of a central grid point. The grid was centered on the centroid of the pyranosic ring of galactose at the 1α or 2γ sites of the ricin structure (accession number 2AAI). The grid spacing was 0.125 angstroms, which led to a grid of 15 × 15 × 15 angstroms. For each molecule, 100 docking runs were performed. The 100 docking poses generated were clustered by root mean square (RMS) difference with a cutoff value of 0.5 angstroms for each case. The top-ranked pose of the most populated clusters was retained and further analyzed with the Discovery Studio Visualizer suite (v21.1.0). β-lactose docking was performed as indicated for D-galactose, but using a grid of 124 × 124 × 124 points and a grid spacing of 0.180 angstroms, resulting in a grid of 22.32 × 22.32 × 22.32 angstroms. The grid was centered on the centroid of the pyranosic rings of lactose at the 1α or 2γ sites of the ricin structure. The 100 docking poses generated were compared with those obtained for D-galactose, and a pose was chosen that (coinciding with the best solution of docking with D-galactose) had a lower binding energy.

5.3.15. Cell Viability and DNA Fragmentation Analyses

Cell viability was determined using a colorimetric assay based on the cleavage of the tetrazolium salt WST-1 to formazan by mitochondrial dehydrogenases in viable cells. First, 3×10^3 COLO 320 or HeLa cells in 0.1 mL of medium were seeded in 96-well plates and incubated at 37 °C under 5% CO_2 in the absence or the presence of RIP for 48 h. Next, the cells were incubated for another 2 h with 10 μL/well of the cell proliferation reagent WST-1 at 37 °C under 5% CO_2. The absorbance of the samples was measured using a microtiter plate reader set at 450 nm with 620 nm as reference (ELISA reader Multiskan). The absorbance of cultures without cells was subtracted as background. For the DNA fragmentation analysis, COLO 320 cells (1×10^6/plate) were incubated for 72 h in the presence of 40 nM nigrin l. After treatment, cells were harvested by centrifugation (1000× *g* for 5 min). The pellets were lysed in 50 mM Tris–HCl, pH 8.0, containing 10 mM EDTA and 0.5% SDS, and the DNA was isolated following the instruction of the Genomic Prep Cells and Tissue DNA Isolation Kit (GE Healthcare, Barcelona, Spain). DNA electrophoresis was carried out in 1.8% agarose gels using TBE buffer (0.089 M Tris, 0.089 M boric acid, 2 mM EDTA, pH 8.0), and 4.0 μg of DNA was electrophoresed and stained with GelRed (Biotium Inc., Hayward, CA, USA) and visualized with an ultraviolet lamp.

Supplementary Materials: The following supporting information can be downloaded at: https://www.mdpi.com/article/10.3390/toxins14090611/s1. Table S1: Amino acid sequences of tryptic peptides from SNAlm obtained by high resolution nanoLC–tandem mass spectrometry and mapped on SNAlm (AC: AAN86132). Table S2: Amino acid sequences of tryptic peptides from SNAld obtained by high-resolution nanoLC–tandem mass spectrometry and mapped on SNAld (AC: AAN86131). Table S3: Amino acid sequences of tryptic peptides from nigrin l obtained by high-resolution nanoLC–tandem mass spectrometry and mapped on nigrin l (AC: AAN86130). Table S4: Amino acid sequences of tryptic peptides from nigrin-RP1 obtained by high-resolution nanoLC–tandem mass spectrometry and mapped on SNLRP2 (AC: AAC49672). Table S5: Amino acid sequences of tryptic peptides from nigrin-RP2 obtained by high-resolution nanoLC–tandem mass spectrometry and mapped on SNLRP2 (AC: AAC49672). Table S6: Amino acid sequences of tryptic peptides from nigrin-RP3 obtained by high-resolution nanoLC–tandem mass spectrometry and mapped on SNLRP1 (AC: AAC49673). Table S7: Inhibition of the hemagglutination activity of nigrin l, SNAlm, and SNAld by sugars compared with the estimated free energy of binding to the sugar binding sites. Figure S1:

Effect of nigrin l, nigrin-RPs 1–4, and nigritin l on protein synthesis. Figure S2: Representative MS/MS fragmentation spectra of SNAlm peptides mapped on the protein SNAlm (AC: AAN86132). Figure S3: Representative MS/MS fragmentation spectra of SNAld peptides mapped on the protein SNAld (AC: AAN86131). Figure S4: Representative MS/MS fragmentation spectra of nigrin l peptides. Figure S5: Representative MS/MS fragmentation spectra of nigrin-RP1 peptides. Figure S6: Representative MS/MS fragmentation spectra of nigrin-RP2 peptides. Figure S7: Representative MS/MS fragmentation spectra of nigrin-RP3 peptides. Figure S8: Comparison of the binding of D-galactose and lactose to the 1α site of nigrin l, SNAlm, and SNAld. Figure S9: Comparison of the binding of D-galactose and lactose to the 2γ site of nigrin l, SNAlm, and SNAld. Figure S10: Structure of the SNAld dimer with D-galactose bound to 1α sites.

Author Contributions: Conceptualization, J.M.F. and L.C.; methodology, R.I., R.R., J.M.F. and L.C.; validation, A.C., J.M.F. and L.C.; formal analysis, R.I., R.R., A.C. and L.C.; investigation, R.I., R.R., N.L., M.V., A.C., A.D.M., J.M.F. and L.C.; resources, R.I., A.C., A.D.M., J.M.F. and L.C.; data curation, R.I., R.R., A.B., J.M.F. and L.C.; writing—original draft preparation, J.M.F. and L.C.; writing—review and editing, R.I., R.R., N.L., A.C., A.B., J.M.F. and L.C.; visualization, J.M.F. and L.C.; supervision, R.I., A.C., A.D.M., A.B., J.M.F. and L.C.; project administration, A.C., A.D.M., A.B., J.M.F. and L.C.; funding acquisition, A.C., A.B., J.M.F. and L.C. All authors have read and agreed to the published version of the manuscript.

Funding: This work was funded by grants BIO39/VA39/14 and BIO/VA17/15 (Consejería de Sanidad, Junta de Castilla y León) to L.C.; grant VA033G19 (Consejería de Educación, Junta de Castilla y León) to the GIR ProtIBio; MISE, project NUTRABEST PON I&C 2014–2020 Prog. n. F/200050/01–03/X45; and the Pallotti Legacies for Cancer Research.

Institutional Review Board Statement: Not applicable.

Informed Consent Statement: Not applicable.

Data Availability Statement: Data are available upon request; please contact the contributing authors.

Conflicts of Interest: The authors declare no conflict of interest.

References

1. Citores, L.; Iglesias, R.; Ferreras, J.M. Ribosome Inactivating Proteins from Plants: Biological Properties and their Use in Experimental Therapy. In *Antitumor Potential and Other Emerging Medicinal Properties of Natural Compounds*; Fang, E.F., Ng, T.B., Eds.; Springer: Dordrecht, The Netherlands, 2013; pp. 127–143. [CrossRef]
2. Di Maro, A.; Citores, L.; Russo, R.; Iglesias, R.; Miguel Ferreras, J.M. Sequence comparison and phylogenetic analysis by the Maximum Likelihood method of ribosome-inactivating proteins from angiosperms. *Plant Mol. Biol.* **2014**, *85*, 575–588. [CrossRef] [PubMed]
3. Ferreras, J.M.; Citores, L.; Iglesias, R.; Jimenez, P.; Girbes, T. Use of Ribosome-Inactivating Proteins from *Sambucus* for the Construction of Immunotoxins and Conjugates for Cancer Therapy. *Toxins* **2011**, *3*, 420–441. [CrossRef]
4. Iglesias, R.; Ferreras, J.M.; Llorente, A.; Citores, L. Ebulin l Is Internalized in Cells by Both Clathrin-Dependent and -Independent Mechanisms and Does Not Require Clathrin or Dynamin for Intoxication. *Toxins* **2021**, *13*, 102. [CrossRef] [PubMed]
5. Zhu, F.; Zhou, Y.K.; Ji, Z.L.; Chen, X.R. The Plant Ribosome-Inactivating Proteins Play Important Roles in Defense against Pathogens and Insect Pest Attacks. *Front. Plant Sci.* **2018**, *9*, 146. [CrossRef] [PubMed]
6. Citores, L.; Iglesias, R.; Ferreras, J.M. Antiviral Activity of Ribosome-Inactivating Proteins. *Toxins* **2021**, *13*, 80. [CrossRef] [PubMed]
7. Polito, L.; Bortolotti, M.; Pedrazzi, M.; Bolognesi, A. Immunotoxins and Other Conjugates Containing Saporin-S6 for Cancer Therapy. *Toxins* **2011**, *3*, 697–720. [CrossRef]
8. De Zaeytijd, J.; Rouge, P.; Smagghe, G.; Van Damme, E.J.M. Structure and Activity of a Cytosolic Ribosome-Inactivating Protein from Rice. *Toxins* **2019**, *11*, 325. [CrossRef] [PubMed]
9. Parente, A.; Chambery, A.; Di Maro, A.; Russo, R.; Severino, V. Ribosome-inactivating Proteins from Phytolaccaceae. In *Ribosome-Inactivating Proteins: Ricin and Related Proteins*; Stirpe, F., Lappi, D., Eds.; Wiley-Blackwell: Oxford, UK, 2014; pp. 28–43. [CrossRef]
10. Domashevskiy, A.V.; Goss, D.J. Pokeweed Antiviral Protein, a Ribosome Inactivating Protein: Activity, Inhibition and Prospects. *Toxins* **2015**, *7*, 274–298. [CrossRef]
11. Ferreras, J.M.; Citores, L.; Iglesias, R.; Jimenez, P.; Girbes, T. Sambucus Ribosome-Inactivating Proteins and Lectins. In *Toxic Plant Proteins–Plant Cell Monographs*, 1st ed.; Lord, J.M., Hartley, M.R., Eds.; Springer: Berlin/Heidelberg, Germany, 2010; Volume 18, pp. 107–131. [CrossRef]

12. Barbieri, L.; Stoppa, C.; Bolognesi, A. Large-scale chromatographic purification of ribosome-inactivating proteins. *J. Chromatograph.* **1987**, *408*, 235–243. [CrossRef]
13. Ferreras, J.M.; Citores, L.; Iglesias, R.; Jimenez, P.; Souza, A.M.; Gayoso, M.J.; Girbes, T. Occurrence and new procedure of preparation of nigrin, an antiribosomal lectin present in elderberry bark. *Food Res. Int.* **2011**, *44*, 2798–2805. [CrossRef]
14. Iglesias, R.; Ferreras, J.M.; Di Maro, A.; Citores, L. Ebulin-RP, a novel member of the *Ebulin* gene family with low cytotoxicity as a result of deficient sugar binding domains. *Biochim. Biophys. Acta Gen. Subj.* **2018**, *1862*, 460–473. [CrossRef]
15. Van Damme, E.J.M.; Barre, A.; Rouge, P.; VanLeuven, F.; Peumans, W. Isolation and molecular cloning of a novel type 2 ribosome-inactivating protein with an inactive B chain from elderberry (*Sambucus nigra*) Bark. *J. Biol. Chem.* **1997**, *272*, 8353–8360. [CrossRef]
16. Iglesias, R.; Citores, L.; Ferreras, J.M. Ribosomal RNA N-glycosylase Activity Assay of Ribosome-inactivating Proteins. *Bio-Protocol* **2017**, *7*, e2180. [CrossRef]
17. Girbes, T.; Barbieri, L.; Ferreras, J.M.; Arias, F.; Rojo, M.; Iglesias, R.; Alegre, C.; Escarmis, C.; Stirpe, F. Effects of ribosome-inactivating proteins on *Escherichia coli* and *Agrobacterium tumefaciens* translation systems. *J. Bacteriol.* **1993**, *175*, 6721–6724. [CrossRef]
18. Iglesias, R.; Citores, L.; Ragucci, S.; Russo, R.; Di Maro, A.; Ferreras, J.M. Biological and antipathogenic activities of ribosome-inactivating proteins from *Phytolacca dioica* L. *Biochim. Biophys. Acta Gen. Subj.* **2016**, *1860*, 1256–1264. [CrossRef]
19. Iglesias, R.; Citores, L.; Di Maro, A.; Ferreras, J.M. Biological activities of the antiviral protein BE27 from sugar beet (*Beta vulgaris* L.). *Planta* **2015**, *241*, 421–433. [CrossRef]
20. Barbieri, L.; Ferreras, J.M.; Barraco, A.; Ricci, P.; Stirpe, F. Some ribosome-inactivating proteins depurinate ribosomal RNA at multiple sites. *Biochem. J.* **1992**, *286*, 1–4. [CrossRef]
21. Barbieri, L.; Valbonesi, P.; Bonora, E.; Gorini, P.; Bolognesi, A.; Stirpe, F. Polynucleotide:adenosine glycosidase activity of ribosome-inactivating proteins: Effect on DNA, RNA and poly(A). *Nucleic Acids Res.* **1997**, *25*, 518–522. [CrossRef]
22. Bortolotti, M.; Maiello, S.; Ferreras, J.M.; Iglesias, R.; Polito, L.; Bolognesi, A. Kirkiin: A New Toxic Type 2 Ribosome-Inactivating Protein from the Caudex of *Adenia kirkii*. *Toxins* **2021**, *13*, 81. [CrossRef]
23. Jumper, J.; Evans, R.; Pritzel, A.; Green, T.; Figurnov, M.; Ronneberger, O.; Tunyasuvunakool, K.; Bates, R.; Zidek, A.; Potapenko, A.; et al. Highly accurate protein structure prediction with Alpha Fold. *Nature* **2021**, *596*, 583–589. [CrossRef]
24. Iglesias, R.; Polito, L.; Bortolotti, M.; Pedrazzi, M.; Citores, L.; Ferreras, J.M.; Bolognesi, A. Primary Sequence and 3D Structure Prediction of the Plant Toxin Stenodactylin. *Toxins* **2020**, *12*, 538. [CrossRef]
25. Maiello, S.; Iglesias, R.; Polito, L.; Citores, L.; Bortolotti, M.; Ferreras, J.M.; Bolognesi, A. Sequence, Structure, and Binding Site Analysis of Kirkiin in Comparison with Ricin and Other Type 2 RIPs. *Toxins* **2021**, *13*, 862. [CrossRef] [PubMed]
26. Rutenber, E.; Robertus, J. Structure of ricin B-chain at 2.5-A resolution. *Proteins* **1991**, *10*, 260–269. [CrossRef] [PubMed]
27. Shang, C.; Van Damme, E.J.M. Comparative analysis of carbohydrate binding properties of *Sambucus nigra* lectins and ribosome-inactivating proteins. *Glycoconj. J.* **2014**, *31*, 345–354. [CrossRef] [PubMed]
28. Mercatelli, D.; Bortolotti, M.; Andresen, V.; Sulen, A.; Polito, L.; Gjertsen, B.; Bolognesi, A. Early Response to the Plant Toxin Stenodactylin in Acute Myeloid Leukemia Cells Involves Inflammatory and Apoptotic Signaling. *Front. Pharmacol.* **2020**, *11*, 630. [CrossRef] [PubMed]
29. Polito, L.; Bortolotti, M.; Battelli, M.G.; Calafato, G.; Bolognesi, A. Ricin: An Ancient Story for a Timeless Plant Toxin. *Toxins* **2019**, *11*, 324. [CrossRef]
30. Saxena, N.; Phatak, P.; Chauhan, V. Differential toxicity of abrin in human cell lines of different organ origin. *Toxicol. In Vitro* **2022**, *78*, 105250. [CrossRef]
31. Battelli, M.; Citores, L.; Buonamici, L.; Ferreras, J.M.; de Benito, F.; Stirpe, F.; Girbes, T. Toxicity and cytotoxicity of nigrin b, a two-chain ribosome-inactivating protein from *Sambucus nigra*: Comparison with ricin. *Arch. Toxicol.* **1997**, *71*, 360–364. [CrossRef]
32. Ito, H.; Hoshi, K.; Osuka, F.; Gotoh, M.; Saito, T.; Hojo, H.; Suzuki, R.; Ohira, H.; Honda, T.; Hashimoto, Y. Lectin inhibits antigen-antibody reaction in a glycoform-specific manner: Application for detecting alpha 2,6sialylated-carcinoembryonic antigen. *Proteomics* **2016**, *16*, 3081–3084. [CrossRef]
33. Hoshi, K.; Hashimoto, Y.; Ito, H. Simple and Rapid Detection of Glycoforms by "Lectin Inhibition" Assay. *Methods Mol. Biol.* **2020**, *2132*, 165–171. [CrossRef]
34. Vernygorodskyi, S.; Shkolnikov, V.; Suhan, D. Lectin binding patterns in normal, dysplastic and *Helicobacter pylori* infected gastric mucosa. *Exp. Oncol.* **2017**, *39*, 138–140. [CrossRef]
35. Ito, H.; Hoshi, K.; Hashimoto, Y.; Honda, T. Glycoform-Specific Visualization in Immunohistochemistry by "Lectin Inhibition". *Methods Mol. Biol.* **2020**, *2132*, 173–181. [CrossRef]
36. Beyer, S.; Kimani, M.; Zhang, Y.; Verhassel, A.; Sternbaek, L.; Wang, T.; Persson, J.; Harkonen, P.; Johansson, E.; Caraballo, R.; et al. Fluorescent Molecularly Imprinted Polymer Layers against Sialic Acid on Silica-Coated Polystyrene Cores-Assessment of the Binding Behavior to Cancer Cells. *Cancers* **2022**, *14*, 1875. [CrossRef]
37. Ogawa, R.; Okimoto, T.; Kodama, M.; Togo, K.; Fukuda, K.; Okamoto, K.; Mizukami, K.; Murakami, K. Changes in Gastric Mucosal Glycosylation Before and After *Helicobacter pylori* Eradication Using Lectin Microarray Analysis. *Turkish J. Gastroenterol.* **2022**, *33*, 88–94. [CrossRef]
38. Kaptan, E.; Sancar-Bas, S.; Sancakli, A.; Bektas, S.; Bolkent, S. The effect of plant lectins on the survival and malignant behaviors of thyroid cancer cells. *J. Cell. Biochem.* **2018**, *119*, 6274–6287. [CrossRef] [PubMed]

39. Chen, Y.; Peumans, W.J.; Van Damme, E.J.M. The *Sambucus nigra* type-2 ribosome-inactivating protein SNA-I′ exhibits in planta antiviral activity in transgenic tobacco. *FEBS Lett.* **2002**, *516*, 27–30. [CrossRef]
40. Vandenbussche, F.; Desmyter, S.; Ciani, M.; Proost, P.; Peumans, W.J.; Van Damme, E.J.M. Analysis of the in planta antiviral activity of elderberry ribosome-inactivating proteins. *Eur. J. Biochem.* **2004**, *271*, 1508–1515. [CrossRef] [PubMed]
41. Shahidi-Noghabi, S.; Van Damme, E.J.M.; Smagghe, G. Carbohydrate-binding activity of the type-2 ribosome-inactivating protein SNA-I from elderberry (*Sambucus nigra*) is a determining factor for its insecticidal activity. *Phytochemistry* **2008**, *69*, 2972–2978. [CrossRef]
42. Shahidi-Noghabi, S.; Van Damme, E.J.M.; Smagghe, G. Expression of *Sambucus nigra* agglutinin (SNA-I′) from elderberry bark in transgenic tobacco plants results in enhanced resistance to different insect species. *Transgen. Res.* **2009**, *18*, 249–259. [CrossRef] [PubMed]
43. Girbes, T.; Citores, L.; Ferreras, J.M.; Rojo, M.; Iglesias, R.; Munoz, R.; Arias, F.; Calonge, M.; Garcia, J.; Mendez, E. Isolation and partial characterization of nigrin-b, a nontoxic novel type-2 ribosome-inactivating protein from the bark of *Sambucus nigra* L. *Plant Mol. Biol.* **1993**, *22*, 1181–1186. [CrossRef]
44. Van Damme, E.J.M.; Barre, A.; Rouge, P.; VanLeuven, F.; Peumans, W. Characterization and molecular cloning of *Sambucus nigra* agglutinin V (nigrin b), a GalNAc-specific type-2 ribosome-inactivating protein from the bark of elderberry (*Sambucus nigra*). *Eur. J. Biochem.* **1996**, *237*, 505–513. [CrossRef]
45. Van Damme, E.J.M.; Roy, S.; Barre, A.; Citores, L.; Mostafapous, K.; Rouge, P.; VanLeuven, F.; Girbes, T.; Goldstein, I.; Peumans, W. Elderberry (*Sambucus nigra*) bark contains two structurally different Neu5Ac(alpha 2,6)Gal/GalNAc-binding type 2 ribosome-inactivating proteins. *Eur. J. Biochem.* **1997**, *245*, 648–655. [CrossRef] [PubMed]
46. de Benito, F.; Citores, L.; Iglesias, R.; Ferreras, J.M.; Camafeita, E.; Mendez, E.; Girbes, T. Isolation and partial characterization of a novel and uncommon two-chain 64-kDa ribosome-inactivating protein from the bark of elder (*Sambucus nigra* L). *FEBS Lett.* **1997**, *413*, 85–91. [CrossRef]
47. Van Damme, E.J.M.; Barre, A.; Rouge, P.; VanLeuven, F.; Peumans, W. The NeuAc(alpha-2,6)Gal/GalNAc-binding lectin from elderberry (*Sambucus nigra*) bark, a type-2 ribosome-inactivating protein with an unusual specificity and structure. *Eur. J. Biochem.* **1996**, *235*, 128–137. [CrossRef]
48. Kaku, H.; Peumans, W.; Goldstein, I. Isolation and characterization of a 2nd lectin (SNA-II) present in elderberry (*Sambucus nigra* L) bark. *Arch. Biochem. Biophys.* **1990**, *277*, 255–262. [CrossRef]
49. de Benito, F.; Iglesias, R.; Ferreras, J.M.; Citores, L.; Camafeita, E.; Mendez, E.; Girbes, T. Constitutive and inducible type 1 ribosome-inactivating proteins (RIPs) in elderberry (*Sambucus nigra* L.). *FEBS Lett.* **1998**, *428*, 75–79. [CrossRef]
50. Citores, L.; de Benito, F.; Iglesias, R.; Ferreras, J.M.; Jimenez, P.; Argueso, P.; Farias, G.; Mendez, E.; Girbes, T. Isolation and characterization of a new non-toxic two-chain ribosome-inactivating protein from fruits of elder (*Sambucus nigra* L.). *J. Exp. Bot.* **1996**, *47*, 1577–1585. [CrossRef]
51. Peumans, W.; Roy, S.; Barre, A.; Rouge, P.; van Leuven, F.; Van Damme, E.J.M. Elderberry (*Sambucus nigra*) contains truncated Neu5Ac(alpha-2,6)Gal/GalNAc-binding type 2 ribosome-inactivating proteins. *FEBS Lett.* **1998**, *425*, 35–39. [CrossRef]
52. Van Damme, E.J.M.; Roy, S.; Barre, A.; Rouge, P.; Van Leuven, F.; Peumans, W. The major elderberry (*Sambucus nigra*) fruit protein is a lectin derived from a truncated type 2 ribosome-inactivating protein. *Plant J.* **1997**, *12*, 1251–1260. [CrossRef]
53. Citores, L.; Iglesias, R.; Munoz, R.; Ferreras, J.M.; Jimenez, P.; Girbes, T. Elderberry (*Sambucus nigra* L.) seed proteins inhibit protein synthesis and display strong immunoreactivity with rabbit polyclonal antibodies raised against the type-2 ribosome-inactivating protein nigrin b. *J. Exp. Bot.* **1994**, *45*, 513–516. [CrossRef]
54. Peumans, W.; Kellens, J.; Allen, A.; Van Damme, E.J.M. Isolation and characterization of a seed lectin from elderberry (*Sambucus nigra* L.) and its relationship to the bark lectins. *Carbohydr. Res.* **1991**, *213*, 7–17. [CrossRef]
55. Karpova, I.; Koretska, N.; Palchykovska, L.; Nehrutska, V. Lectins from *Sambucus nigra* L. inflorescences: Isolation and investigation of biological activity using procaryotic test-systems. *Ukr. Biokhimicheskii Zhurnal* **1999**, *79*, 8.
56. Karpova, I.; Lylo, V.; Macewicz, L.; Kotsarenko, K.; Palchkovska, L.; Ruban, T.; Lukash, L. Lectins of *Sambucus nigra* as biologically active and DNA-protective substances. *Acta Hortic.* **2015**, *1061*, 93–102. [CrossRef]
57. Broekaert, W.; Nsimbalubaki, M.; Peeters, B.; Peumans, W. A lectin from elder (*Sambucus nigra* L.) bark. *Biochem. J.* **1984**, *221*, 163–169. [CrossRef]
58. Rojo, M.; Citores, L.; Arias, F.; Ferreras, J.M.; Jimenez, P.; Girbes, T. cDNA molecular cloning and seasonal acumulation of an ebulin 1-related dimeric lectin of dwarf elder (*Sambucus ebulus* L.) leaves. *Int. J. Biochem. Cell Biol.* **2003**, *35*, 1061–1065. [CrossRef]
59. Citores, L.; de Benito, F.; Iglesias, R.; Ferreras, J.M.; Argueso, P.; Jimenez, P.; Mendez, E.; Girbes, T. Presence of polymerized and free forms of the non-toxic type 2 ribosome-inactivating protein ebulin and a structurally related new homodimeric lectin in fruits of *Sambucus ebulus* L. *Planta* **1998**, *204*, 310–317. [CrossRef]
60. Iglesias, R.; Citores, L.; Ferreras, J.M.; Perez, Y.; Jimenez, P.; Gayoso, M.J.; Olsnes, S.; Tamburino, R.; Di Maro, A.; Parente, A.; et al. Sialic acid-binding dwarf elder four-chain lectin displays nucleic acid N-glycosidase activity. *Biochimie* **2010**, *92*, 71–80. [CrossRef]
61. Too, P.; Ma, M.; Mak, A.; Wong, Y.; Tung, C.; Zhu, G.; Au, S.; Wong, K.; Shaw, P. The C-terminal fragment of the ribosomal P protein complexed to trichosanthin reveals the interaction between the ribosome-inactivating protein and the ribosome. *Nucleic Acids Res.* **2009**, *37*, 602–610. [CrossRef]
62. Zhou, Y.; Li, X.; Chen, B.; Tumer, N. Ricin uses arginine 235 as an anchor residue to bind to P-proteins of the ribosomal stalk. *Sci. Rep.* **2017**, *7*, 42912. [CrossRef]

63. Ruggiero, A.; Di Maro, A.; Severino, V.; Chambery, A.; Berisio, R. Crystal Structure of PD-L1, a Ribosome Inactivating Protein from *Phytolacca dioica* L. Leaves with the Property to Induce DNA Cleavage. *Biopolymers* **2009**, *91*, 1135–1142. [CrossRef]
64. Baek, M.; DiMaio, F.; Anishchenko, I.; Dauparas, J.; Ovchinnikov, S.; Lee, G.; Wang, J.; Cong, Q.; Kinch, L.; Schaeffer, R.; et al. Accurate prediction of protein structures and interactions using a three-track neural network. *Science* **2021**, *373*, 871–876. [CrossRef]
65. Pascal, J.; Day, P.; Monzingo, A.; Ernst, S.; Robertus, J.; Iglesias, R.; Perez, Y.; Ferreras, J.M.; Citores, L.; Girbes, T. 2.8-angstrom crystal structure of a nontoxic type-II ribosome-inactivating protein, ebulin 1. *Proteins* **2001**, *43*, 319–326. [CrossRef] [PubMed]
66. Girbes, T.; Citores, L.; Iglesias, R.; Ferreras, J.M.; Munoz, R.; Rojo, M.; Arias, F.; Garcia, J.; Mendez, E.; Calonge, M. Ebulin l, a nontoxic novel type-2 ribosome-inactivating protein from *Sambucus ebulus* L. leaves. *J. Biol. Chem.* **1993**, *268*, 18195–18199. [CrossRef]
67. Kalb, V.J.; Bernlohr, R. A new spectrophotometric assay for protein in cell extracts. *Anal. Biochem.* **1977**, *82*, 362–371. [CrossRef]
68. Laemmli, U. Cleavage of structural proteins during the assembly of the head of bacteriophage T4. *Nature* **1974**, *227*, 680–685. [CrossRef]
69. Di Maro, A.; Chambery, A.; Daniele, A.; Casoria, P.; Parente, A. Isolation and characterization of heterotepalins, type 1 ribosome-inactivating proteins from *Phytolacca heterotepala* leaves. *Phytochemistry* **2007**, *68*, 767–776. [CrossRef]
70. Schneidman-Duhovny, D.; Inbar, Y.; Nussinov, R.; Wolfson, H. PatchDock and SymmDock: Servers for rigid and symmetric docking. *Nucleic Acids Res.* **2005**, *33*, W363–W367. [CrossRef]
71. Kim, S.; Chen, J.; Cheng, T.; Gindulyte, A.; He, J.; He, S.; Li, Q.; Shoemaker, B.; Thiessen, P.; Yu, B.; et al. PubChem in 2021: New data content and improved web interfaces. *Nucleic Acids Res.* **2021**, *49*, D1388–D1395. [CrossRef]
72. Morris, G.; Huey, R.; Lindstrom, W.; Sanner, M.; Belew, R.; Goodsell, D.; Olson, A. AutoDock4 and AutoDockTools4: Automated Docking with Selective Receptor Flexibility. *J. Comput. Chem.* **2009**, *30*, 2785–2791. [CrossRef]

toxins

Article

Characterization of Lung Injury following Abrin Pulmonary Intoxication in Mice: Comparison to Ricin Poisoning

Anita Sapoznikov *, Yoav Gal, Ron Alcalay, Yentl Evgy, Tamar Sabo, Chanoch Kronman and Reut Falach *

Department of Biochemistry and Molecular Genetics, Israel Institute for Biological Research, Ness-Ziona 74100, Israel
* Correspondence: anitas@iibr.gov.il (A.S.); reutf@iibr.gov.il (R.F.); Tel.: +972-89381847 (A.S.); +972-89381522 (R.F.)

Abstract: Abrin is a highly toxic protein obtained from the seeds of the rosary pea plant *Abrus precatorius*, and it is closely related to ricin in terms of its structure and chemical properties. Both toxins inhibit ribosomal function, halt protein synthesis and lead to cellular death. The major clinical manifestations following pulmonary exposure to these toxins consist of severe lung inflammation and consequent respiratory insufficiency. Despite the high similarity between abrin and ricin in terms of disease progression, the ability to protect mice against these toxins by postexposure antibody-mediated treatment differs significantly, with a markedly higher level of protection achieved against abrin intoxication. In this study, we conducted an in-depth comparison between the kinetics of in vivo abrin and ricin intoxication in a murine model. The data demonstrated differential binding of abrin and ricin to the parenchymal cells of the lungs. Accordingly, toxin-mediated injury to the nonhematopoietic compartment was shown to be markedly lower in the case of abrin intoxication. Thus, profiling of alveolar epithelial cells demonstrated that although toxin-induced damage was restricted to alveolar epithelial type II cells following abrin intoxication, as previously reported for ricin, it was less pronounced. Furthermore, unlike following ricin intoxication, no direct damage was detected in the lung endothelial cell population following abrin exposure. Reduced impairment of intercellular junction molecules following abrin intoxication was detected as well. In contrast, similar damage to the endothelial surface glycocalyx layer was observed for the two toxins. We assume that the reduced damage to the lung stroma, which maintains a higher level of tissue integrity following pulmonary exposure to abrin compared to ricin, contributes to the high efficiency of the anti-abrin antibody treatment at late time points after exposure.

Keywords: abrin; ricin; intranasal; lungs; alveolar epithelial type II cells; neutrophils; alveolar–capillary barrier; junction proteins; glycocalyx

Key Contribution: Pulmonary exposure of mice to a lethal dose of abrin induces less pronounced damage to the lung stroma and a reduced impairment of intercellular junction molecules in comparison to ricin.

Citation: Sapoznikov, A.; Gal, Y.; Alcalay, R.; Evgy, Y.; Sabo, T.; Kronman, C.; Falach, R. Characterization of Lung Injury following Abrin Pulmonary Intoxication in Mice: Comparison to Ricin Poisoning. *Toxins* **2022**, *14*, 614. https://doi.org/10.3390/toxins14090614

Received: 10 August 2022
Accepted: 31 August 2022
Published: 2 September 2022

1. Introduction

The family of ribosome-inactivating proteins (RIPs) groups all enzymes (EC.3.2.2.22) with a so-called RIP domain which comprises N-glycosylase activity and enables these proteins to catalytically inactivate ribosomes. The highest number of plant RIPs has been found in angiosperm plants [1–4], including fungi [3,5,6], algae [7] and bacteria [8]. Structurally, plant RIPs can be divided into two main groups, depending on the presence or absence of a quaternary structure. Type 1 RIPs (~30 kDa) are single-chain proteins with enzymatic action, whereas type 2 RIPs (~60 kDa) consist of an enzymatically active A-chain linked to a B-chain with lectinic properties through a disulfide bridge. The B-chain binds to carbohydrates on the cell surface, allowing A-chain cell internalization. The absence of a lectinic chain prevents type 1 RIPs from binding to the cell, which are consequently less toxic with respect to type 2 RIPs, due to the difficulty of entering the cell [3]. Moreover, a third group of RIPs, known as type 3 RIPs, consists of a type-1-like N-terminal domain with N-glycosylase activity, covalently linked to a C-terminal domain with an unknown function [9].

The toxins abrin and ricin produced from the seeds of *Abrus precatorius* and *Ricinus communis*, respectively, are classified as type II RIPs and consist of an enzymatically active A-chain disulfide linked to a B-chain. The B-chain is a galactose-specific lectin that is responsible for the binding of toxins to glycoproteins or glycolipids on the surface of cells to promote endocytosis of the toxin [10]. Receptor-dependent internalization of the toxins involves retrograde transport to the endoplasmic reticulum, where the disulfide bond connecting the A and B subunits is reduced [11], allowing the release of the catalytically active A-chain into the cytoplasm [12]. The A subunit of both toxins is an RNA N-glycosylase that catalyzes the site-specific release of an essential adenine moiety, located in a highly conserved stem–loop within the small subunit of the ribosome 28S rRNA [13–15]. The irreversible depurination of the stem–loop by the A subunit prevents the binding of elongation factors to ribosomes, thereby inhibiting protein synthesis and eventually causing cell death [16,17].

Abrin's potential use as a chemical weapon stems from its high toxicity, together with the fact that it can be isolated from jequirity beans at a low cost by a relatively simple procedure. The relatively low-scale cultivation of jequirity plants compared with *Ricinus* (castor oil) plants would suggest a smaller and more focused terrorist-type chemical attack with abrin. One possible scenario is that once isolated, abrin could be aerosolized as a dry powder. Studies have found that the overall pattern and time course of damage following inhalation were similar for ricin and abrin and characterized by rapidly progressive and overwhelming pulmonary edema accompanied by acute destructive alveolitis and necrosis/apoptosis of the lower respiratory tract epithelium accounting for the majority of deaths [18–21]. The clinical manifestations following pulmonary (intranasal) exposure to these toxins were entirely restricted to the lungs and characterized by severe pulmonary edematous inflammation, neutrophil recruitment and development of a proinflammatory cytokine storm [22,23]. Pulmonary (intranasal) ricin and abrin intoxication in mice are similar with regard to pathological features and kinetics. However, despite their resemblance, the ability to protect mice against ricin and abrin intoxication by postexposure antibody-mediated treatment differs drastically. Rabbit-derived polyclonal anti-ricin antibody-based treatment showed almost complete protection a few hours after exposure to a lethal ricin dose; however, when this treatment was delayed to 24 h after intoxication, only one-third of the mice survived [22]. In contrast, the intranasal administration of polyclonal anti-abrin antibodies to mice even as late as 72 h postexposure to a lethal dose of abrin conferred exceedingly high-level protection [23]. Interestingly, the efficient protection by polyclonal anti-abrin antibodies cannot be attributed to the specific neutralization of a particular A or B subunit of the toxin, as antibodies raised against chimeric toxins of either an $A_{abrin}B_{ricin}$ or $A_{ricin}B_{abrin}$ structure conferred exceptionally high protection levels to mice following intranasal exposure to a lethal dose of abrin [24].

In view of these findings, we characterized and quantified the cellular and molecular changes in murine lung tissue following pulmonary exposure to abrin, as compared to ricin intoxication, to delineate toxin-specific patho-physiologic factors that may play a role in determining the differential ability to protect against abrin and ricin by postexposure antibody administration.

2. Results

2.1. *Differential Binding of Abrin to Lung Cell Populations*

Previously, we showed that following intranasal intoxication of mice with a lethal dose of ricin, the toxin binds to alveolar macrophages (AMs) and dendritic cells (DCs) of the hematopoietic compartment, as well as to the lung parenchyma, epithelial and endothelial cells [25]. We examined, in a similar system, the interactions between abrin and lung cells after intranasal intoxication of mice with fluorescently labeled abrin (abrin Alexa Fluor 488 (abrin AF488)) at a lethal dose of $2LD_{50}$. Abrin-associated cells could be visualized by flow cytometry 3 h following exposure to the toxin (Figure 1A). To determine the kinetics of toxin binding to individual cell populations in the lung, mice were intoxicated with abrin or ricin, and lung cells isolated at different time points thereafter were analyzed for toxin binding. In the hematopoietic compartment ($CD45^+$ cells), peak binding of both toxins was detected as early as 3 h after intoxication, and at later time points, fewer cells were detected in association with the toxins. The kinetics of $CD45^+$ cell binding and the proportion of toxin-bound cells at all examined time points was similar for both toxins (Figure 1B). Among cells of hematopoietic origin, abrin exhibited prompt binding to AMs and DCs, with peak binding at 3 h after intoxication (Figure 1C). To analyze the correlation between the binding of abrin and its ability to eliminate cells by inhibiting protein synthesis, we quantified the number of AMs and DCs at different time points following intoxication. In contrast to ricin pulmonary intoxication, where AMs were significantly reduced 3 h after exposure [25], AMs were heavily reduced 6 h after abrin intoxication, and their numbers stayed low at later time points (24–72 h postexposure, Figure 1D). At 24 h after either ricin or abrin intoxication, the population of AMs comprised only ~40–50% of the initial population of AMs observed in naïve mice (Figure 1E). In a similar manner, the DC population was reduced starting from 6 h post abrin exposure (Figure 1F), in contrast to the significant reduction in these cells already at 3 h after ricin intoxication [25]. One day after intoxication with either of the toxins, the DC population in the lung consisted of only ~60–70% of the initial population measured in non-intoxicated mice (Figure 1G).

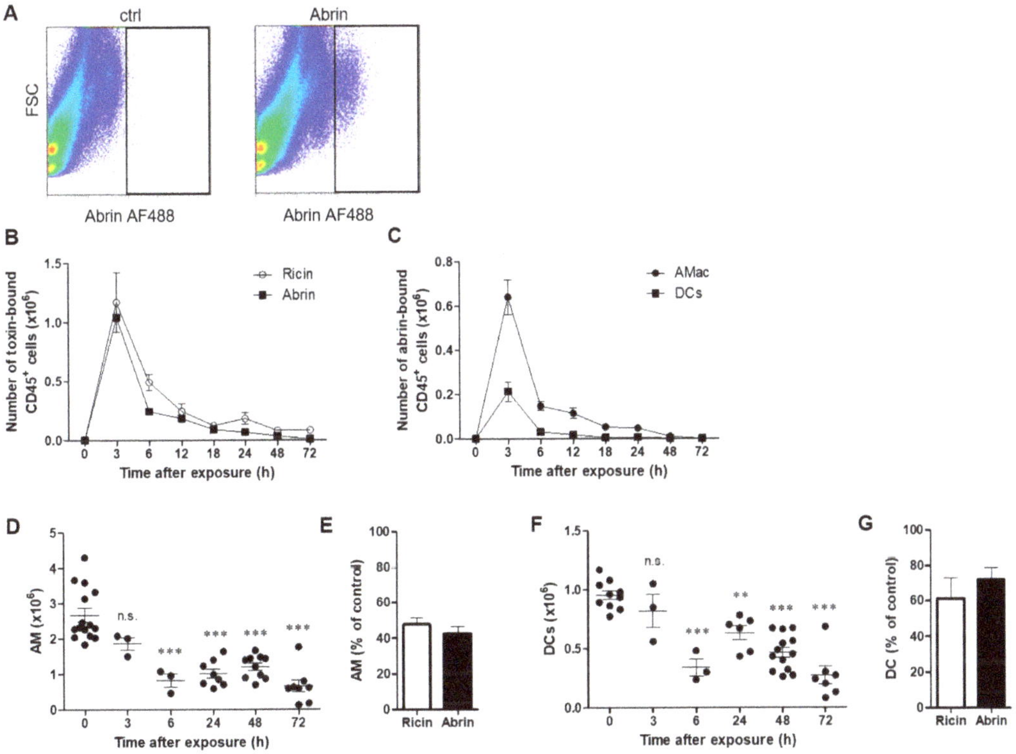

Figure 1. Kinetics of abrin binding to hematopoietic cells and alteration in cell populations in the lung. Mice were intranasally exposed to fluorescent abrin or ricin AF488 (20 or 14 μg/kg body weight, respectively),

and lung cells were isolated at 3, 6, 12, 18, 24, 48 and 72 h after exposure and analyzed by flow cytometry for toxin binding by detection of AF488$^+$ cells and different cell population counts in the lungs. (**A**) Dot plots represent abrin AF488 staining in lung cells isolated 3 h after abrin intoxication or in cells isolated from control mice. (**B**) Quantification of toxin-bound CD45$^+$ cells at different time points following intoxication. A comparison between ricin and abrin intoxications (n = 3–9 mice in each group). (**C**) Quantification of abrin-bound AMs and DCs at different time points following abrin intoxication (n = 3–6 mice in each group). Quantification of AM (**D**) and DC (**F**) population sizes at different time points following abrin intoxication (10 µg/kg, n = 3–15 mice in each group; each point indicates individual mice). The results are depicted as the means ± SEMs. ** $p < 0.01$, *** $p < 0.001$ in comparison to nonintoxicated mice; n.s., not significant. Comparison between abrin and ricin [25] AMs (**E**) and DCs (**G**) (% of control) at 24 h postexposure to toxins.

Next, we analyzed the binding of both toxins to the parenchymal cell populations of the lung (CD45$^-$ cells). Interestingly, we found that binding to CD45$^-$ cells was considerably more pronounced in the case of ricin exposure. Thus, at 3 h after intoxication, twice as many CD45$^-$ cells were bound to ricin than to abrin, while at 6 h postexposure, almost 6-fold more cells were associated with ricin than with abrin. Higher levels of toxin binding to lung parenchymal cells following ricin exposure were further observed at all later time points (12–72 h postexposure, Figure 2A). Within the parenchymal compartment, we distinguished between vascular endothelial cells and alveolar epithelial cells. Surprisingly, examination of the endothelial cell population following abrin intoxication did not result in observable damage to the cells, as evidenced by the preserved cell numbers at all tested time points postexposure (Figure 2B). This was in contrast to ricin intoxication, where a ~25% reduction in endothelial cells was detected 48 h postexposure (Figure 2C). Profiling of epithelial cells demonstrated a significant reduction in these cells from 48 h to 72 h after abrin intoxication (Figure 2D). A comparison of ricin and abrin at the same time point showed that the damage to epithelial cells after ricin intoxication was more pronounced, displaying a loss of ~60% of the population, while a reduction in these cells following abrin pulmonary intoxication was no more than ~40% (Figure 2E). Examination of subsets of the epithelial cells demonstrated that although no change was found in alveolar epithelial type I cells after pulmonary abrin intoxication (Figure 2F), the number of alveolar epithelial type II cells decreased significantly by 48 h postexposure (Figure 2G). The reduction in the alveolar epithelial type II population was also manifested with immunohistochemistry by specific labeling of pro-surfactant C (pro-SPC). Immunostaining of the lungs 48 h after abrin intoxication for alveolar epithelial type I cells (anti-T1a) and endothelial cells (anti-CD31) further confirmed no injury to these populations, as comparable staining levels were detected both in nonintoxicated (control) and intoxicated lungs (Figure 2H).

2.2. *Pulmonary Exposure to Abrin and Ricin Induces Comparable Neutrophil Influx to the Lungs Accompanied by Lung Hyperpermeability*

One prominent hallmark of ricin-mediated pulmonary intoxication is the rapid and massive influx of neutrophils to the lungs, where they contribute to the developing inflammation yet may also cause tissue damage, thereby promoting ricin-mediated morbidity [26–28]. These neutrophils are refractive to ricin binding [25]. Examination of the binding of abrin to neutrophils in the lungs following intranasal exposure of mice demonstrated that this cell type did not bind abrin at any time point (3–72 h) tested after intoxication (Figure 3A,B). There were no abrin AF488-positive neutrophils in the lungs, which could be appreciated by the low mean fluorescent intensity (MFI) of fluorescent abrin in these cells at all time points after intoxication (Figure 3C). These results are in sharp contrast to AMs, which readily bound abrin, as shown by the increased abrin AF488 intensity (Figure 3B). Next, we monitored neutrophil numbers at different time points after abrin intoxication. Neutrophil counts that comprised ~2 × 106 cells in healthy mice were raised to 26 ± 4 × 106 cells and 50 ± 11 × 106 cells at 24 h and 72 h, respectively, after exposure to abrin (Figure 3D). Alignment between the elevation of neutrophils after ricin and abrin intoxications showed equal recruitment of these cells to the lungs 24–72 h postexposure

(Figure 3E). Since uncontrolled massive recruitment of neutrophils may cause tissue damage and promote permeability and edema, we measured lung permeability by the Evans blue dye (EBD) extravasation assay. To this end, mice were intravenously injected with EBD at different time points after intranasal exposure to abrin, lungs were harvested, and EBD was extracted and quantified. Pulmonary EBD levels were found to be elevated significantly at 48 and 72 h post abrin exposure (Figure 3F). These results indicate that similar to ricin pulmonary intoxication, exposure to abrin promotes comparable neutrophil influx into the lungs, which is accompanied by lung hyperpermeability.

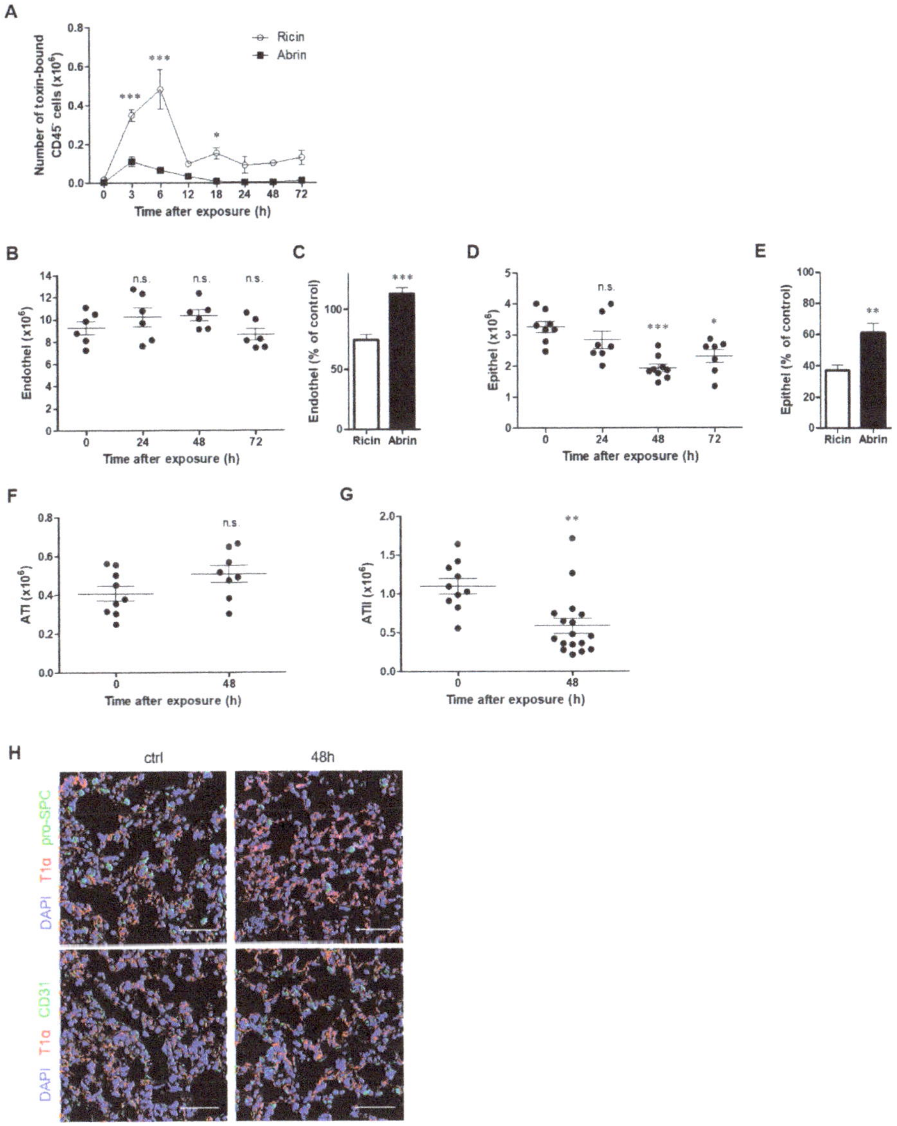

Figure 2. Kinetics of abrin binding to parenchymal cells and alteration in cell populations in the lung. Mice were intranasally exposed to fluorescent abrin or ricin AF488 (20 or 14 μg/kg body weight, respectively), and lung cells were isolated at 24, 48 and 72 h and analyzed by flow cytometry for toxin

binding by detection of AF488⁺ cells and for different cell population counts in the lungs. (**A**) Quantification of toxin-bound CD45⁻ cells at different time points following intoxication. Comparison between ricin and abrin intoxications (n = 3 mice per group). (**B**) Mice were intranasally exposed to abrin (10 μg/kg), lungs were removed at indicated time points, and endothelial and epithelial (**D**) numbers were determined by flow cytometry (n = 6–9 mice in each group). Comparison between the percent of endothelial (**C**) or epithelial (**E**) cells at 48 h post abrin exposure to the percent of these cells at the same time point after ricin intoxication [25]. (**F**) Quantification of alveolar epithelial type I (ATI) and alveolar epithelial type II (ATII) (**G**) cell populations at 48 h post abrin exposure (n = 8–17 mice in each group). (**H**) Immunofluorescence analysis of ATI (T1α, red) and ATII (pro-SPC, green) or endothelial cell (CD31, green) staining of lung tissue in nonintoxicated mice (control) versus abrin 48 h postexposed mice (blue, 4′,6-diamidino-2-phenylidole (DAPI) staining of nuclei). Scale bar: 50 μm. The results are depicted as the means ± SEMs. * $p < 0.05$, ** $p < 0.01$, *** $p < 0.001$; n.s., not significant. (**A**,**C**,**E**) Comparison between intoxications at each time point; (**B**,**D**,**F**,**G**) comparison with nonintoxicated mice.

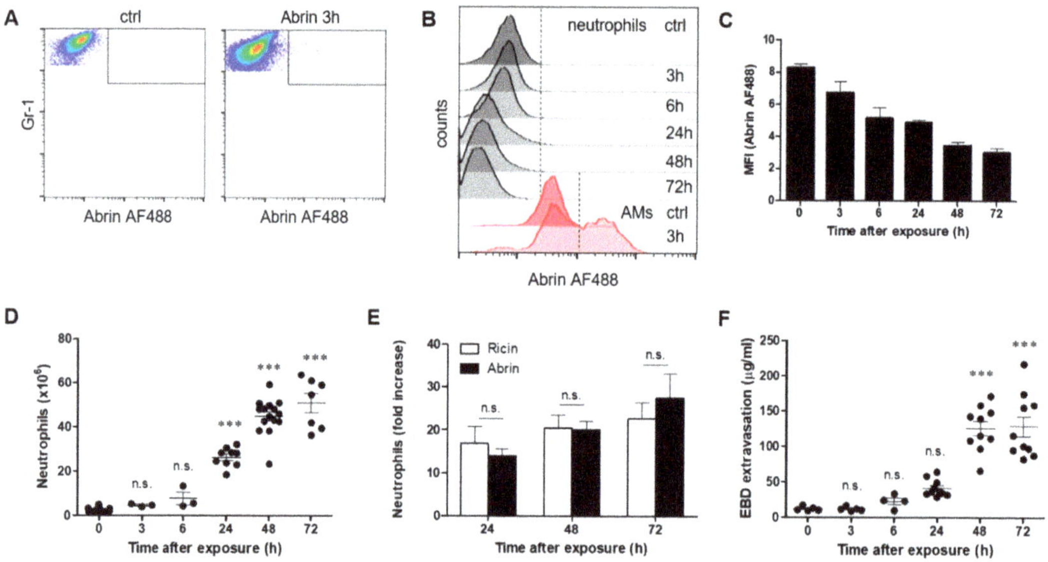

Figure 3. Effect of exposure to abrin on neutrophils and lung permeability. Mice were intranasally exposed to fluorescent abrin AF488 (20 μg/kg body weight), and lung cells were isolated at 3, 6, 24, 48 and 72 h and analyzed for neutrophils by flow cytometry. (**A**) Dot plots represent abrin AF488 staining in neutrophils isolated 3 h after abrin intoxication or in cells isolated from control mice. (**B**) Abrin AF488 binding to neutrophils (black histograms) and AMs (red histograms) at different time points following abrin intoxication. (AMs are autofluorescent in the lungs and exhibit high background in control mice.) (**C**) MFI of abrin AF488 binding to neutrophils (n = 3–7 mice in each group). (**D**) Neutrophil count in the lungs at different time points following abrin intoxication. (**E**) Comparison between the increase in neutrophils after abrin and ricin [25] intoxication (n = 3–15 mice in each group). (**F**) Lung EBD extravasation following abrin intoxication. Control or abrin-intoxicated mice were intravenously injected with 50 mg/kg EBD at the indicated time points, and lungs were monitored for EBD content (n = 4–10 mice in each group). The results are depicted as the means ± SEMs. *** $p < 0.001$; n.s., not significant. In (**D**,**F**), comparison with nonintoxicated mice.

2.3. Pulmonary Exposure to Abrin Leads to Inferior Impairment of Junction Proteins in the Lungs in Comparison to Ricin

The integrity of the alveolar wall barrier depends on the intercellular junctions of the alveolar epithelial and capillary endothelial cells. Junction protein complexes formed by tight junctions (TJs), adherens junctions (AJs) and gap junctions (GJs) stabilize the connections between contiguous cells. Disruption of the integrity of these complexes results in increased permeability and the formation of lung edema [29–31]. Indeed, we have previously shown that in the case of ricin pulmonary intoxication, disruption of differential intercellular junction proteins leads to impairment of the alveolar–capillary barrier and to the development of lung edema, which in turn results in the impairment of oxygenation [27,32]. Occludin, a TJ protein, is expressed in the alveoli, bronchial epithelial cells and endothelial cells of blood vessels in healthy lungs. We did not detect any change in occludin expression at 24 and 48 h after abrin intoxication (Figure 4A,B), which was in contrast to ricin intoxication which triggered a two-fold reduction in occludin in the lungs (Figure 4C). Next, we examined the expression of VE-cadherin, which is essential for endothelial barrier integrity. The VE-cadherin level was significantly decreased following abrin exposure (Figure 4D,E), although in comparison to ricin intoxication, which brought nearly complete elimination of VE-cadherin at 3–6 h postexposure, the damage to this protein by abrin was only moderate, even at 24 h post abrin exposure (Figure 4F). Similar to VE-cadherin, claudin 18, which is expressed by alveolar epithelial cells, was also diminished after abrin intoxication (Figure 4G,H), but once again, its reduction was less pronounced than that recorded after pulmonary exposure to ricin (Figure 4I).

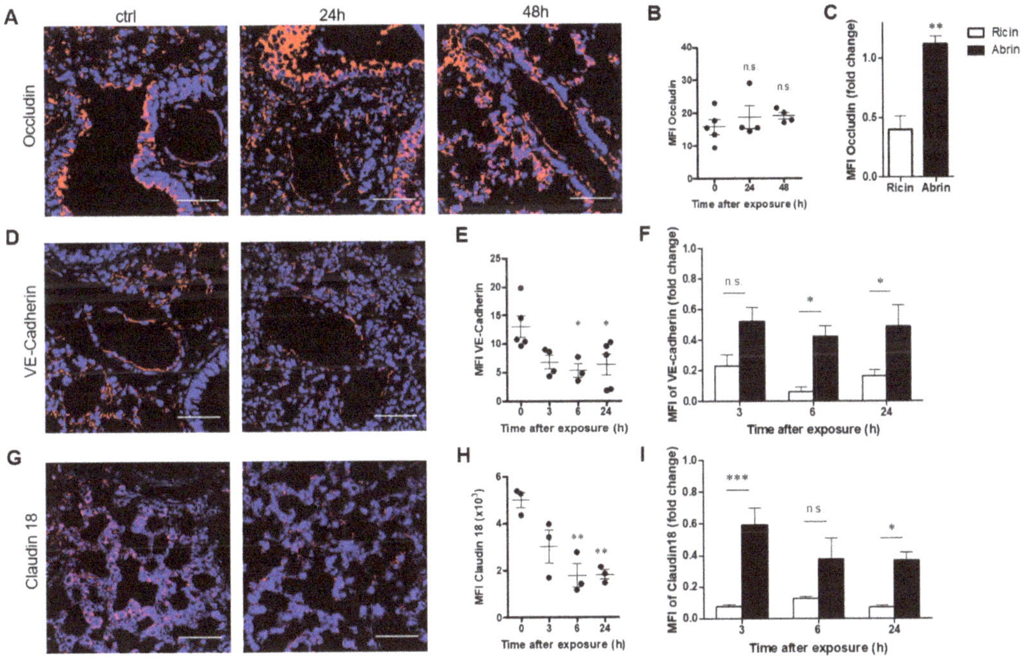

Figure 4. Alterations in occludin, VE-cadherin and claudin 18 in the lungs of abrin-intoxicated mice. Lungs of abrin-intoxicated (10 µg/kg body weight) mice were harvested at the indicated time points, and junction proteins were quantified by immunohistochemical analysis of lung sections. (**A**,**D**,**G**) Confocal microscopy scans of lung sections stained for occludin, VE-cadherin and claudin 18 (red), respectively, and identification of nuclei by DAPI (blue). (**B**,**E**,**H**) Scatterplots represent the immunofluorescence staining intensities of occludin, VE-cadherin and claudin 18, expressed as MFI (n = 3–5 mice in each group). Scale bar: 50 µm. (**C**,**F**,**I**) Comparison between the abrin- and ricin-induced reduction in occludin (at 48 h), VE-cadherin and claudin 18 [27]. The results are depicted as the means ± SEMs. * $p < 0.05$, ** $p < 0.01$, *** $p < 0.001$; n.s., not significant. In (**B**,**E**,**H**), comparison with nonintoxicated mice.

Examination of connexin 43, a GJ protein, showed that its level was heavily decreased at 6 and 24 h post abrin intoxication (Figure 5A,B). This pronounced decrease in connexin 43 correlated with its elimination after ricin intoxication (Figure 5C). Finally, we examined the expression of claudin 5, which is expressed in both endothelial and airway epithelial cells. The level of claudin 5 was significantly diminished 3 to 24 h after abrin intoxication (Figure 5D,E). As in the connexin 43 case, the reduction in claudin 5 following abrin intoxication was comparable to the level of the protein at the corresponding time points after ricin intoxication, as demonstrated by MFI (Figure 5F).

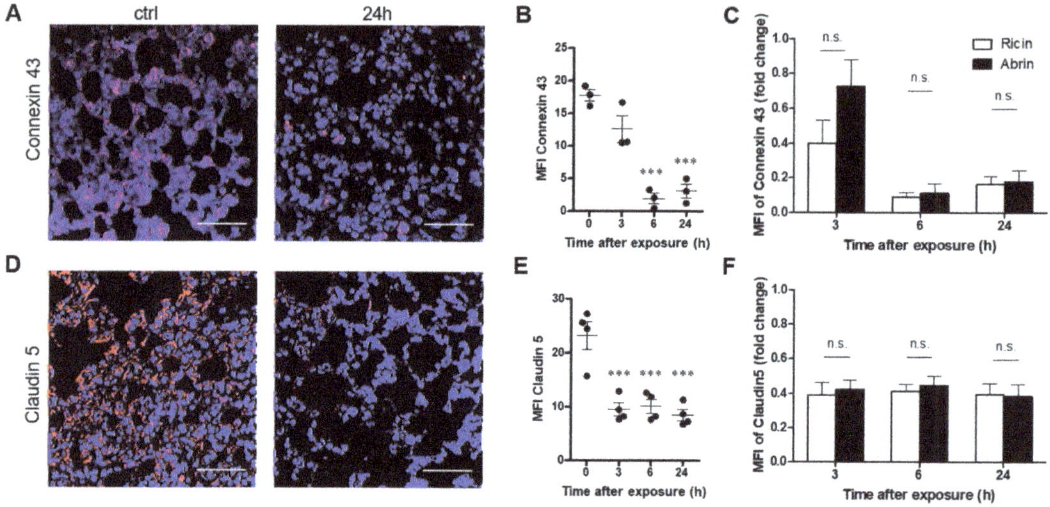

Figure 5. Alterations in connexin 43 and claudin 5 in the lungs of abrin-intoxicated mice. Lungs of abrin-intoxicated (10 µg/kg body weight) mice were harvested at the indicated time points, and junction proteins were quantified by immunohistochemical analysis of lung sections. (**A**,**D**) Confocal microscopy scans of lung sections stained for connexin 43 and claudin 5 (red) and identification of nuclei by DAPI (blue). (**B**,**E**) Scatterplots represent the immunofluorescence staining intensities of connexin 43 and claudin 5, expressed as MFI (n = 3–4 mice in each group). Scale bar: 50 µm. (**C**,**F**) Comparison between the abrin- and ricin-induced reduction in connexin 43 and claudin 5 [27]. The results are depicted as the means ± SEMs. *** $p < 0.001$. In (**B**,**E**), comparison with nonintoxicated mice; n.s., not significant.

2.4. Pulmonary Exposure to Abrin- and Ricin-Induced Comparable Damage to the Endothelial Glycocalyx

The endothelial glycocalyx is a complex layer of glycoproteins, proteoglycans and glycosaminoglycans that coat the luminal surface of the vascular endothelium. Hydrated glycosaminoglycans form a thick and rigid endothelial surface layer (ESL) that plays a key role in limiting vascular permeability and regulating leukocyte adhesion [33,34]. Because shedding of the ESL results in hyperpermeability and inappropriate leukocyte adhesion [35], we decided to evaluate the integrity of the ESL after pulmonary exposure to abrin in comparison to ricin intoxication. Mice were intranasally exposed to abrin or ricin, and soluble shed compounds of glycocalyx were analyzed in bronchoalveolar fluid (BALF) harvested at different time points. The detection of soluble glycocalyx compounds, which are indicative of endothelial glycocalyx degradation, negatively correlates with ESL thickness and positively correlates with vascular permeability [30,36]. Syndecan-1, heparan sulfate and hyaluronic acid are the main components whose shedding has been claimed to represent the endothelial glycocalyx state of health. An analysis of the hyaluronic acid levels, a ubiquitous glycosaminoglycan of the ECL, revealed elevated levels of this compound in the BAL of ricin- and abrin-intoxicated mice at 24 h postexposure, which continued to increase at later time points, 48–72 h postexposure. No difference was found in the intensity of the shedding of hyaluronic acid between abrin- and ricin-intoxicated mice (Figure 6A). Monitoring of heparan sulfate, the predominant glycosaminoglycan, demonstrated a marked release of this component at 72 h post abrin and ricin exposures. As in the case of hyaluronic acid, heparan sulfate shedding was similar in response to both toxins at all indicated time points (Figure 6B). Despite the degradation of hyaluronic acid and heparan sulfate, we did not detect shedding of the transmembrane core protein of the glycocalyx, syndecan-1, following either abrin or ricin pulmonary exposure (Figure 6C).

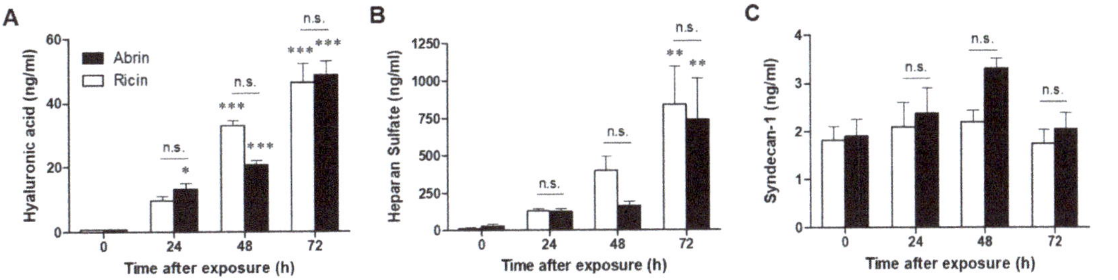

Figure 6. Degradation of the glycocalyx in abrin- and ricin-exposed mice. Levels of soluble hyaluronic acid (**A**), heparan sulfate (**B**) and syndecan-1 (**C**) were determined in the BALF collected from abrin- and ricin (10 or 7 μg/kg body weight, respectively)-exposed mice at the indicated time points (n = 4–9 mice in each group). The results are depicted as the means ± SEMs. * $p < 0.05$, ** $p < 0.01$, *** $p < 0.001$; n.s., not significant. The comparison of each column with nonintoxicated mice and between abrin and ricin at each time point.

3. Discussion

The toxicity of abrin and ricin depends on the route of exposure, with inhalatory exposures considered the most fatal [16]. The clinical manifestation following intranasal exposure of mice to these toxins is the onset of localized yet severe pulmonary edematous inflammation, which is accompanied by massive recruitment of neutrophils to the lungs and onset of a turbulent proinflammatory cytokine storm within this organ [22,37]. Despite the similarity in morbidity and mortality, following pulmonary abrin and ricin intoxications in mice, the ability to protect mice against ricin and abrin intoxications by postexposure antibody-mediated treatment differs radically. In the case of lethal ricin intoxication, rabbit-derived polyclonal anti-ricin antibody-based treatment of the mice at 24 h postexposure resulted in 34% survival rates [22]. When antibody treatment was administered at 48 h postexposure to ricin, protection was no more than marginal [38]. In sharp contrast, the administration of polyclonal anti-abrin antibodies to mice intranasally exposed to a lethal dose of abrin led to very high survival rates (~70–80%), even when the antibodies were applied as late as 72 h after intoxication [22]. This efficient protection by polyclonal anti-abrin antibodies could not be attributed to the neutralization of a single subunit because specific antibodies against the A or B subunits of abrin were equally effective in protecting mice against pulmonary intoxication with chimeric reciprocal toxins harboring one of the subunits of abrin and the other of ricin [24]. These observations indicated that the difference in the protection conferred by anti-abrin and anti-ricin antibodies against abrin and ricin intoxications, respectively, is not related to the difference in the quality of the two antibody preparations. Therefore, in this study, we dissected the differences in abrin and ricin lung pathology following pulmonary exposure of mice to either of the toxins. We found that both toxins bound similarly to hematopoietic cells, especially to AMs and DCs, and triggered their early and persistent elimination from the lungs. These results are in agreement with an earlier study which showed that macrophages are the most sensitive cells to RIPs [39]. In opposition to cells of hematopoietic origin, the toxins differed in their efficiency of binding to parenchymal cells of the lungs. The binding of abrin to $CD45^-$ cells was considerably less effective at all tested time points following intoxication. Consequently, following abrin intoxication, there was no direct damage found in the endothelial cell compartment, and the epithelial cell damage that, as in the case of ricin, was mostly confined to the alveolar epithelial type II cells was significantly lower than that observed following ricin intoxication. Supporting these results, we have previously shown that following intranasal intoxication, the in vivo catalytic performance of abrin, i.e., ribosomal depurination of pulmonary tissue, is significantly lower than that observed following ricin intoxication. In particular, the depurination levels of endothelial cells and pulmonary epithelial cells were markedly lower following abrin intoxication in comparison to ricin intoxication [40]. Furthermore, the lesser damage to lung parenchymal cells is in line with an older study that described the histopathology of the lungs in rats and showed that the appearance of apoptosis in the alveolar epithelium was far more marked following inhalation of ricin than abrin [19]. Interestingly, examination of the effect of antibody treatment against the two toxins on lung cell composition following exposure of mice to abrin and ricin showed significant reversion in the cells of hematopoietic origin, neutrophils and macrophages, after both intoxications [23]. However, neither the antibody-based treatment against abrin nor that against ricin conferred any beneficial influence on epithelial cells (data not shown). Since we did not find any repair of the epithelial population in the near term after antibody treatment, we estimate that the intensity of the epithelial damage has a direct effect on the survival of intoxicated and treated mice. As the epithelial damage in the lung is much more extensive after exposure to ricin compared to abrin, the protection ability of anti-ricin antibody treatment is limited.

One prominent hallmark of ricin-mediated pulmonary intoxication is the rapid and massive influx of neutrophils to the lungs [26–28]. This uncontrolled recruitment of neutrophils and the overwhelming activation in sterile inflammation, such as in the case of abrin or ricin intoxication, contributes to tissue damage by the release of proteinases, cationic polypeptides, cytokines and reactive oxygen species [41,42]. It has been previously shown that ricin does not bind neutrophils [25,40]. Similarly, we show in this study that infiltrating neutrophils, unlike other cells of hematopoietic origin, did not bind abrin following intoxication.

The dramatic influx over time of toxin-nonbinding neutrophils occurs mainly in the small capillaries spanning the alveolar network [43] and induces indirect lung damage by compromising the permeability of the alveolar–capillary barrier [44]. The kinetics and the extent of neutrophil influx to the lungs were found to be similar following abrin and ricin pulmonary intoxications. However, while the alveolar–capillary barrier integrity was compromised at early stages following ricin intoxication [27], lung permeability following abrin exposure was significantly increased only at later time points (48–72 h). This finding may stem from the relatively reduced levels of irreversible cellular damage inflicted by abrin. Alveolar–capillary barrier permeability is tightly regulated by the molecular interplay of intercellular junction molecules that span the gap between neighboring epithelial and endothelial cells. These junction molecules cooperate in maintaining tissue integrity to limit epithelial and endothelial permeability and to allow for just minimal leakage of fluids into the interstitial compartment [30,45]. We show in this study that in addition to reduced cellular damage, the insult to junction proteins, such as VE-cadherin and claudin 18, following abrin intoxication was less prominent, both at early and late stages, than the damage observed following ricin exposure. Moreover, the tight junction protein occludin, which is reduced at a later stage (48 h) following ricin intoxication [27], was not impaired at all post-abrin-exposure time points. These data may also account for the delayed hyperpermeability of the lungs post abrin exposure in comparison to ricin. However, the pronounced lung hyperpermeability at later time points after exposure to abrin may stem from robust recruitment of neutrophils at these time points and marked impairment in connexin 43 and claudin 5. In addition, although in contrast to ricin, no direct damage to endothelial cells was observed following abrin intoxication, collateral damage to these cells was discerned. In fact, we show in this study that the intensity of the damage to the pulmonary vascular endothelial glycocalyx was comparable following intoxication with the two toxins.

In summary, we propose that the relatively superior performance of the anti-abrin antibody-based treatment and the ability to protect against abrin lethality, but not against ricin, when administered lately following intranasal intoxication in mice is due to the differences in the timing and intensity of the damage to the lung stroma inflicted by abrin and ricin.

4. Materials and Methods

4.1. Animals

Experiments were performed in accordance with Israeli law and approved by the Institutional Animal Care and Use Committee (IACUC) of the Israel Institute for Biological Research (Ness-Ziona, Iarael). Treatment of animals was in accordance with regulations outlined in the USDA Animal Welfare Act and the conditions specified in the National Institute of Health Guide for Care and Use of Laboratory Animals. Female CD-1 mice (27–32 g) were purchased from Charles River Laboratories Ltd., Margate, UK. Mice were housed in filter-top cages in an environmentally controlled room and maintained at 21 ± 2 °C and 55 ± 10% humidity. Lighting was set to mimic a 12/12 h dawn-to-dusk cycle. Mice were housed in a purpose-built animal holding facility for 4–8 days prior to the beginning of the experiment. Animals were allowed access to water ad libitum and 4% body weight food per day.

4.2. Fluorescent Toxin Labeling and Intoxication

Abrin and ricin were purified as previously described [22,23] and conjugated (1 mg) with the Alexa Fluor™ 488 protein labeling kit (Molecular probes, Thermo Fisher Scientific, Waltham, MA, USA) according to the manufacturer's instructions. The cytotoxicity of labeled toxin (~5 dye/protein (mol/mol)) was determined in a cell-based assay developed in the past [46]. Briefly, labeled abrin or ricin was added to HEK-293 cell cultures, which secrete the enzyme acetylcholinesterase (AChE) in a constitutive manner. Secreted AChE was measured in the cell growth medium at 18 h postexposure, and activity levels were compared to those measured for nonlabeled toxin. We determined a ~2-fold reduction in the toxicity of the labeled toxins. Intranasal intoxication with toxin at a $2LD_{50}$ dose (unlabeled and labeled ricin or abrin, 7 and 14 μg/kg or 10 and 20 μg/kg, respectively) was applied (2 × 25 μL).

4.3. Flow Cytometry

Lungs were minced into small pieces and subjected to enzymatic digestion with 4 mg/mL collagenase D (Roche, Mannheim, Germany) for 2 h at 37 °C. The tissues were then meshed through a 40 μm cell strainer, and red blood cells were lysed with red blood cell lysis buffer (Sigma–Aldrich, Rehovot, Israel). For staining, cell suspensions were stained with CD45 (clone 30-F11), CD11b (M1/70), Ly6G (1A8), CD11c (N418), MHC class II (M5/114), Siglec F (S17007L), CD31 (390), CD326 (G8.8), T1α (8.1.1) and proSPC (Millipore, Temecula, CA, USA) followed by allophycocyanin (APC) donkey anti-rabbit IgG (Jackson ImmunoResearch, West Grove, PA, USA). All antibodies were purchased from Biolegend (San Diego, CA, USA) unless otherwise indicated. Neutrophils were identified as $CD45^{high}$, $Ly6G^{high}$ and $CD11b^{high}$; AMs as $CD45^{high}$, autofluorescent, Siglec F^{high} and $CD11c^{high}$; DCs as $CD45^{high}$, Siglec F^{neg} and $CD11c^{high}$ and MHC class II^{high}; endothelial cells as $CD45^{neg}$ and $CD31^{high}$; epithelial cells as $CD45^{neg}$ and $CD326^{high}$; alveolar epithelial cells type I as $CD45^{neg}$ and $CD31^{neg}$, $CD326^{high}$ and $T1\alpha^{high}$; epithelial cells as $CD45^{neg}$ and $CD326^{high}$; and alveolar epithelial cells type II as $CD45^{neg}$ and $CD31^{neg}$, $CD326^{high}$ and $proSPC^{high}$ cells. Flow cytometry was performed on a FACSCalibur (BD Biosciences, San Jose, CA, USA) and analyzed using FlowJo software v.10.8.0 (Tree Star, Ashland, OR, USA).

4.4. Immunohistochemistry

Lungs were collected and fixed in 4% buffered formaldehyde in PBS pH 7.2–7.4 (Bio Lab, Jerusalem, Israel) for 2 weeks. Sections of 5 μm were prepared after paraffin embedding using an RM 2255 microtome (Leica, Nussloch, Germany). Antigen retrieval was performed by incubation in Target Retrieval Solution (S1700, DAKO, Carpinteria, CA, USA, 30 min, 95 °C). After blocking in 5% BSA in PBS, slides were incubated (overnight, 4 °C) with purified anti-CD31 (390, Biolegend, San Diego, CA, USA), proSPC (Millipore, Temecula, CA, USA), podoplanin (T1α, 8.1.1, Biolegend), VE-cadherin (ab33168), claudin 5 (ab15106), connexin 43 (ab117843), occludin (ab31721) or claudin 18 (ab203563) (Abcam, Cambridge, MA, USA). Alexa Fluor 594- or 488-coupled donkey anti-rabbit or Alexa Fluor 594-coupled goat anti-Armenian hamster antibodies were used for detection (Molecular probes®, Thermo Fisher Scientific, Carlsbad, CA, USA). For nuclear staining, slides were mounted with Prolong® Gold antifade reagent containing DAPI (Molecular probes®, Thermo Fisher Scientific, Carlsbad, CA, USA). Analysis was performed using an LSM 710 confocal scanning microscope (Zeiss, Jena, Germany) equipped with the following lasers: argon multiline (458/488/514 nm), diode 405 nm, DPSS 561 nm and helium-neon 633 nm.

4.5. Permeability Analysis

Lung permeability was determined by the Evans blue dye (EBD) extravasation method as follows: EBD (7.5 mg/mL, Sigma–Aldrich, Rehovot, Israel) was injected intravenously into mice at a dose of 50 mg/kg and allowed to circulate for 1 h. Mice were then anesthetized, and the lungs were perfused by cutting the left atrium and flushing with 5 mL PBS through the right ventricle. The lungs were removed, and EBD was extracted by incubation of the tissues in 0.5 mL of formamide (Sigma–Aldrich, Rehovot, Israel) at 60 °C for 24 h. The EBD optical density in the supernatant was measured at 620 nm in a spectrophotometer (Molecular Devices, Sunnyvale, CA, USA), and the total amount of dye was calculated by means of a standard calibration curve.

4.6. Analysis of Glycocalyx Shedding

BALF was performed by flushing the lungs with 1 mL of PBS using a tracheal cannula. The BALF was centrifuged at 950× *g* at 4 °C for 10 min, and the supernatants were collected and tested for soluble heparan sulfate, hyaluronic acid and syndecan-1 levels using an LSBio (Seattle, WA, USA) Mouse Heparan Sulfate ELISA kit (LS-F39210), R&D Systems (Abingdon, UK) Porcine/Mouse Quantikine Hyaluronan Immunoassay kit (DHYAL0) and Diaclone (Besancon Cedex, France) Murine sCD138 (Syndecan-1) ELISA kit (860.090.096) according to the manufacturer's instructions.

4.7. Statistical Analysis

All statistical analyses were conducted with GraphPad Prism software (version 5.01, GraphPad Software Inc., La Jolla, CA, USA, 2007). Data are presented as the means ± SEMs. Significance was assessed by Student's t test, and for multiple comparisons, one-way analysis of variance (ANOVA) followed by Tukey's multiple comparisons test or two-way ANOVA with Bonferroni correction was used for planned comparisons. Differences were considered significant at $p < 0.05$. Points in graphs indicate individual mice.

Author Contributions: Conceptualization, A.S., T.S., C.K., Y.G. and R.F.; Formal analysis, A.S., C.K., Y.G. and R.F.; Investigation, A.S., Y.G., R.A., T.S., C.K. and R.F.; Data curation, A.S., Y.G., R.A., Y.E., T.S., C.K. and R.F.; Methodology, A.S., Y.G., R.A., Y.E., T.S., C.K. and R.F.; Visualization, A.S. and Y.E.; Validation, A.S., Y.G., T.S., C.K. and R.F.; Writing—original draft preparation, A.S. and R.F.; Writing—review and editing, A.S., Y.G., R.A., Y.E., T.S., C.K. and R.F.; Supervision, T.S. and C.K. All authors have read and agreed to the published version of the manuscript.

Funding: This study was supported by the Israel Institute for Biological Research.

Institutional Review Board Statement: The study was conducted in accordance with Israeli law and was approved by the Ethics Committee for Animal Experiments of the Israel Institute for Biological Research (project identification codes M-42-13, M-63-13, M-09-15 and M-44-15, approval dates 2 July 2013, 22 December 2013, 22 January 2015 and 23 June 2015, respectively).

Informed Consent Statement: Not applicable.

Data Availability Statement: Not applicable.

Acknowledgments: We are grateful to the staff of the animal facilities of the Israel Institute of Biological Research.

Conflicts of Interest: The authors declare no conflict of interest.

References

1. Di Maro, A.; Citores, L.; Russo, R.; Iglesias, R.; Ferreras, J.M. Sequence comparison and phylogenetic analysis by the Maximum Likelihood method of ribosome-inactivating proteins from angiosperms. *Plant Mol. Biol.* **2014**, *85*, 575–588. [CrossRef] [PubMed]
2. Landi, N.; Ruocco, M.R.; Ragucci, S.; Aliotta, F.; Nasso, R.; Pedone, P.V.; Di Maro, A. Quinoa as source of type 1 ribosome inactivating proteins: A novel knowledge for a revision of its consumption. *Food Chem.* **2021**, *342*, 128337. [CrossRef] [PubMed]
3. Stirpe, F.; Gilabert-Oriol, R. *Ribosome-Inactivating Proteins: An Overview*; Plant Toxins; Carlini, C.R., Ligabue-Braun, R., Eds.; Springer: Berlin/Heidelberg, Germany, 2015; pp. 153–182.
4. Girbés, T.; Ferreras, J.M.; Arias, F.J.; Stirpe, F. Description, distribution, activity and phylogenetic relationship of ribosome-inactivating proteins in plants, fungi and bacteria. *Mini Rev. Med. Chem.* **2004**, *4*, 461–476. [CrossRef] [PubMed]
5. Ng, T. Peptides and proteins from fungi. *Peptides* **2004**, *25*, 1055–1073. [CrossRef]
6. Pizzo, E.; Di Maro, A. A new age for biomedical applications of Ribosome Inactivating Proteins (RIPs): From bioconjugate to nanoconstructs. *J. Biomed. Sci.* **2016**, *23*, 54. [CrossRef]
7. Liu, R.S.; Yang, J.H.; Liu, W.Y. Isolation and enzymatic characterization of lamjapin, the first ribosome-inactivating protein from cryptogamic algal plant (Laminaria japonica A). *Eur. J. Biochem.* **2002**, *269*, 4746–4752. [CrossRef]
8. O'Loughlin, E.V.; Robins-Browne, R.M. Effect of Shiga toxin and Shiga-like toxins on eukaryotic cells. *Microbes Infect.* **2001**, *3*, 493–507. [CrossRef]
9. Lapadula, W.J.; Sanchez Puerta, M.V.; Juri Ayub, M. Revising the taxonomic distribution, origin and evolution of ribosome inactivating protein genes. *PLoS ONE* **2013**, *8*, e72825. [CrossRef]
10. Sandvig, K.; Olsnes, S.; Pihl, A. Kinetics of binding of the toxic lectins abrin and ricin to surface receptors of human cells. *J. Biol. Chem.* **1976**, *251*, 3977–3984. [CrossRef]
11. Spooner, R.A.; Smith, D.C.; Easton, A.J.; Roberts, L.M.; Lord, M.J. Retrograde transport pathways utilised by viruses and protein toxins. *Virol. J.* **2006**, *3*, 26. [CrossRef]
12. Sandvig, K.; Van Deurs, B. Endocytosis, intracellular transport, and cytotoxic action of Shiga toxin and ricin. *Physiol. Rev.* **1996**, *76*, 949–966. [CrossRef] [PubMed]
13. Endo, Y.; Tsurugi, K. RNA N-glycosidase activity of ricin A-chain. Mechanism of action of the toxic lectin ricin on eukaryotic ribosomes. *J. Biol. Chem.* **1987**, *262*, 8128–8130. [CrossRef]
14. Endo, Y.; Tsurugi, K. The RNA N-glycosidase activity of ricin A-chain. The characteristics of the enzymatic activity of ricin A-chain with ribosomes and with rRNA. *J. Biol. Chem.* **1988**, *263*, 8735–8739. [CrossRef]
15. Endo, Y.; Mitsui, K.; Motizuki, M.; Tsurugi, K. The mechanism of action of ricin and related toxic lectins on eukaryotic ribosomes. The site and the characteristics of the modification in 28 S ribosomal RNA caused by the toxins. *J. Biol. Chem.* **1987**, *262*, 5908–5912. [CrossRef]
16. Audi, J.; Belson, M.; Patel, M.; Schier, J.; Osterloh, J. Ricin poisoning: A comprehensive review. *JAMA* **2005**, *294*, 2342–2351. [CrossRef] [PubMed]
17. Montanaro, L.; Sperti, S.; Testoni, G.; Mattioli, A. Effect of elongation factor 2 and of adenosine diphosphate-ribosylated elongation factor 2 on translocation. *Biochem. J.* **1976**, *156*, 15–23. [CrossRef]
18. Bhaskaran, M.; Didier, P.J.; Sivasubramani, S.K.; Doyle, L.A.; Holley, J.; Roy, C.J. Pathology of lethal and sublethal doses of aerosolized ricin in rhesus macaques. *Toxicol. Pathol.* **2014**, *42*, 573–581. [CrossRef]
19. Griffiths, G.D.; Rice, P.; Allenby, A.C.; Bailey, S.C.; Upshall, D.G. Inhalation toxicology and histopathology of ricin and abrin toxins. *Inhal. Toxicol.* **1995**, *7*, 269–288. [CrossRef]
20. Pincus, S.H.; Bhaskaran, M.; Brey, R.N., III; Didier, P.J.; Doyle-Meyers, L.A.; Roy, C.J. Clinical and pathological findings associated with aerosol exposure of macaques to ricin toxin. *Toxins* **2015**, *7*, 2121–2133. [CrossRef]
21. Wilhelmsen, C.; Pitt, M. Lesions of acute inhaled lethal ricin intoxication in rhesus monkeys. *Vet. Pathol.* **1996**, *33*, 296–302. [CrossRef]
22. Gal, Y.; Mazor, O.; Alcalay, R.; Seliger, N.; Aftalion, M.; Sapoznikov, A.; Falach, R.; Kronman, C.; Sabo, T. Antibody/doxycycline combined therapy for pulmonary ricinosis: Attenuation of inflammation improves survival of ricin-intoxicated mice. *Toxicol. Rep.* **2014**, *1*, 496–504. [CrossRef] [PubMed]

23. Sabo, T.; Gal, Y.; Elhanany, E.; Sapoznikov, A.; Falach, R.; Mazor, O.; Kronman, C. Antibody treatment against pulmonary exposure to abrin confers significantly higher levels of protection than treatment against ricin intoxication. *Toxicol. Lett.* **2015**, *237*, 72–78. [CrossRef] [PubMed]
24. Gal, Y.; Sapoznikov, A.; Falach, R.; Mazor, O.; Alcalay, R.; Elhanany, E.; Aftalion, M.; Ehrlich, S.; Kronman, C.; Sabo, T. Equal Neutralization Potency of Antibodies Raised against Abrin Subunits. *Antibodies* **2020**, *9*, 4. [CrossRef] [PubMed]
25. Sapoznikov, A.; Falach, R.; Mazor, O.; Alcalay, R.; Gal, Y.; Seliger, N.; Sabo, T.; Kronman, C. Diverse profiles of ricin-cell interactions in the lung following intranasal exposure to ricin. *Toxins* **2015**, *7*, 4817–4831. [CrossRef]
26. Lindauer, M.L.; Wong, J.; Iwakura, Y.; Magun, B.E. Pulmonary inflammation triggered by ricin toxin requires macrophages and IL-1 signaling. *J. Immunol.* **2009**, *183*, 1419–1426. [CrossRef]
27. Sapoznikov, A.; Gal, Y.; Falach, R.; Sagi, I.; Ehrlich, S.; Lerer, E.; Makovitzki, A.; Aloshin, A.; Kronman, C.; Sabo, T. Early disruption of the alveolar-capillary barrier in a ricin-induced ARDS mouse model: Neutrophil-Dependent and-independent impairment of junction proteins. *Am. J. Physiol. Lung Cell. Mol. Physiol.* **2019**, *316*, L255–L268. [CrossRef]
28. Wong, J.; Korcheva, V.; Jacoby, D.B.; Magun, B. Intrapulmonary delivery of ricin at high dosage triggers a systemic inflammatory response and glomerular damage. *Am. J. Pathol.* **2007**, *170*, 1497–1510. [CrossRef]
29. Harris, E.S.; Nelson, W.J. VE-Cadherin: At the front, center, and sides of endothelial cell organization and function. *Curr. Opin. Cell Biol.* **2010**, *22*, 651–658. [CrossRef]
30. Herrero, R.; Sanchez, G.; Lorente, J.A. New insights into the mechanisms of pulmonary edema in acute lung injury. *Ann. Transl. Med.* **2018**, *6*, 32. [CrossRef]
31. Ohta, H.; Chiba, S.; Ebina, M.; Furuse, M.; Nukiwa, T. Altered expression of tight junction molecules in alveolar septa in lung injury and fibrosis. *Am. J. Physiol. Lung Cell. Mol. Physiol.* **2012**, *302*, L193–L205. [CrossRef]
32. Katalan, S.; Falach, R.; Rosner, A.; Goldvaser, M.; Brosh-Nissimov, T.; Dvir, A.; Mizrachi, A.; Goren, O.; Cohen, B.; Gal, Y. A novel swine model of ricin-induced acute respiratory distress syndrome. *Dis. Models Mech.* **2017**, *10*, 173–183. [CrossRef]
33. Schmidt, E.P.; Lee, W.L.; Zemans, R.L.; Yamashita, C.; Downey, G.P. On, around, and through: Neutrophil-Endothelial interactions in innate immunity. *Physiology* **2011**, *26*, 334–347. [CrossRef] [PubMed]
34. Yang, Y.; Schmidt, E.P. The endothelial glycocalyx: An important regulator of the pulmonary vascular barrier. *Tissue Barriers* **2013**, *1*, 1217–1223. [CrossRef] [PubMed]
35. Curry, F.; Adamson, R. Endothelial glycocalyx: Permeability barrier and mechanosensor. *Ann. Biomed. Eng.* **2012**, *40*, 828–839. [CrossRef] [PubMed]
36. Torres Filho, I.P.; Torres, L.N.; Salgado, C.; Dubick, M.A. Plasma syndecan-1 and heparan sulfate correlate with microvascular glycocalyx degradation in hemorrhaged rats after different resuscitation fluids. *Am. J. Physiol. Heart Circ. Physiol.* **2016**, *310*, H1468–H1478. [CrossRef]
37. Lindauer, M.; Wong, J.; Magun, B. Ricin toxin activates the NALP3 inflammasome. *Toxins* **2010**, *2*, 1500–1514. [CrossRef]
38. Gal, Y.; Sapoznikov, A.; Falach, R.; Ehrlich, S.; Aftalion, M.; Kronman, C.; Sabo, T. Total body irradiation mitigates inflammation and extends the therapeutic time window for anti-ricin antibody treatment against pulmonary ricinosis in mice. *Toxins* **2017**, *9*, 278. [CrossRef]
39. Barbieri, L.; Battelli, M.G.; Stirpe, F. Ribosome-inactivating proteins from plants. *Biochim. Biophys. Acta (BBA)-Rev. Biomembr.* **1993**, *1154*, 237–282. [CrossRef]
40. Falach, R.; Sapoznikov, A.; Gal, Y.; Israeli, O.; Leitner, M.; Seliger, N.; Ehrlich, S.; Kronman, C.; Sabo, T. Quantitative profiling of the in vivo enzymatic activity of ricin reveals disparate depurination of different pulmonary cell types. *Toxicol. Lett.* **2016**, *258*, 11–19. [CrossRef]
41. Aulakh, G.K. Neutrophils in the lung: "The first responders". *Cell Tissue Res.* **2018**, *371*, 577–588. [CrossRef]
42. Grommes, J.; Soehnlein, O. Contribution of neutrophils to acute lung injury. *Mol. Med.* **2011**, *17*, 293–307. [CrossRef] [PubMed]
43. Doerschuk, C.M. Leukocyte trafficking in alveoli and airway passages. *Respir. Res.* **2000**, *1*, 4. [CrossRef] [PubMed]
44. Yang, S.-C.; Tsai, Y.-F.; Pan, Y.-L.; Hwang, T.-L. Understanding the role of neutrophils in acute respiratory distress syndrome. *Biomed. J.* **2021**, *44*, 439–446. [CrossRef] [PubMed]
45. Shen, L. Tight junctions on the move: Molecular mechanisms for epithelial barrier regulation. *Ann. New York Acad. Sci. USA* **2012**, *1258*, 9–18. [CrossRef]
46. Cohen, O.; Mechaly, A.; Sabo, T.; Alcalay, R.; Aloni-Grinstein, R.; Seliger, N.; Kronman, C.; Mazor, O. Characterization and epitope mapping of the polyclonal antibody repertoire elicited by ricin holotoxin-based vaccination. *Clin. Vaccine Immunol.* **2014**, *21*, 1534–1540. [CrossRef]

toxins

MDPI

Article

LRP1-Mediated Endocytosis May Be the Main Reason for the Difference in Cytotoxicity of Curcin and Curcin C on U2OS Osteosarcoma Cells

Siying Qin [1,2], Xueying Wang [1], Pan Han [1], Zhiping Lai [1], Yingying Ren [1], Rui Ma [1], Cheng Cheng [1], Ting Wang [1] and Ying Xu [1,*]

1 Key Laboratory of Bio-Resources and Eco-Environment of Ministry of Education, College of Life Sciences, Sichuan University, Chengdu 610041, China
2 The First Affiliated Hospital of Chengdu Medical College, School of Clinical Medicine, Chengdu Medical College, Chengdu 610500, China
* Correspondence: xuying@scu.edu.can

Abstract: Curcin and Curcin C, both of the ribosome-inactivating proteins of *Jatropha curcas*, have apparent inhibitory effects on the proliferation of osteosarcoma cell line U20S. However, the inhibitory effect of the latter is 13-fold higher than that of Curcin. The mechanism responsible for the difference has not been studied. This work aimed to understand and verify whether there are differences in entry efficiency and pathway between them using specific endocytosis inhibitors, gene silencing, and labeling techniques such as fluorescein isothiocyanate (FITC) labeling. The study found that the internalization efficiency of Curcin C was twice that of Curcin for U2OS cells. More than one entering pathway was adopted by both of them. Curcin C can enter U2OS cells through clathrin-dependent endocytosis and macropinocytosis, but clathrin-dependent endocytosis was not an option for Curcin. The low-density lipoprotein receptor-related protein 1 (LRP1) was found to mediate clathrin-dependent endocytosis of Curcin C. After LRP1 silencing, there was no significant difference in the 50% inhibitory concentration (IC_{50}) and endocytosis efficiency between Curcin and Curcin C on U2OS cells. These results indicate that LRP1-mediated endocytosis is specific to Curcin C, thus leading to higher U2OS endocytosis efficiency and cytotoxicity than Curcin.

Keywords: ribosome-inactivating proteins; Curcin; Curcin C; osteosarcoma U20S cells; endocytosis

Key Contribution: LRP1 was found to be the endocytic receptor of Curcin C. One of the main reasons for the difference in the activity of Curcin and Curcin C was discovered.

Citation: Qin, S.; Wang, X.; Han, P.; Lai, Z.; Ren, Y.; Ma, R.; Cheng, C.; Wang, T.; Xu, Y. LRP1-Mediated Endocytosis May Be the Main Reason for the Difference in Cytotoxicity of Curcin and Curcin C on U2OS Osteosarcoma Cells. *Toxins* **2022**, *14*, 771. https://doi.org/10.3390/toxins14110771

Received: 9 October 2022
Accepted: 5 November 2022
Published: 8 November 2022

Publisher's Note: MDPI stays neutral with regard to jurisdictional claims in published maps and institutional affiliations.

1. Introduction

Osteosarcoma is the most common primary bone malignant tumor in children and adolescents [1]. Since the 1970s, the treatment received by osteosarcoma patients has not changed [2,3]. In recent decades, many clinical trials of new drugs and research on alternative strategies of standard chemotherapy have not successfully improved the prognosis of osteosarcoma patients [3]. Therefore, exploring new efficient medicines has become a new research focus in osteosarcoma. Scientists have paid more attention to macromolecular drugs than traditional small-molecule drugs because of their efficacy and high specificity [4]. In recent years, biological macromolecular drugs accounted for about 30% of innovative drugs approved by the U.S. Food and Drug Administration, and the ratio is still rising [5].

RIPs are rRNA N-glycosylases (EC 3.2.2.22) that catalyze the elimination of a specific adenine (A4324 in rat ribosomes or the equivalent in other organisms) located in the sarcin–ricin loop (SRL) of animal 28S ribosomal RNA [6,7]. Most type I RIPs have antitumor activity and lower toxicity to intact normal cells, being the potential cancer drug candidates [8]. As early as the 1990s, TCS was found toxic to leukemia/lymphoma cells in vitro [9]. MAP30, a type I ribosome-inactivating protein extracted from *Momordica charantia*, can inhibit the proliferation of MDA MB 231 breast cancer cells that are insensitive to chemotherapy due to overexpression of HER2 [10]. Pachyerosin is a new type I ribosome-inactivating protein extracted from the seeds of *Rhamnus rhamnoides*. MTT test shows it can inhibit the proliferation of liver cancer cell HUH 7 [11]. α-MMC can block the cell cycle and inhibit the growth of lung cancer cells [12]. Curcin and Curcin C are type I RIPs found in *Jatropha curcas*. The latter has a powerful inhibitory effect on the growth of osteosarcoma U20S cells, even ten times that of Curcin [13].

Entry efficiency is one of the critical factors affecting the curative effect. Type I RIPs generally have low internalization efficiency and cytotoxicity due to the lack of a B chain with lectin activity. Moreover, their internalization pathways in mammalian cells are still less studied compared with ricin and other type II RIPs. Only a few mammalian internalization pathways of type I RIPs, such as saporin and trichosanthin, have been reported [14–19]. There is still no related report of Curcin and Curcin C. Therefore, this work tried to establish in vitro tracer systems of Curcin and Curcin C by three methods, including FITC labeling, linking eGFP tags and immunofluorescence tags and exploring their internalization efficiency, and internalization pathways and related receptors of Curcin and Curcin C into osteosarcoma U2OS cells.

2. Results

2.1. FITC-Curcin C Entered U2OS Cells More Efficiently Than FITC-Curcin

2.1.1. Comparison of Three Methods for Labeling Curcin and Curcin C

To explore how Curcin and Curcin C enter U2OS cells, we first tried to establish an in vitro tracking system of Curcin and Curcin C by three methods: FITC labeling and eGFP through prokaryotic expression and immunofluorescence labeling.

FITC labeling is a commonly used method for in vitro protein tracking, which is simple and has little effect on the structure of the target protein [20,21]. When the protein concentration was 10 mg/mL, the labeling ratio was FITC:C/CC = 70:1 and, incubated at 4 °C overnight in the dark for 12 h, the F/P value can be greater than 3, which meets the requirements of subsequent experiments. However, it is worth noting that the cytotoxicity of labeled FITC-Curcin and FITC-Curcin C were reduced. The IC_{50} values of them were 1.65 μM and greater than 2 μM (Table 1), respectively, which were much larger than those of unlabeled proteins (15.04 nM and 1.05 nM) (Figure 1A). It may be due to the presence of FITC-labeled target arginine (Arg212) in the highly conserved RNA N-glycosylase active site of Curcin and Curcin C.

Table 1. IC_{50} [1] of C [2], CC [3], C, and CC with different tags.

Protein	Curcin	Curcin C	FITC-C	FITC-CC	His-eGFP-C	His-eGFP-CC
IC_{50} (nM)	15.04	1.05	1650.24	>2000.00	216.28	209.32

[1] IC—inhibitory concentration; [2] C—Curcin; [3] CC—Curcin C.

Subsequently, we tried to generate His-eGFP-Curcin and His-eGFP-Curcin C using a prokaryotic expression system so that Curcin and Curcin C could be traced by eGFP tag. Although the strain expressing the recombinant vector emits specific green fluorescence of GFP, both His-eGFP-Curcin and His-eGFP-Curcin C were only obtained as inclusion bodies. Their antitumor activities after renaturation decreased by 14.38-fold and 199.35-fold, respectively (Figure 1B and Table 1). Moreover, at the experimental concentration, the fluorescence intensities of His-eGFP-Curcin and His-eGFP-Curcin C could not reach the required intensity (Figure 1C,D).

In addition, we also tried to trace His-tagged proteins (rCurcin and rCurcin C) in vitro with His-tag primary antibody and FITC-conjugated fluorescent secondary antibody. However, after many attempts, the fluorescence produced by this method was also very low and prone to false positives. Although the three in vitro tracing methods tried were not perfect, the FITC labeling method was still used in the subsequent analysis.

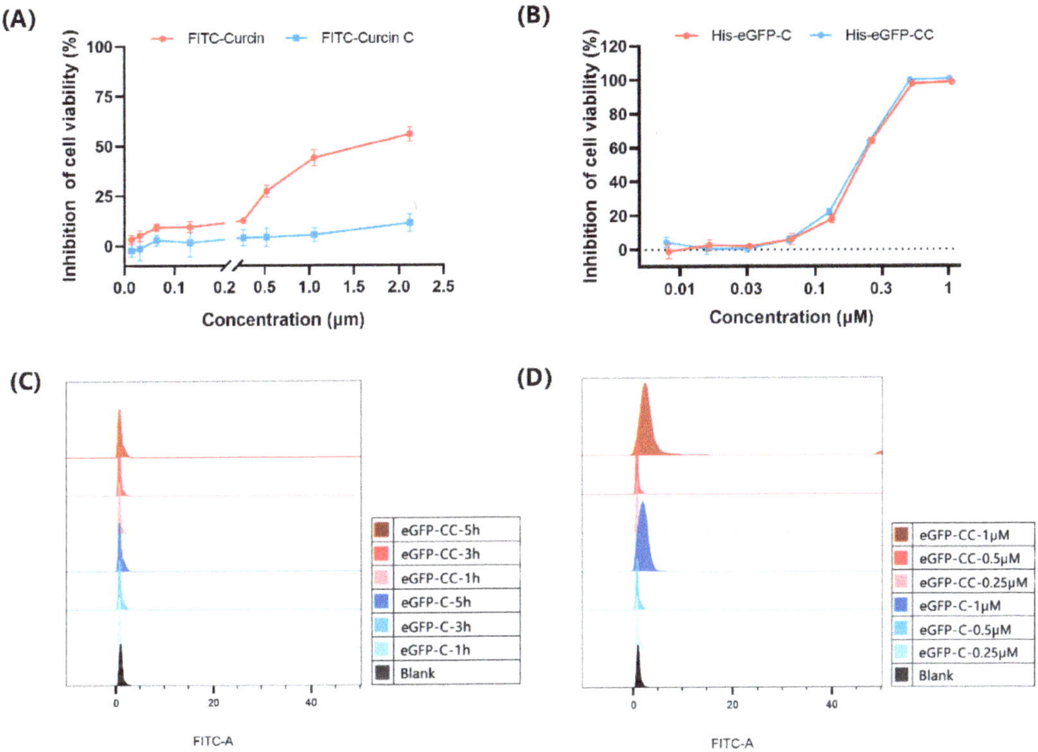

Figure 1. Cytotoxicity and intracellular fluorescence intensity of FITC-Curcin (FITC-C), FITC-Curcin C (FITC-CC), His-eGFP-Curcin (eGFP-C), and His-eGFP-Curcin C (eGFP-CC). (**A**) Viability of U2OS cells after 48 h FITC-C and FITC-CC treatment. (**B**) Viability of U2OS cells after 48 h eGFP-C and eGFP-CC treatment. (**C**) Flow cytometry histograms of intracellular fluorescence intensity after 1, 3, and 5 h incubation with 0.5 μM eGFP-C and eGFP-CC. (**D**) Flow cytometry histograms of intracellular fluorescence intensity after 4 h incubation with 0.25 μM, 0.5 μM, and 1 μM eGFP-C and eGFP-CC. All data are the mean ± SD of three independent experiments.

2.1.2. FITC-Curcin C Entered U2OS Cells More Efficiently Than FITC-Curcin

To visually observe whether Curcin and Curcin C can enter the cells, U2OS cells were incubated with 0.5 μM FITC-Curcin and FITC-Curcin C for 4 h, respectively. When observed under a fluorescence microscope, it is easy to detect that green fluorescence did appear inside the cells, mainly distributing in the cytoplasmic region. The results suggested Curcin and Curcin C entered the cells (Figure 2A). Meanwhile, FITC-labeled bovine serum albumin (FITC-BSA) was used as a control and was not observed entering U2OS cells.

When the concentration of FITC-Curcin and FITC-Curcin C was 0.5 μM, the amount of FITC-Curcin and FITC-Curcin C entering U2OS cells showed time-dependent laziness with the prolongation of incubation time (Figure 2B,C). When the incubation time was 12 h, the entry of FITC-Curcin C was 50% higher than that of FITC-Curcin. Then, the incubation time was fixed (12 h), and it was found that the amount of FITC-Curcin and FITC-Curcin C entering U2OS cells also showed a dose-dependent manner. When the concentration reached 1 μM, the influx of FITC-Curcin C was almost twice that of FITC-Curcin (Figure 2D,E).

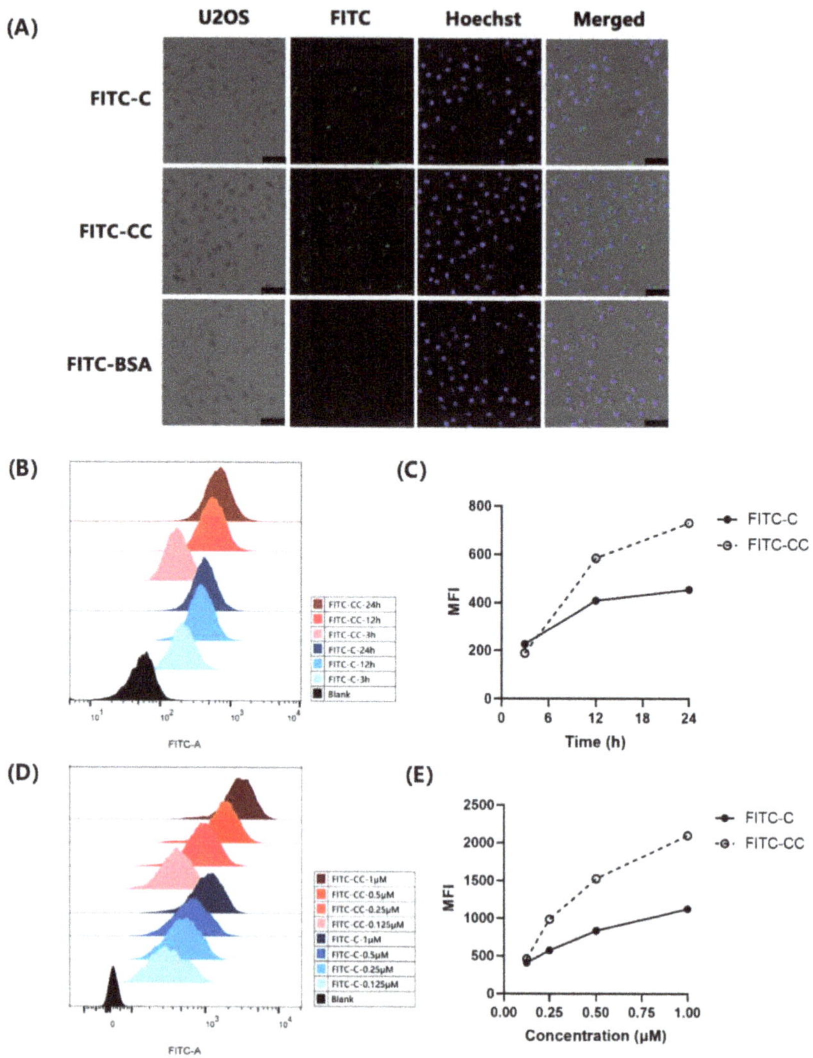

Figure 2. FITC-C and FITC-CC entered U2OS cells in a time- and concentration-dependent manner. (**A**) Observation of FITC-C-, FITC-CC-, and FITC-labeled bovine serum albumin (FITC-BSA) in U2OS cells under a fluorescence microscope. The nucleus was stained with DNA-specific fluorescent dye Hoechst 33258. Scale bar, 75 μm. (**B**) Flow cytometry histograms of 0.5 μM FITC-Curcin and FITC-Curcin C fluorescence intensity in U2OS cells after 3 h, 12 h, and 24 h incubation. (**C**) Mean fluorescence intensity (MFI) of 0.5 μM FITC-Curcin and FITC-Curcin C in U2OS cells after 3 h, 12 h, and 24 h incubation. (**D**) Flow cytometry histograms of FITC-Curcin and FITC-Curcin C fluorescence intensity in U2OS cells after 12 h incubation with various concentrations (0.125 μM, 0.25 μM, 0.5 μM, and 1 μM). (**E**) The MFI of FITC-Curcin and FITC-Curcin C in U2OS cells after 12 h incubation with various concentrations (0.125 μM, 0.25 μM, 0.5 μM, and 1 μM).

2.2. *FITC-Curcin and FITC-Curcin C Have Different Entry Pathways*

The effects of six different endocytosis inhibitors on the cytotoxicity of Curcin and Curcin C were determined further to understand the reasons for the difference in entry efficiency.

To assess whether the entry of Curcin and Curcin C into U2OS cells depends on pH-dependent endocytic pathways, we first studied the effects of two lysosomotropic agents, chloroquine (CQ) and NH_4Cl [22]. Compared with the control, the treatment significantly reduced the proliferation inhibition rates of both Curcin and Curcin C, resulting in a significant increase in their IC_{50}, suggesting that endocytosis was indeed their entry point (Figure 3).

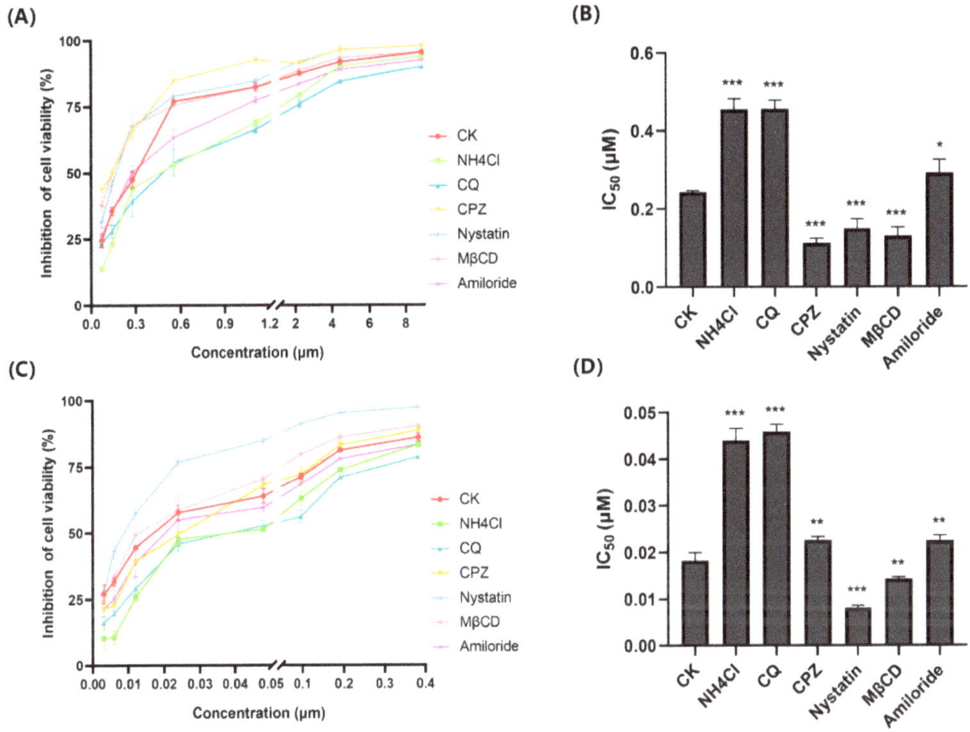

Figure 3. Effects of different endocytosis inhibitors on the Curcin and Curcin C cytotoxicity against U2OS cells. Control check (CK); chloroquine (CQ); chlorpromazine (CPZ); methyl-β-cyclodextrin (MβCD). (**A**) Proliferation inhibition rate of different endocytosis inhibitor-pretreated U2OS cells after 48 h Curcin treatment. (**B**) IC_{50} of Curcin after 48 h Curcin treatment on different endocytosis inhibitor-pretreated U2OS cells. (**C**) Proliferation inhibition rate of different endocytosis inhibitor-pretreated U2OS cells after 48 h Curcin C treatment. (**D**) IC_{50} of Curcin C after 48 h Curcin C treatment on different endocytosis inhibitor-pretreated U2OS cells. All data are presented as the mean ± SD of three independent experiments (***, $p < 0.001$; **, $p < 0.01$; * $p < 0.05$).

Chlorpromazine (CPZ), a cationic amphiphilic drug, inhibits the clathrin-dependent endocytic pathway [23]. CPZ pretreatment also significantly increased the IC_{50} of Curcin C compared with the control group without inhibitor treatment. However, the increase was still lower than that in the NH_4Cl and chloroquine (CQ) pretreatment groups (Figure 3D), suggesting the clathrin-dependent endocytosis pathway was only one of the entry pathways of Curcin C. Especially, when the concentration of Curcin C increased, the inhibition rate of cell proliferation in the CPZ pretreatment group was slightly higher than that in the control group. It might be because the clathrin-dependent endocytic pathway may be the preferred choice for Curcin C. However, when the amount of Curcin C exceeds the carrying capacity of the pathway, cells enable other pathways to transport Curcin C. Interestingly, after incubation with CPZ, the proliferation inhibition rate of Curcin at each concentration was higher than that of the control group (Figure 3A), and the IC_{50} of Curcin decreased obviously (Figure 3B), indicating that Curcin should not enter U2OS cells through a clathrin-dependent endocytic pathway. The decreased IC_{50} may be because cells' normal growth and development are affected, and the resistance to Curcin is reduced after the clathrin-dependent endocytic pathway is inhibited.

The clathrin-independent endocytosis pathways include caveolin-dependent endocytosis, caveolin-independent endocytosis, and micropinocytosis [24]. The boundaries between the first two pathways are not very clear. Nystatin and methyl-β-cyclodextrin (MβCD) can prevent cholesterol, sphingolipids, and caveolin from invaginating on the cell membrane by removing cholesterol from the cell surface and inhibiting caveolin-dependent endocytosis [22,25]. Compared with the control, the IC_{50}s of Curcin and Curcin C were significantly decreased after pretreatment of U2OS cells with Nystatin (Nystatin) and methyl-β-cyclodextrin (MβCD), indicating that the caveolin-dependent endocytosis is not the way for them to enter the cells.

Amiloride can inhibit the macropinocytosis-mediated endocytosis pathway by inhibiting Na+/H+ exchange and lowering the submembrane pH [26]. Compared with the control group, Amiloride treatment resulted in a significant increase in the IC_{50} of Curcin and Curcin C, but lower than those of the NH_4Cl and chloroquine (CQ) pretreatment groups, which proved that macropinocytosis-mediated endocytosis is also one of the pathways by which they enter cells. However, it is worth noting that when the concentration of Curcin was lower than 0.6 μM, the proliferation inhibition rate of the Amiloride treatment group was slightly lower than that of the control group. Only when the concentration was higher than this, the inhibition rate of proliferation was lower than that of the control group, suggesting that Curcin may not enter cells through macropinocytosis at lower concentrations. Another unknown pathway might have mediated its entry.

2.3. LRP1 Is One of Curcin C Endocytic Receptors

2.3.1. Correlation Analysis of LRP1 Abundance with Curcin and Curcin C Cytotoxicities

The above studies show that the clathrin-dependent pathway significantly differed between the Curcin and Curcin C endocytic pathways. The low-density lipoprotein receptor family is one of the most representative receptor families in the clathrin-dependent pathway [27]. Many studies have shown that the low-density lipoprotein receptor family, especially the LRP1 receptor, mediates the endocytosis of type I RIPs [14–18,28,29]. To understand whether LRP1 is also involved in the transport of Curcin and Curcin C, an analysis was first performed using databases TIMER2.0 and GEPIA2 to analyze the abundances of seven central LDL family members in 30 tumor cells and adjacent normal tissues (Figure 4A). The endocytic receptor LRP1 was found to have the highest expression in human sarcoma tumor cells to which U2OS cells belong (Figure 4B). Moreover, its levels in various tumor cells were found to correlate with the IC50s of Curcin C (Figure 4C). That is, the higher the level of LRP1, the lower the IC_{50} of Curcin C. However, this correlation was not observed when it concerned Curcin (data not shown). Thus, it may be speculated that LRP1 may only be involved in the endocytosis of Curcin C and is the main reason for the difference in entry efficiency between the two (IC_{50} data are derived from previous laboratory research data [13,30]).

2.3.2. LRP1 Gene Silencing Partially Reduces the Cytotoxicity of Curcin C to U2OS Cells

To verify whether LRP1 was involved in the transport of Curcin C, siRNA transfection was used to silence LRP1 of U2OS cells. Western Blot detection showed that there was still a tiny amount of LRP1 24 h after transfection, but no LRP1 protein could be detected when transfected for 48 h to 96 h (Figure 5A).

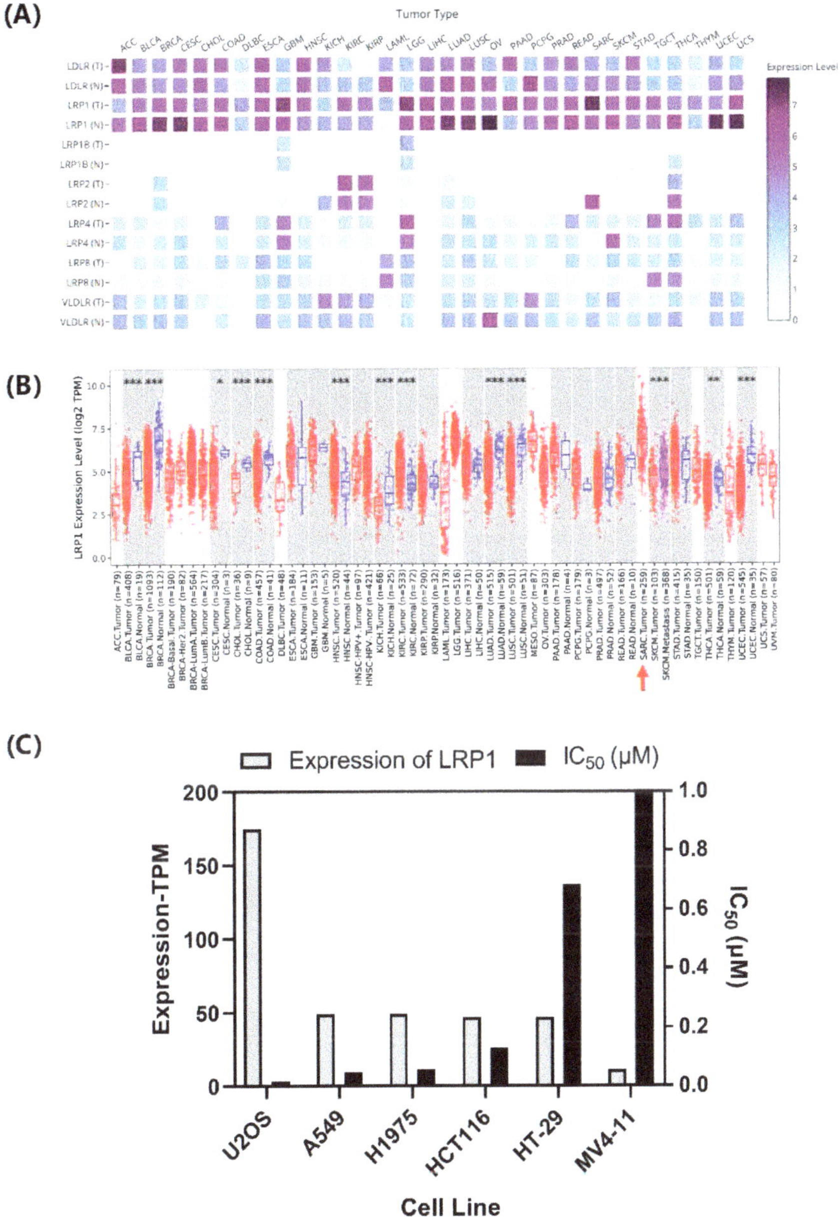

Figure 4. Expression of LDL family and LRP1 in tumor tissues and the relationship between LRP1 and cytotoxicity of Curcin C. (**A**) Heat map of LDL receptor family expression in 30 types of tumors and adjacent normal tissues, T—tumor tissues; N—normal tissues. (**B**) Expression of LRP1 in 35 tumor types and their adjacent normal tissues. The red arrow indicates the expression of LRP1 in sarcoma. (***, $p < 0.001$; **, $p < 0.01$; * $p < 0.05$). (**C**) Expression of LRP1 in 6 cell lines and IC_{50} of Curcin C for their corresponding cell lines. These 6 cell lines belong to 4 tumors (SARC: U2OS; LUAD: A549, H1975; COAD: HCT116, HT-29; LAML: MV4-11). IC_{50} data from Zhang et al., 2017 and Wang et al., 2022 [13,30]; the average of three independent experiments.

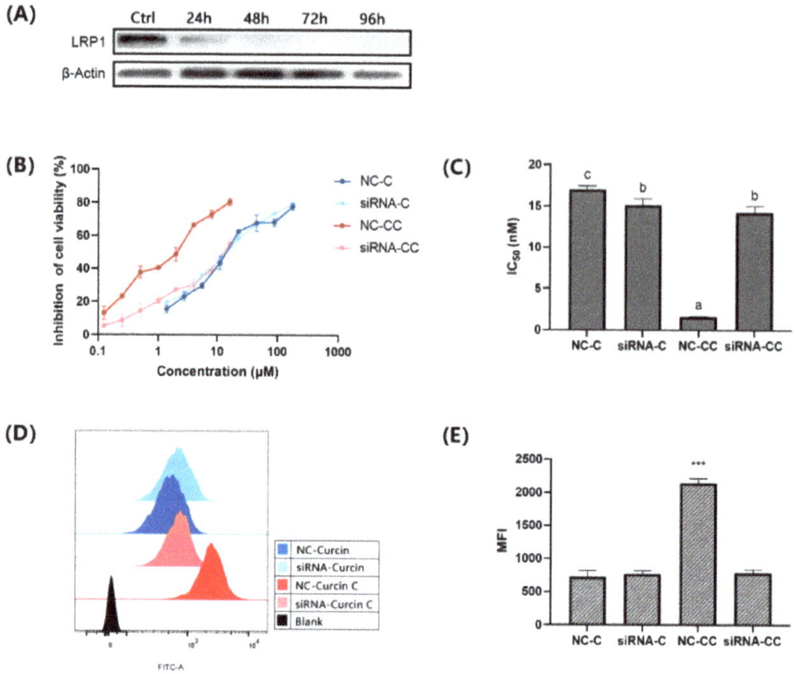

Figure 5. LRP1 is one of the Curcin C endocytic receptors. (**A**) Western Blot analysis of LRP1 expression levels at different time points after siRNA silencing. (**B**) Proliferation inhibition rate of siRNA-treated cells vs. untreated cells after 48 h treatment with different concentrations of Curcin and Curcin C. The "siRNA-C" and "siRNA-CC" indicate that cells were pretransfected with LRP1-siRNA for 48 h and then treated with C/CC for 48 h, and "NC-C" and "NC-CC" represent cells as the negative control, which are pre transfected with NC-siRNA for 48 h, and then treated with C/CC for 48 h. (**C**) IC_{50} of Curcin and Curcin C after 48 h of Curcin and Curcin C treatment in U2OS cells with different concentrations. (**D**) Flow cytometry was used to detect the fluorescence intensity of FITC-Curcin and FITC-Curcin C after LRP1 silencing in U2OS cells. Cells not treated with Curcin and Curcin C protein as a blank group. (**E**) The MFI of FITC-Curcin and FITC-Curcin C after LRP1 silencing vs. NC-C and NC-CC. All data are the mean ± SD of three independent experiments (***, $p < 0.001$). Different letters (a–c) indicate significant differences ($p < 0.05$; one-way ANOVA with Tukey's honestly significant difference test).

The toxicity of different concentrations of Curcin and Curcin C to U2OS cells 48 h after transfection was determined. It was found that the IC_{50} of Curcin C increased significantly. After 48 h of incubation, there was an approximately 9.13-fold increase. In contrast, the IC_{50} value of Curcin decreased slightly, indicating silencing the LRP1 even enhanced the toxicity of Curcin to U2OS cells. A notable phenomenon is that after the LRP1 was silenced, the IC_{50} of Curcin and Curcin C were very close to U2OS cells, and there was no significant difference (Figure 5B,C).

2.3.3. Silencing LRP1 Partially Inhibits the Entry of FITC-Curcin C into U2OS Cells

Flow cytometry was used to measure the fluorescence after U2OS cells incubated with 1 μM FITC-Curcin and FITC-Curcin C for 12 h to explore the effect of LRP1 silencing on the entry of FITC-Curcin and FITC-Curcin C into U2OS cells (Figure 5D). In the negative control group, the average fluorescence intensity of FITC-Curcin C was 2.97 times that of FITC-Curcin, indicating that the amount of FITC-Curcin C entering cells was higher than that of FITC-Curcin. When the LRP1 of U2OS cells was silenced, the mean fluorescence intensity of FITC-Curcin C decreased significantly but was not substantially different from that of FITC-Curcin (Figure 5E). It indicates that LRP1 silencing prevented FITC-Curcin C from entering U2OS cells. Considering there was no effect on the entry of FITC-Curcin into U2OS cells, the result suggests that the difference in the endocytosis receptors LRP1 may be one of the main reasons for the difference in endocytosis efficiency between Curcin and Curcin C.

3. Discussion

The amino acid composition and three-dimensional structure of Curcin and Curcin C are similar, but their cytotoxicity to osteosarcoma cell U2OS is 14 times different [13]. The in vitro tracer system and endocytosis inhibitors demonstrated that there might be multiple pathways involved in their transport and that the LRP1-mediated endocytic pathway might be the main reason for the difference in cytotoxicity to osteosarcoma U2OS.

Internalization efficiency primarily affects the efficacy of molecular drugs [14,15,29,31]. Due to the presence of B-chain lectins, type II RIPs possess high cell entry efficiency, leading to a dramatic increase in toxicity [32,33]. Conese et al. also found evidence that LRP1 can mediate saporin endocytosis in U937 cells, and downregulation of LRP1 can reduce the sensitivity of U937 cells to urokinase–saporin conjugates [14]. Our results also proved that though both FITC-Curcin and FITC-Curcin C could enter U2OS cells in a time- and concentration-dependent manner, the endocytosis efficiency of FITC-Curcin C was significantly higher than that of FITC-Curcin. These results suggest that it is likely that the difference in endocytosis efficiency is responsible for such significant activity between the two.

There are multiple routes for protein drugs to enter cells. For the entry of RIPs, research showed that there is also more than one pathway [34]. For example, in the endocytosis of ricin, the involved pathways include clathrin-dependent and clathrin-independent pathways. The ricin can bind to multiple receptors before it enters cells [35–37]. Shiga toxin can enter the cell by inducing invaginations at the cell surface [38]. Although lacking the B chain's help, Type I RIPs such as trichosanthin can still be transported into trophoblasts and proximal tubule epithelial cells through LRP1-mediated and megalin-mediated pathways, respectively [29,31]. It is reported that the membrane insertion pathway mediated by low pH also plays a role in the intracellular transport of trichosanthin [39]. The results in this study also support that LRP1 is one of the significant pathways that mediate the entry of Curcin C into cells. Moreover, as the cytotoxicity and entry of Curcin and Curcin C were not significantly different when LRP 1 was silenced in U2OS cells, it indicates that LRP1 plays an essential role in the efficient entry of Curcin C into U2OS cells and is most likely the main reason for the difference in the activity of Curcin C and Curcin. However, the silencing of LRP1 did not reduce the entry efficiency of Curcin, and the activity of Curcin C was not completely inhibited at the same time, suggesting that there must be other endocytic pathways involved in the entry of them in addition to the LRP1-mediated pathway.

Some RIPs such as ricin enter the endocytic pathway and travel backward from the Golgi complex to the ER, which is thought to parasitize the ER-associated degradation (ERAD) pathway [32]. CQ is a weak base that accumulates inside acidic subcellular compartments, e.g., endosomes, lysosomes, and Golgi vesicles [40]. It remains trapped in a protonated state, causing an increase in pH, and thereby inhibiting the functions of these cellular compartments. We found that the CQ treatment affected the cytotoxicity of Curcin and Curcin C, implying that the intracellular transport of Curcin and Curcin C may depend on the Golgi apparatus [15]. This situation is opposite to saporin [15] but similar to ricin [32,41], which might be related to the closer evolutionary relationship between Curcin and ricin.

LRP1 not only mediates endocytosis but also plays a role in signaling. The downstream proteins include JNK pathway-related proteins, human amyloid beta (Aβ), precursor protein-binding family B member 1, Disabled-1 protein (DAB1), dense postsynaptic zone protein-954, and other proteins [42]. These proteins regulate cell proliferation and apoptosis through the mitogen-activated protein kinases (MAPK) pathway [43]. Several RIPs, including ricin A, Korean mistletoe lectin II, TCS, and Shiga toxin, induce apoptosis by activating the MAPK pathways (SAPK/JNK) [43]. Our previous study also found that Curcin C can induce apoptosis in U2OS cells by activating JNK and inhibiting the ERK signaling pathway [30]. However, further experimental evidence is needed to confirm whether Curcin C activates the JNK signaling pathway through the LRP1 receptor.

In summary, there should be multiple pathways involved in the endocytosis of Curcin and Curcin C, and LRP1 might be the main reason for the difference in activity between the two. However, we also need to provide direct evidence for the interaction between Curcin C and LRP1 through bio-layer interferometry (BLI), surface plasmon resonance (SPR) or other protein interaction technologies. At the same time, the possible endocytic receptors that mediate the entry of Curcin and Curcin C into U2OS cells via macropinocytosis are also worthy of further investigation. Moreover, the entry amount of FITC-Curcin C is only about twice that of FITC-Curcin, which is insufficient to explain the activity difference on the order of magnitude. Different entry pathways trigger various downstream events, so these mechanisms may have efficiency differences. Therefore, understanding the intracellular events triggered differently can provide more valuable clues for answering questions.

4. Materials and Methods

4.1. Materials and Cell Lines

The mature seeds of *Jatropha curcas* used in this work were collected from Panzhihua City, Sichuan Province and cultivated in our laboratory (Sichuan University, Chengdu, China). They were soaked in distilled water for 2 h and germinated in phytotron under 16/8 h day/night photoperiods at 30 °C with 1:1 mix of coconut soil and nutrient soil until the cotyledons were expanded totally, and quantitative distilled water was added daily to keep seedlings growing.

Human osteosarcoma cells U2OS were purchased from the National Collection of Authenticated Cell Cultures, and the cells were maintained in McCoy's 5A medium (Gibco, Australia) supplemented with 10% fetal bovine serum (Cellmax Co., Ltd. (Beijing, China) and cultured in a CO_2 incubator (Shanghai Lishen Scientific Instrument Co., Ltd. (Shanghai, China)) at 37 °C and 5% CO_2.

4.2. FITC Labeling of Curcin, Curcin C, and BSA

Curcin, Curcin C, and BSA solutions were prepared by dissolving in the reaction solution (100 mM $NaHCO_3$, 10 mM Na_2CO_3, 125 mM NaCl, and pH = 9.8) to achieve a concentration of 10 mg/mL. A total of 10 mg/mL of FITC working solution was prepared in DMSO. The volume of FITC solution was determined by the following substance ratio:FITC:BSA = 10:1 and FITC:Curcin/Curcin C = 70:1. The FITC solution was added to the protein solution slowly, while shaking to mix the solution after every 10 μL of FITC was added until all the required FITC solution was added. The resulting solution was further mixed by shaking for 12 h at 4 °C in dark. The reaction was terminated by adding 1% of the total volume of termination solution (5M NH_4Cl) to the reaction system and then shaking for 2 h at 4 °C in dark. The solution was subsequently ultrafiltered 8 times using 10 kDa ultrafiltration tubes to remove unbound FITC from the solution. Protein concentration was determined by the Bradford protein assay kit. F/P was calculated by the following equation: F/P = A495 * Dilution factor * Path length correction/68,000 * Protein concentration (M).

4.3. Induced Expression of His-eGFP-Curcin and His-eGFP-Curcin C in Prokaryotes

The amino acid sequences of eGFP, Curcin, and Curcin C were obtained from the NCBI database. Codon optimization was carried out according to the genetic codon preference of *Escherichia coli*. The whole gene synthesis was carried out by Zoonbio Biotechnology. Briefly, His-eGFP-Curcin and His-eGFP-Curcin C genes were cloned into expression vectors pET30a and pCZN1 after codon optimization. The expression plasmids were transformed into the expression strains BL21 (DE3), Arctic Express, Rosetta, Shuffle T7-B, and Shuffle T7-K12. The expression was induced by different concentrations of IPTG, different temperatures, and different induction times, and the expression results were validated by SDS-PAGE. After successful induction, the expressed protein of interest was extracted using Ni column, and the supernatant was directly dialyzed into 1 × PBS (pH 7.4) at 4 °C. After dialysis, the supernatant protein was sterilized by filtering against a 0.22 μM filter head, and the resulting protein solution was aliquoted and stored at −80 °C. The inclusion bodies were dialyzed into inclusion renaturation buffer (1 × PBS (pH 7.4), 4 mM GSH, 0.4 mM GSSG, 0.4 M L-Arginine, 1 M Urea) at 4 °C for renaturation. After renaturation, the protein was dialyzed into 1 × PBS (pH 7.4). After dialysis, the protein was sterilized by filtering through the 0.22 μM filter heads. The protein was then aliquoted and stored at −80 °C.

4.4. Fluorescent Microscopy Observation of FITC-Curcin/Curcin C and His-eGFP-Curcin/Curcin C Entry into Cells

U2OS cells at logarithmic growth stage were seeded at 6×10^4 cells/mL in a 6-well plate (2 mL per well) and incubated overnight at 37 °C with 5% CO_2 to allow for the cells to adhere. The cells were then incubated with FITC-Curcin/ Curcin C and His-eGFP-Curcin/ Curcin C medium at the corresponding concentrations for a certain period of time, respectively. After incubation, the medium was removed, and the cells were washed 3 times with PBS. The nuclei were stained with Hoechst 33258 (Beyotime Biotech (Shanghai, China) Co., Ltd.) for 10 min at 37 °C. The cells were then washed 3 times with PBS, and then the complete medium was added. The cells were visualized using an inverted fluorescent microscope (Leica Microsystems (Wezlaer, Germany) Co., Ltd., Leica DMi 8).

4.5. Flow Cytometry of FITC-Curcin/Curcin C and His-eGFP-Curcin/Curcin C in Cells

U2OS cells at logarithmic growth stage were seeded at 1×10^5 cells/mL in a 6-well plate (2 mL per well) and incubated overnight at 37 °C with 5% CO_2 to allow for the cells to adhere. The cells were then incubated with FITC-Curcin/ Curcin C and His-eGFP-Curcin/ Curcin C medium at the corresponding concentrations for a certain period of time, respectively. After incubation, the medium was removed, and the cells were washed 3 times with PBS. The cells were then dissociated with 0.25% trypsin (Hyclone (Logan, UT, USA) Co., Ltd.) at 37 °C for 3 min (250 μL per well). The digestion was stopped by adding an equal amount of complete medium. The cells were dislodged by blowing, and the dislodged cell suspension was transferred into 1.5 mL EP tubes and centrifuged at 800 rpm for 3 min. The supernatant was discarded, and the cells were resuspended with 0.5 mL PBS for flow cytometry (BD (New York, NY, USA) Co., Ltd., BD LSRfortessa).

4.6. U2OS Cell Endocytic Inhibitors Pretreatment

U2OS cells at logarithmic growth stage were seeded at 6×10^4 cells/mL in a 96-well plate (100 μL per well) and incubated overnight at 37 °C with 5% CO_2 for cell adhesion. The cells were then pretreated for 2 h with media supplemented with 1 M MβCd, 100 mM NH_4Cl, 50 μM CPZ, 250 mM CQ, 125 μM Amiloride, and 100X diluted Nystatin (Sigma (Shanghai, China) Co., Ltd., Topscience (Shanghai, China) Co., Ltd.). Media without any inhibitor supplemented were used as control group. The cells were subsequently treated with different concentrations of Curcin C for 48 h. Cell viability was determined by CCK-8 assay.

4.7. CCK8 Assay for Cell Viability Characterization

U2OS cells at logarithmic growth stage were seeded at 6×10^4 cells/mL in a 96-well plate (100 μL per well) and incubated overnight at 37 °C with 5% CO_2 for cell adhesion. Eight concentration gradients of Curcin and Curcin C protein were introduced to the cells. In addition, a blank group containing only medium and a control group without Curcin or Curcin C were set up; 5 wells of cells were used in each treatment condition. After

48 h of drug treatment, CCK8-assay was performed by culturing the cells with 10% CCK8 (Dingyou Biotech Co., LTD (Chengdu, China) at 37 °C for 1–2 h until the A450 of the control group was about 1.0. The absorbance value (A) of each group at 450 nm was measured by a microplate reader (proliferation inhibition rate = ((1 − (A of administration group—A of blank group))/(A of control group—A of blank group)) × 100%). The IC_{50} of the drug was calculated by Quest Graph™ IC_{50} Calculator.

4.8. LRP1 Gene Silencing in U2OS Cells by siRNA

U2OS cells at logarithmic growth stage were seeded at 6 × 10^4 cells/mL in a 6-well plate (2 mL per well) and incubated overnight at 37 °C with 5% CO_2 for cell adhesion. For each well, the transfection reagent was prepared by adding 100 pmol siRNA (Sangon Biotech Co., Ltd. (Shanghai, China)) into 125 μL serum-free 1640 media. Then, 3 μL Lipo8000™ transfection reagent (Beyotime Biotech Co., Ltd. (Shanghai, China)) was added. The mixture was gently mixed and incubated for 20 min at 25 °C. A total of 125 μL of the transfection reagent was added into each well, and the plate was shaken briefly to mix the reagent. After 6–8 h of culture, the medium was changed to McCoy's 5A medium with 1 × antibiotics and 10% serum. After culture of 24 h, 48 h, 72 h, and 96 h, total protein was extracted and Western Blot was used to verify the silencing effect of LRP1. Three siRNAs were designed for desired silencing efficiency (Table 2).

Table 2. siRNA [1] sequences.

Number	Gene Name	Sequence (5′-3′)
1	hLRP1-3430	CCUGCAACAAUGGCAGAUGUATT
2	hLRP1-9620	GUCCAACUACACGUUACUUAATT
3	hLRP1-4970	GCGAACAAACACACUGGCUAATT

[1] siRNA was prepared by Sangon Biotechnology (Shanghai, China) Co., LTD.

4.9. Western Blotting for LRP1 Silencing Efficiency Characterization in U2OS Cells

After transfecting cells with siRNA in 6-well plates for 24 h, 48 h, 72 h, and 96 h, the cells were digested with 0.25% trypsin, and the digestion was stopped with an equal volume of complete medium. The cell suspension was centrifuged at 1000 rpm for 5 min at 4 °C, and the supernatant was discarded. The cells were then resuspended in 1 mL PBS. The resulting cell suspension was centrifuged again, and supernatant was discarded. For each well, the cells were lysed by adding 100 μL cell lysis buffer, 1 μL phosphatase inhibitor, and 1 μL protease inhibitor (Sigma (Shanghai, China) Co., Ltd.) and incubated at 4 °C for 10 min, followed by shaking for 15 s on ice 3 times. The cell lysates were centrifuged at 1500 RPM at 4 °C for 30 min, and the supernatant was collected and measured by Bradford protein detection kit (Sangon Biotech Co., Ltd. (Shanghai, China)). Protein concentration from each sample was normalized and separated by SDS-PAGE. The protein was transferred to PVDF membrane at 100 V for 90 min, blocked with rapid blocking solution (Beyotime Biotech (Shanghai, China) Co., Ltd.) at 25 °C for 30 min, and washed with 1X TBST for 10 min 3 times. Primary antibody was diluted 1:1000 in 5% skim milk powder and incubated overnight at 4 °C (primary antibody was LRP1, β-actin, Abways Tech Co., Ltd. (Shanghai, China)). The membrane was washed with 1× TBST for 10 min 3 times. The secondary antibody labeled with horseradish peroxidase was diluted 1:5000 in 5% skim milk powder and then incubated for 1–2 h at 25 °C. The membrane was washed with 1X TBST for 10 min × 3 times. The bands were visualized by ECL.

4.10. Data Analysis and Processing

All experiments were performed in biological triplicates, and the results were plotted as the mean value. GraphPad Prism 8, IBM SPSS Statistics 24, Excel, and Photoshop 2019 were used for data plotting and analysis. By one-way analysis of variance, $p < 0.05$ was defined as significant difference and marked with "*", $p < 0.01$ was defined as highly significant difference and marked with "**", and $p < 0.001$ was marked with "***".

Author Contributions: Conceptualization, S.Q.; data curation, R.M.; formal analysis, Z.L.; funding acquisition, Y.X.; investigation, Y.R.; methodology, S.Q.; project administration, Y.X.; resources, S.Q.; software, X.W.; supervision, Y.X.; validation, X.W. and P.H.; visualization, C.C. and T.W.; writing—original draft, S.Q. and X.W.; writing—review and editing, S.Q. All authors have read and agreed to the published version of the manuscript.

Funding: This work is supported by the National Nature Science Foundation, China (No. 31870315).

Institutional Review Board Statement: Not applicable.

Informed Consent Statement: Not applicable.

Data Availability Statement: No new data were created or analyzed in this study. Data sharing is not applicable to this article.

Conflicts of Interest: The authors declare no conflict of interest.

References

1. Lindsey, B.A.; Markel, J.E.; Kleinerman, E.S. Osteosarcoma Overview. *Rheumatol. Ther.* **2017**, *4*, 25–43. [CrossRef] [PubMed]
2. Harrison, D.J.; Geller, D.S.; Gill, J.D.; Lewis, V.O.; Gorlick, R. Current and future therapeutic approaches for osteosarcoma. *Expert Rev. Anti-Infect. Ther.* **2017**, *18*, 39. [CrossRef] [PubMed]
3. Saraf, A.J.; Fenger, J.M.; Roberts, R.D. Osteosarcoma: Accelerating Progress Makes for a Hopeful Future. *Front. Oncol.* **2018**, *8*, 4. [CrossRef] [PubMed]
4. Crommelin, D.; Storm, G.; Verrijk, R.; Leede, L.D.; Jiskoot, W.; Hennink, W.E. Shifting paradigms: Biopharmaceuticals versus low molecular weight drugs. *Int. J. Pharm.* **2003**, *266*, 3–16. [CrossRef]
5. Milletti, F. Cell-penetrating peptides: Classes, origin, and current landscape. *Drug Discov. Today* **2012**, *17*, 850–860.
6. Barbier, J.; Gillet, D. Ribosome Inactivating Proteins: From Plant Defense to Treatments against Human Misuse or Diseases. *Toxins* **2018**, *10*, 160. [CrossRef]
7. Rust, A.; Partridge, L.J.; Davletov, B.; Hautbergue, G.M. The Use of Plant-Derived Ribosome Inactivating Proteins in Immunotoxin Development: Past, Present and Future Generations. *Toxins* **2017**, *9*, 344. [CrossRef]
8. Lin, J.Y.; Tserng, K.Y.; Chen, C.C.; Lin, L.T.; Tung, T.C. Abrin and ricin: New anti-tumour substances. *Nature* **1970**, *227*, 292–293.
9. Sha, O.; Niu, J.; Ng, T.B.; Cho, E.Y.-P. Anti-tumor action of trichosanthin, a type 1 ribosome-inactivating protein, employed in traditional Chinese medicine: A mini review. *Cancer Chemother. Pharmacol.* **2013**, *71*, 1387–1393. [CrossRef]
10. Lee-Huang, S.; Huang, P.L.; Sun, Y.; Chen, H.C.; Kung, H.F.; Huang, P.L.; Murphy, W.J. Inhibition of MDA-MB-231 human breast tumor xenografts and HER2 expression by anti-tumor agents GAP31 and MAP30. *Anticancer Res.* **2000**, *20*, 653–659.
11. Guo, J.L.; Cheng, Y.L.; Qiu, Y.; Shen, C.H.; Yi, B.; Peng, C. Purification and characterization of a novel type i ribosome inactivating protein, pachyerosin, from Pachyrhizus erosus seeds, and preparation of its immunotoxin against human hepatoma cells. *Planta Med.* **2014**, *80*, 896–901. [CrossRef] [PubMed]
12. Fan, X.; He, L.; Meng, Y.; Li, G.; Li, L.; Meng, Y. A-MMC and MAP30, two ribosome-inactivating proteins extracted from Momordica charantia, induce cell cycle arrest and apoptosis in A549 human lung carcinoma cells. *Mol. Med. Rep.* **2015**, *11*, 3553–3558. [CrossRef] [PubMed]
13. Zhang, Y.; Yang, Q.; Li, C.; Ding, M.; Lv, X.; Tao, C.; Yu, H.; Chen, F.; Xu, Y. Curcin C, a novel type I ribosome-inactivating protein from the post-germinating cotyledons of Jatropha curcas. *Amino Acids* **2017**, *49*, 1619–1631. [CrossRef] [PubMed]
14. Conese, M.; Cavallaro, U.; Sidenius, N.; Olson, D.; Soria, M.R.; Blasi, F. PMA-induced down-regulation of the receptor for α2-macroglobulin in human U937 cells. *FEBS Lett.* **1995**, *358*, 73–78. [CrossRef]
15. Vago, R.; Marsden, C.J.; Lord, J.M.; Ippoliti, R.; Flavell, D.J.; Flavell, S.U.; Ceriotti, A.; Fabbrini, M.S. Saporin and ricin A chain follow different intracellular routes to enter the cytosol of intoxicated cells. *Febs J.* **2010**, *272*, 4983–4995.
16. Jiao, Y.Z.; Liu, W.Y. Low-density lipoprotein receptor-related protein 1 is an essential receptor for trichosanthin in 2 choriocarcinoma cell lines. *Biochem. Biophys. Res. Commun.* **2010**, *391*, 1579–1584. [CrossRef]
17. Fabbrini, M.S.; Carpani, D.; Bello-Rivero, I.; Soria, M.R. The amino-terminal fragment of human urokinase directs a recombinant chimeric toxin to target cells: Internalization is toxin mediated. *Faseb J.* **1997**, *11*, 1169–1176.
18. Fabbrini, M.S.; Rappocciolo, E.; Carpani, D.C.; Solinas, M.; Valsasina, B.; Breme, U.; Cavallaro, U.; Nykjaer, A.; Rovida, E.; Legname, G. Characterization of a saporin isoform with lower ribosome-inhibiting activity. *Biochem. J.* **1997**, *322 Pt 3*, 719–727. [CrossRef]
19. Bolognesi, A.; Polito, L.; Scicchitano, V.; Orrico, C.; Pasquinelli, G.; Musiani, S.; Santi, S.; Riccio, M.; Bortolotti, M.; Battelli, M.G. Endocytosis and intracellular localisation of type 1 ribosome-inactivating protein saporin-s6. *J. Biol. Regul. Homeost. Agents* **2012**, *26*, 97–109.
20. Schnaible, V.; Przybylski, M. Identification of fluorescein-5′-isothiocyanate-modification sites in proteins by electrospray-ionization mass spectrometry. *Bioconjugate Chem.* **1999**, *10*, 861–866.
21. Lin, S.H.; Faller, L.D. Preparation of Na, K-ATPase specifically modified on the anti-fluorescein antibody-inaccessible site by fluorescein 5′-isothiocyanate. *Anal. Biochem.* **2000**, *287*, 303–312. [PubMed]

22. Shi, B.-J.; Liu, C.-C.; Zhou, J.; Wang, S.-Q.; Gao, Z.-C.; Zhang, X.-M.; Zhou, B.; Chen, P.-Y. Entry of Classical Swine Fever Virus into PK-15 Cells via a pH-, Dynamin-, and Cholesterol-Dependent, Clathrin-Mediated Endocytic Pathway That Requires Rab5 and Rab7. *J. Virol.* **2016**, *90*, 9194–9208. [CrossRef] [PubMed]
23. Nawa, M.; Takasaki, T.; Yamada, K.-I.; Kurane, I.; Akatsuka, T. Interference in Japanese encephalitis virus infection of Vero cells by a cationic amphiphilic drug, chlorpromazine. *J. Gen. Virol.* **2003**, *84 Pt 7*, 1737–1741. [PubMed]
24. Mayor, S.; Parton, R.G.; Donaldson, J.G. Clathrin-Independent Pathways of Endocytosis. *Cold Spring Harb. Perspect. Biol.* **2014**, *6*, a016758. [CrossRef] [PubMed]
25. Li, Z.; Zhao, K.; Lan, Y.; Lv, X.; Hu, S.; Guan, J.; Lu, H.; Zhang, J.; Shi, J.; Yang, Y.; et al. Porcine Hemagglutinating Encephalomyelitis Virus Enters Neuro-2a Cells via Clathrin-Mediated Endocytosis in a Rab5-, Cholesterol-, and pH-Dependent Manner. *J. Virol.* **2017**, *91*, e01083-17. [CrossRef]
26. Wang, B.M.; Zhuang, X.Y.; Deng, Z.B.; Jiang, H.; Mu, J.Y.; Wang, Q.L.; Xiang, X.Y.; Guo, H.X.; Zhang, L.F.; Dryden, G.; et al. Targeted Drug Delivery to Intestinal Macrophages by Bioactive Nanovesicles Released from Grapefruit. *Mol. Ther.* **2014**, *22*, 522–534. [CrossRef]
27. Lin, L.; Shi, A.B. Endocytic recycling pathways and the regulatory mechanisms. *Yi Chuan Hered.* **2019**, *41*, 451–468.
28. Maria, F.; Miku, K.; Ikuhiko, N.; Riccardo, V. Plant Ribosome-Inactivating Proteins: Progesses, Challenges and Biotechnological Applications (and a Few Digressions). *Toxins* **2017**, *9*, 314.
29. Tang, N.; Chan, W.L.; Ke, Y.B.; Mak, M.; Lai, F.M.; Tam, S.C. Acute Renal Failure and Proximal Tubule Lesions after Trichosanthin Injection in Rats. *Exp. Mol. Pathol.* **1997**, *64*, 78–89. [CrossRef]
30. Wang, F.; Wu, P.; Qin, S.; Deng, Y.; Han, P.; Li, X.; Fan, C.; Xu, Y. Curcin C inhibit osteosarcoma cell line U_2OS proliferation by ROS induced apoptosis, autophagy and cell cycle arrest through activating JNK signal pathway. *Int. J. Biol. Macromol.* **2022**, *195*, 433–439. [CrossRef]
31. Chan, W.L.; Shaw, P.C.; Tam, S.C.; Jacobsen, C.; Gliemann, J.; Nielsen, M.S. Trichosanthin Interacts with and Enters Cells via LDL Receptor Family Members. *Biochem. Biophys. Res. Commun.* **2000**, *270*, 453–457. [PubMed]
32. Sandvig, K.; Bo, V.D. Transport of protein toxins into cells: Pathways used by ricin, cholera toxin and Shiga toxin. *FEBS Lett.* **2002**, *529*, 49–53. [CrossRef]
33. Sandvig, K.; Deurs, B.V. Entry of ricin and Shiga toxin into cells: Molecular mechanisms and medical perspectives. *Embo J.* **2000**, *19*, 5943–5950. [PubMed]
34. Sandvig, K.; Kavaliauskiene, S.; Skotland, T. The Protein Toxins Ricin and Shiga Toxin as Tools to Explore Cellular Mechanisms of Internalization and Intracellular Transport. *Toxins* **2021**, *13*, 377. [CrossRef] [PubMed]
35. Roberts, L.M.; Lord, J.M. Ribosome-Inactivating Proteins: Entry into Mammalian Cells and Intracellular Routing. *Mini Rev. Med. Chem.* **2004**, *4*, 505–512. [CrossRef]
36. Sandvig, K.; Torgersen, M.L.; Engedal, N.; Skotland, T.; Iversen, T.G. Protein toxins from plants and bacteria: Probes for intracellular transport and tools in medicine. *Febs Lett.* **2010**, *584*, 2626–2634. [CrossRef] [PubMed]
37. Seaman, M.; Peden, A.A. Ricin Toxin Hits a Retrograde Roadblock. *Cell* **2010**, *141*, 222–224. [CrossRef]
38. Römer, W.; Berland, L.; Chambon, V.; Gaus, K.; Windschiegl, B.; Tenza, D.; Aly, M.R.; Fraisier, V.; Florent, J.C.; Perrais, D.; et al. Shiga toxin induces tubular membrane invaginations for its uptake into cells. *Nature* **2007**, *450*, 670–675. [CrossRef]
39. Xia, X.F.; Sui, S.F. The membrane insertion of trichosanthin is membrane-surface-pH dependent. *Biochem. J.* **2000**, *349 Pt 3*, 835.
40. Al-Bari, M.A.A. Targeting endosomal acidification by chloroquine analogs as a promising strategy for the treatment of emerging viral diseases. *Pharmacol. Res. Perspect.* **2017**, *5*, e00293.
41. Rapak, A.; Falnes, P.O.; Olsnes, S. Retrograde transport of mutant ricin to the endoplasmic reticulum with subsequent translocation to cytosol. *Proc. Natl. Acad. Sci. USA* **1997**, *94*, 3783–3788. [PubMed]
42. García-Fernández, P.; Üçeyler, N.; Sommer, C. From the low-density lipoprotein receptor-related protein 1 to neuropathic pain: A potentially novel target. *Pain Rep.* **2021**, *6*, e898. [PubMed]
43. Wang, L.; Shen, F.; Zhang, M.; He, Q.; Zhao, H.; Yu, X.; Yang, S.; Liu, Y.; Deng, N.; Zheng, J.; et al. Cytotoxicity mechanism of α-MMC in normal liver cells through LRP1 mediated endocytosis and JNK activation. *Toxicology* **2016**, *357–358*, 33–43. [CrossRef] [PubMed]

MDPI
St. Alban-Anlage 66
4052 Basel
Switzerland
Tel. +41 61 683 77 34
Fax +41 61 302 89 18
www.mdpi.com

Toxins Editorial Office
E-mail: toxins@mdpi.com
www.mdpi.com/journal/toxins

www.ingramcontent.com/pod-product-compliance
Lightning Source LLC
Chambersburg PA
CBHW061521210926
51850CB00008B/1990
9783036568225